Special Issue on Nucleon Resonances

Elena Santopinto · Ralf Gothe
Viktor I. Mokeev
Editors

Special Issue on Nucleon Resonances

Previously published as Topical Collection in the journal
Few-Body Systems, volume 57, 2016

Springer

Editors
Elena Santopinto
Department of Physics, INFN, sezione di
 Genova
University of Genova
Genoa
Italy

Viktor I. Mokeev
Jefferson Lab
Newport News, VA
USA

Ralf Gothe
Department of Physics and Astronomy
University of South Carolina
Columbia, SC
USA

ISBN 978-3-7091-4870-9

Library of Congress Control Number: 2017955849

Printed on acid-free paper

This Springer imprint is published by Springer Nature
The registered company is Springer-Verlag GmbH Austria
The registered company address is: Prinz-Eugen-Str. 8-10, 1040 Wien, Austria

Contents

Part I Intro

Nucleon resonances: From Photoproduction to High Photon Virtualities 3
Ralf W. Gothe, Viktor Mokeev and Elena Santopinto

Part II Review

Nucleon Resonance Physics 9
Volker D. Burkert

Baryons and the Borromeo 19
Craig D. Roberts and Jorge Segovia

Part III Advances in Experimental Studies of the N* Spectrum and Structure

**Impact of the $\gamma_v NN^*$ electrocoupling parameters at high photon
virtualities and preliminary cross sections off the neutron** 31
Ralf W. Gothe and Ye Tian

**Updates on the Studies of N^* Structure with CLAS and the Prospects
with CLAS12** 39
V.I. Mokeev

Nucleon Resonance Structure Studies Via Exclusive KY Electroproduction 47
Daniel S. Carman

**Exclusive Single Pion Electroproduction off the Proton:
Results from CLAS** 55
Kijun Park

Baryon spectroscopy with polarization observables from CLAS 61
Steffen Strauch

Exotic baryons from a heavy meson and a nucleon 69
Yasuhiro Yamaguchi

**Part IV Advances and Prospects in Extraction of Resonance
                  Parameters from the Data**

**Light-quark baryon spectroscopy within ANL-Osaka dynamical
coupled-channels approach** 79
Hiroyuki Kamano

Hard exclusive pion leptoproduction 87
Peter Kroll

Electromagnetic transition form factor of nucleon resonances   97
Toru Sato

Pion Electroproduction and Siegert's Theorem   103
Lothar Tiator

Part V   Nucleon Resonances and Strong QCD Dynamics from Dyson Schwinger Equation Approaches

Excited hadrons and the analytical structure of bound-state interaction kernels   113
Bruno El-Bennich, Gastão Krein, Eduardo Rojas and Fernando E. Serna

Progress in the calculation of nucleon transition form factors   123
Gernot Eichmann

$\gamma v NN^*$ Electrocouplings in Dyson–Schwinger Equations   133
Jorge Segovia

Comments on Formulating Meson Bound-State Equations Beyond Rainbow-Ladder Approximation   141
Si-xue Qin

Part VI   Nucleon Resonances from Lattice QCD

Hadron wave functions from lattice QCD   151
V.M. Braun

Meson electro-/photo-production from QCD   159
Raúl A. Briceño

Light-cone sum rule approach for Baryon form factors   167
Nils Offen

N^* Resonances in Lattice QCD from (mostly) Low to (sometimes) High Virtualities   177
David G. Richards

Part VII   Nucleon Resonance Studies in Quark Models

Open flavor strong decays   187
H. García-Tecocoatzi, R. Bijker, J. Ferretti, G. Galatà and E. Santopinto

High Q^2 helicity amplitudes in the hypercentral Constituent Quark Model   195
M.M. Giannini

The Spectroscopy and Form Factors of Nucleon Resonances from Superconformal Quantum Mechanics and Holographic QCD   205
Guy F. de Téramond

Nucleon resonance electrocouplings from light-front quark models at Q^2 up to 12 GeV^2   213
Igor T. Obukhovsky, Amand Faessler, Thomas Gutsche and Valery E. Lyubovitskij

Effective degrees of freedom in baryon spectroscopy   221
E. Santopinto and J. Ferretti

A Relativistic Model for the Electromagnetic Structure of Baryons from the 3rd Resonance Region   229
G. Ramalho

Few-Body Syst (2016) 57:869–871
DOI 10.1007/s00601-016-1152-7

Ralf W. Gothe · Viktor Mokeev · Elena Santopinto

Nucleon Resonances: From Photoproduction to High Photon Virtualities

Received: 23 August 2016 / Accepted: 24 August 2016 / Published online: 22 September 2016
© Springer-Verlag Wien 2016

The topical workshop "Nucleon Resonances: From Photoproduction to High Photon Virtualities" took place at the European Center for Theoretical Studies in Nuclear Physics and Related Areas in Trento, Italy from October 12–16, 2015. The organizing committee consisted of R.W. Gothe (Chair, USC), V.I. Mokeev (Jefferson Lab), and E. Santopinto (INFN).

The Workshop had around 40 participants and was supported in part also by generous contributions from JSA/Jefferson Lab, the University of South Carolina, and INFN.

The studies of the excited nucleon N* spectrum and the structure of these states offer unique information on many facets of the non-perturbative strong interaction in the generation of excited nucleon states of different quantum numbers. The workshop was devoted to the study of the spectrum and structure of N* states, as they become accessible through the electromagnetic excitation of the nucleon in exclusive meson production. It was focused on: (a) the baryon spectrum in exclusive meson photoproduction, (b) the search for new baryon states in the combined studies of exclusive photo- and electroproduction at small and moderate photon virtualities (Q^2), (c) electro-excited N* states and their structure for Q^2 up to 5 GeV^2, and (d) the extension of these studies in the future experiments with CLAS12 at Jefferson Lab up to 12 GeV^2.

The current status of the studies of the excited nucleon spectrum and structure from exclusive meson photo- and electroproduction data, as well as the challenges and prospects in this field, such as the search for new states of baryon matter, the so-called hybrid baryons, with glue as an extra structural component, were reviewed in a keynote talk by V.D. Burkert (JLab), which helped to shape the Workshop discussions and outcome.

The experimental studies of exclusive meson photoproduction off the nucleon at the JLab/CLAS, ELSA, and MAMI facilities continue their rapid progress. The results were presented in talks by S. Strauch (USC) and A. D'Angelo (Rome U.). These experiments provided detailed information on all exclusive meson photo-production channels relevant in the resonance region, including differential cross sections, and single, double, and triple polarization asymmetries. The data on exclusive photoproduction off the nucleon with more than one meson in the final states keeps growing. The status and prospects of the advanced reaction models for the extraction of the resonance parameters from these data were presented in talks by L. Tiator (Mainz U.),

---

This preface belongs to the special issue "Nucleon Resonances".

Ralf W. Gothe
Department of Physics and Astronomy, University of South Carolina, 712 Main Street, Columbia, SC 29208, USA
E-mail: gothe@sc.edu

Viktor Mokeev
Jefferson Lab' 1200 Jefferson Ave Suite 5, Newport News, VA 23606, USA
E-mail: mokeev@jlab.org

Elena Santopinto (✉)
Istituto Nazionale di Fisica Nucleare, via Dodecaneso 33, 16146 Genova, Italy
E-mail: elena.santopinto@ge.infn.it

I. Danilkin (JPAC at JLab), H. Haberzettl (GWU), J. Nys (Ghent U.), A. Sarantsev (Bonn U. and Petersburg Nucl. Phys. Inst.), and H. Kamano (Osaka U.). Analyses of the aforementioned experimental data considerably extended our knowledge of the N* spectrum. Several candidate N* states were included to the 2014 edition of the PDG as an outcome of these efforts with decisive contribution of the KY photoproduction data from Jlab/CLAS and ELSA carried out within the framework of global multi-channel Bonn-Gatchina approach. The search for manifestations of these new N* states in the future exclusive electroproduction data at low Q^2 will be the critical step to prove or rule out their existence.

The CLAS detector at JLab has produced the dominant part of the available worldwide data on all relevant meson electroproduction channels off the nucleon in the resonance region for Q^2 up to 5.0 GeV^2. The recent results from these studies were presented in talks by K. Park (JLab) and D.S. Carman (JLab). Analyses of the data on N π, N η, and $\pi^+ \pi^-$ p exclusive electroproduction off the proton have provided the only worldwide results available on the Q^2 evolution of the helicity amplitudes for the transitions from the initial photon-proton to the final N* states, the so-called γ_v NN* electrocouplings, which allow us to explore the N* internal structure. The efforts on the extraction of the γ_v NN* electrocouplings were reviewed in the talk by V.I. Mokeev (JLab). γ_v NN* electrocouplings have become available for most excited nucleon states in the mass range up to 1.8 GeV at photon virtualities up to 5.0 GeV^2 (up to 7.5 GeV^2 for $\Delta(1232)3/2^+$ and 7.0 GeV^2 for N(1535)1/2$^-$). The current results on the resonance electrocouplings have been collected at: https://userweb. jlab.org/~mokeev/resonance_electrocouplings/. Physics analyses of these results have revealed the structure of N* states for $Q^2 < 5.0$ GeV^2 as a complex interplay between the inner core of three dressed quarks and the external meson-baryon cloud.

For the first time results on exclusive π^- p electroproduction off bound neutrons corrected for kinematical final-state-interactions and Fermi-motion have become available and were presented in the talk by R.W. Gothe (USC). The impressive progress in advanced reaction models allowing us to account for the π^- p final-state-interactions and Fermi-motion of the neutron-target inside the deuteron, presented in the talk by T-S.H. Lee (ANL), opens up the prospects for extracting the γ_v NN* electrocouplings off the neutron.

The development of theoretical approaches capable of relating the non-perturbative strong interaction mechanisms behind the formation of N* states to the results on the N* spectrum and the γ_v NN* electrocouplings, reviewed in the colloquium talk by C.D. Roberts (ANL) and the presentation by V.M. Braun (Regensburg U.), represent a key part in the synergetic efforts between experimentalists and theorists in the studies of N* states.

The Dyson-Schwinger Equations of QCD (DSEQCD) successfully reproduce the data on elastic and transition N $\to \Delta(1232)3/2^+$, N $\to$ N(1440)1/2$^+$ form factors for $Q^2 > 2.5$ GeV^2 with the same dressed quark mass function, demonstrating the relevance of dressed quarks in the structure of the ground and excited nucleons and the capability of accessing this fundamental quantity from the data on the nucleon elastic and transition N $\to$ N* form factors. Impressive developments in the studies of the N* spectrum and structure within DSEQCD were presented in the talk by J. Segovia (Technical U of Munich), G. Eichmann (Giessen U.), S. Qin (ANL), B. El-Bennich (Sao Paulo U.), A. Bashir (Michoacán U.), P. Rodriguez-Quintero (Huelva U.), and D. Binosi (ECT*).

The novel approach presented in the talk by N. Offen (Regensburg U.) provides insight into the N* partonic structure, allowing us to constrain the quark distribution amplitudes of the N(1535)1/2$^-$ resonance, relating them to the resonance electrocouplings determined by employing Light Cone Sum Rules. The moments of the quark distribution amplitudes can be computed from the QCD Lagrangian within LQCD and confronted to those derived from the γ_v NN* electrocoupling values offering an additional promising avenue in relating empirical resonance electrocoupling values from experiment to the first principles of QCD.

Plans for the evaluation of the γ_v NN* electrocouplings at intermediate photon virtualities from the QCD Lagrangian within the LQCD framework were presented in talks by D.G. Richards (JLab) and R. Briceno (JLab).

The constituent quark models remain the only available tool in the studies of the N* structure over the full N* spectrum. The recent results of the advanced quark models were presented in talks by G. de Teramond (Univ. of Costa Rica), I.T. Obukhovsky (Moscow State U.), G. Ramalho (Univ. of Rio Grande), H. Garcia (INFN), J. Feretti (INFN), Y. Yamaguchi (INFN), M.M. Giannini (Univ. of Genova), and E. Santopinto (INFN). New results on resonance electrocouplings for $Q^2 < 12$ GeV^2 from these promising approaches are urgently needed.

After completion of the Jefferson Lab 12 GeV Upgrade, CLAS12 will be the only facility foreseen worldwide capable of exploring the γ_v NN* electrocouplings at the smallest 0.05 GeV^2 < Q^2 < 0.3 GeV^2 and at the highest Q^2 up to 12 GeV^2 ever achieved in exclusive reactions. The search for new types of baryon matter, the so-called hybrid baryons with glue as a structural component, represents a new flagship experiment in the N* studies at small/intermediate Q^2 with CLAS12. The plans for the preparation of this experiment recently approved (JLab PAC44) experimental program with the CLAS12 were presented in the talk by L. Lanza (INFN). The three other approved experiments aimed toward obtaining the $\gamma_v NN^*$ electrocouplings of most N* states in mass range up to 3.0 GeV and at $2.0\,\mathrm{GeV}^2 < Q^2 < 12\,\mathrm{GeV}^2$ from exclusive $N\pi$, $\pi^+\pi^- p$, and KY electroproduction data will start in the first run period in 2017 and possibly 2018. Further development of the reaction models incorporating the quark degrees of freedom is urgently needed to facilitate the extraction of the resonance electrocouplings at high Q^2. The prospects for the development of such an approach were outlined by P. Kroll (University of Wuppertal and University of Regensburg).

The resonance electroexcitation will be explored in these experiments at the distances where the quark core dominates, for the first time offering direct access to dressed quarks and their non-perturbative strong interaction. The dressed quark mass function will be probed at the distance scales where the transition from the quark–gluon confinement to pQCD regimes takes place, addressing the most challenging and still open problems of the Standard Model on the nature of the dominant part of the hadron mass, quark–gluon confinement, and the emergence of hadrons from QCD.

The reader can find more details about the mentioned presentations and many other interesting contributions in this peer-reviewed special issue of these Proceedings. The detailed program and all the presentations can be found in: http://boson.physics.sc.edu/~gothe/ect*-15/program.html.

Few-Body Syst (2016) 57:873–882
DOI 10.1007/s00601-016-1121-1

Volker D. Burkert

Nucleon Resonance Physics

Received: 14 February 2016 / Accepted: 25 April 2016 / Published online: 25 July 2016
© Springer-Verlag Wien 2016

Abstract Recent results of meson photo-production at the existing electron machines with polarized real photon beams and the measurement of polarization observables of the final state baryons have provided high precision data that led to the discovery of new excited nucleon and Δ states using multi-channel partial wave analyses procedures. The internal structure of several prominent excited states has been revealed employing meson electroproduction processes. On the theoretical front, lattice QCD is now predicting the baryon spectrum with very similar characteristics as the constituent quark model, and continuum QCD, such as is represented in the Dyson–Schwinger equations approach and in light front relativistic quark models, describes the non-perturbative behavior of resonance excitations at photon virtuality of $Q^2 > 1.5\,\mathrm{GeV}^2$. In this talk I discuss the need to continue a vigorous program of nucleon spectroscopy and the study of the internal structure of excited states as a way to reveal the effective degrees of freedom underlying the excited states and their dependence on the distance scale probed.

1 Introduction

The excited states of the nucleon have been studied experimentally since the 1950's [1]. They contributed to the discovery of the quark model in 1964 by Gell-Mann and Zweig [2,3], and were critical for the discovery of "color" degrees of freedom as introduced by Greenberg [4]. The quark structure of baryons resulted in the prediction of a wealth of excited states with underlying spin-flavor and orbital symmetry of $SU(6) \otimes O(3)$, and led to a broad experimental effort to search for these states. Most of the initially observed states were found with hadronic probes. However, of the many excited states predicted in the quark model, only a fraction have been observed to date. Search for the "missing" states and detailed studies of the resonance structure are now mostly carried out using electromagnetic probes and have been a major focus of hadron physics for the past decade [5]. A broad experimental effort is currently underway with measurements of exclusive meson photoproduction and electroproduction reactions, including many polarization observables. Precision data and the development of multi-channel partial wave analysis procedures have resulted in the discovery of several new excited states of the nucleon, which have been entered in the Review of Particle Physics [6].

A quantitative description of baryon spectroscopy and the structure of excited nucleons must eventually involve solving QCD for a complex strongly interacting multi-particle system. Recent advances in Lattice QCD led to predictions of the nucleon spectrum in QCD with dynamical quarks [7], albeit with still large pion masses of 396 MeV. Lattice prediction can therefore only be taken as indicative of the quantum numbers of excited states and not of the masses of specific states. In parallel, the development of dynamical coupled

---

This article belongs to the special issue "Nucleon Resonances".

V. D. Burkert (✉)
Jefferson Laboratory, 12000 Jefferson Avenue, Newport News, VA 23606, USA
E-mail: burkert@jlab.org

channel models is being pursued with new vigor. The EBAC group at JLab has shown [8] that dynamical effects can result in significant mass shifts of the excited states. As a particularly striking result, a very large shift was found for the Roper resonance pole mass to 1365 MeV downward from its bare core mass of 1736 MeV. This result has clarified the longstanding puzzle of the incorrect mass ordering of $N(1440)\frac{1}{2}^{+}$ and $N(1535)\frac{1}{2}^{-}$ resonances in the constituent quark model. Developments on the phenomenological side go hand in hand with a world-wide experimental effort to produce high precision data in many different channel as a basis for a determination of the light-quark baryon resonance spectrum. On the example of experimental results from CLAS, the strong impact of precise meson photoproduction data is discussed. Several reviews have recently been published on this and related subjects [9–13].

It is interesting to point out recent findings that relate the observed baryon spectrum of different quark flavors with the baryon densities in the freeze out temperature in heavy ion collisions, which show evidence for missing baryons in the strangeness and the charm baryon sector [14,15]. These data hint that an improved baryon model including further unobserved light quark baryons may resolve the current discrepancy between lattice QCD results and the results obtained using a baryon model that includes only states listed by the PDG. A complete accounting of excited baryon states of all flavors seems essential for a quantitative description of the occurrence of baryons in the evolution of the microsecond old universe.

Accounting for the complete excitation spectrum of the nucleon (protons and neutrons) and understanding the effective degrees of freedom is perhaps the most important and certainly the most challenging task of hadron physics. The experimental N* program currently focusses on the search for new excited states in the mass range from 2 to 2.5 GeV using energy-tagged photon beams in the few GeV range, and the study of the internal structure of prominent resonances in meson electroproduction.

2 Establishing the N* Spectrum

The complex structure of the light-quark (u and d quarks) baryon excitation spectrum complicates the experimental search for individual states. As a result of the strong interaction, resonances are wide, often 200 to 400 MeV, and are difficult to uniquely identify when only differential cross sections are measured. Most of the excited nucleon states listed in the Review of Particle Properties prior to 2012 have been observed in elastic pion scattering $\pi N \to \pi N$. However there are important limitations in the sensitivity to the higher mass nucleon states that may have small $\Gamma_{\pi N}$ decay widths. The extraction of resonance contributions then becomes exceedingly difficult in πN scattering. Estimates for alternative decay channels have been made in quark model calculations [16] for various channels. This has led to a major experimental effort at Jefferson Lab, ELSA, GRAAL, and MAMI to chart differential cross sections and polarization observables for a variety of meson photoproduction channels. At JLab with CLAS, several final states have been measured with high precision [17–27] that are now employed in multi-channel analyses.

2.1 New States from Open Strangeness Photoproduction

Here one focus has recently been on measurements of $\gamma p \to K^{+}\Lambda$, using a polarized photon beam several polarization observables can be measured by analyzing the parity violating decay of the recoil $\Lambda \to p\pi^{-}$. It is well known that the energy-dependence of a partial-wave amplitude for one particular channel is influenced by other reaction channels due to unitarity constraints. To fully describe the energy-dependence of an amplitude one has to include other reaction channels in a coupled-channel approach. Such analyses have been developed by the Bonn–Gatchina group [28], at JLab [29], at Jülich [30] and other groups.

The data sets with the highest impact on resonance amplitudes in the mass range above 1.7 GeV have been kaon-hyperon production using a spin-polarized photon beam and where the polarization of the Λ or Σ° is also measured. The high precision cross section and polarization data [23–27] provide nearly full polar angle coverage and span the $K^{+}\Lambda$ invariant mass range from threshold to 2.9 GeV, hence covering the full nucleon resonance domain where new states might be discovered.

The backward angle $K^{+}\Lambda$ data in Fig. 1 show clear resonance-like structures at 1.7 and 1.9 GeV that are particularly prominent and well-separated from other structures at backward angles, while at more forward angles (not shown) t-channel processes become prominent and dominate the cross section. The broad enhancement at 2.2 GeV may also indicate resonant behavior although it is less visible at more central angles with larger background contributions. The $K^{+}\Sigma$ channel also indicates significant resonant behavior as seen in

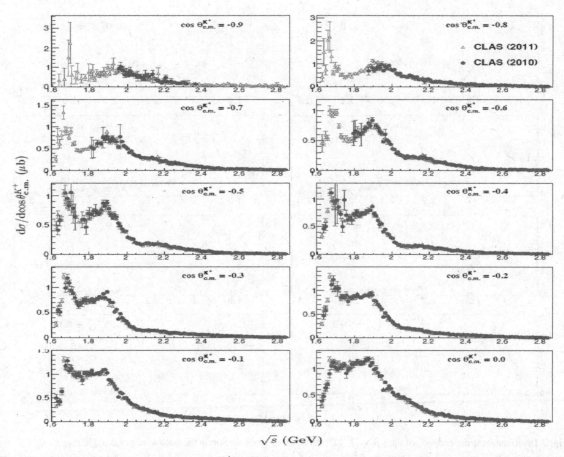

Fig. 1 Invariant mass dependence of the $\gamma p \rightarrow K^+\Lambda$ differential cross section in the backward polar angle range. There are 3 structure visible that may indicate resonance excitations, at 1.7, 1.9, and 2.2 GeV. The *blue full circles* are based on the topology $K^+ p\pi^-$, the *red open triangles* are based on topology $K^+ p$ or $K^+\pi^-$, which extended coverage towards lower W at backward angles and allows better access to the resonant structure near threshold (color figure online)

Fig. 2. The peak structure at 1.9 GeV is present at all angles with a maximum strength near 90°, consistent with the behavior of a $J^P = \frac{3}{2}^+$ p-wave. Other structures near 2.2 to 2.3 GeV are also visible. Still, only a full partial wave analysis can determine the underlying resonances, their masses and spin-parity. The task is somewhat easier for the $K\Lambda$ channel, as the iso-scalar nature of the Λ selects isospin-$\frac{1}{2}$ states to contribute to the $K\Lambda$ final state, while both isospin-$\frac{1}{2}$ and isospin-$\frac{3}{2}$ states can contribute to the $K\Sigma$ final state.

These cross section data together with the Λ and Σ recoil polarization and polarization transfer data to the Λ and Σ had strong impact on the discovery of several new nucleon states. They also provided new evidence for several candidate states that had been observed previously but lacked confirmation as shown in Fig. 3. It is interesting to observe that five of the observed nucleon states have nearly degenerate masses near 1.9 GeV. Similarly, the new Δ state appears to complete a mass degenerate multiplet near 1.9 GeV as well. There is no obvious mechanism for this apparent degeneracy. Nonetheless, all new states may be accommodated within the symmetric constituent quark model based on $SU(6) \otimes O(3)$ symmetry group as far as quantum numbers are concerned. As discussed in Sect. 1 for the case of the Roper resonance $N(1440)\frac{1}{2}^+$, the masses of all pure quark model states need to be corrected for dynamical coupled channel effects to compare them with observed resonances. The same applies to the recent Lattice QCD predictions [31] for the nucleon and Delta spectrum.

2.2 Vectormeson Photoproduction

In the mass range above 2.0 GeV resonances tend to decouple from simple final states like $N\pi$, $N\eta$, and $K\Lambda$. We have to consider more complex final states with multi-mesons or vector mesons, such as $N\omega$, $N\phi$, and

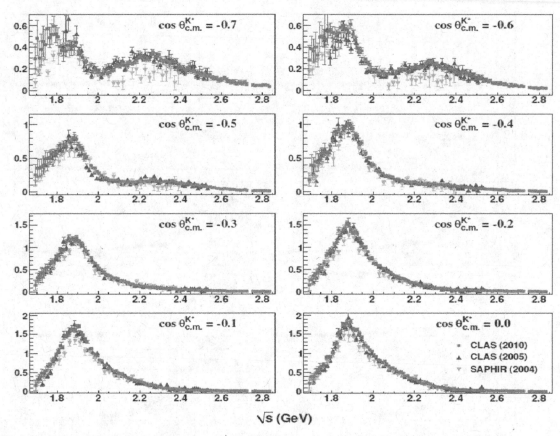

Fig. 2 Invariant mass dependence of the $\gamma p \to K^+ \Sigma^\circ$ differential cross section in the backward polar angle range

$K^* \Sigma$. The study of such final states adds significant complexity as more amplitudes can contribute to spin-1 mesons photoproduction compared to pseudo-scalar meson production. As is the case for $N\eta$ production, the $N\omega$ channel is selective to isospin $\frac{1}{2}$ nucleon states only. CLAS has collected a tremendous amount of data in the $p\omega$ [21,22] and $p\phi$ [32,33] final states on differential cross sections and spin-density matrix elements that are now entering into the more complex multi-channel analyses such as Bonn–Gatchina [34]. The CLAS collaboration performed a single channel event-based analysis, whose results are shown in Fig. 4, and provide further evidence for the $N(2000)\frac{5}{2}^+$.

Photoproduction of ϕ mesons is also considered a potentially rich source of new excited nucleon states in the mass range above 2 GeV. Some lower mass states such as $N(1535)\frac{1}{2}^-$ may have significant $s\bar{s}$ components [35]. Such components may result in states coupling to $p\phi$ with significant strength above threshold. Differential cross sections and spin-density matrix elements have been measured for $\gamma p \to p\phi$ in a mass range up to nearly 3 GeV. In Fig. 5 structures are seen near 2.2 GeV in the forward most angle bins and at very backward angles for both decay channels $\phi \to K^+K^-$ and $\phi \to K_l^0 K_s^0$, and with the exception of the smallest forward angle bin the structures are more prominent at backward angles. Only a multi-channel partial wave analysis will be able to pull out any significant resonance strength. Fig. 6 shows the differential cross section $d\sigma/dt$ of the most forward angle bin. A broad structure at 2.2 GeV is present, but does not show the typical Breit–Wigner behavior of a single resonance. It also does not fit the data in a larger angle range, which indicates that contributions other than genuine resonances may be significant. The forward and backward angle structures may also hint at the presence of dynamical effects possibly due to molecular contributions such as diquark-anti-triquark contributions [36], the strangeness equivalent to the recently observed hidden charm P_c^+ states.

Another process that has promise in the search for new excited baryon states, including those with isospin-$\frac{3}{2}$ is $\gamma p \to K^* \Sigma$ [34]. In distinction to the vector mesons discussed above, diffractive processes do not play a role in this channel, which then should allow better direct access to s-channel resonance production.

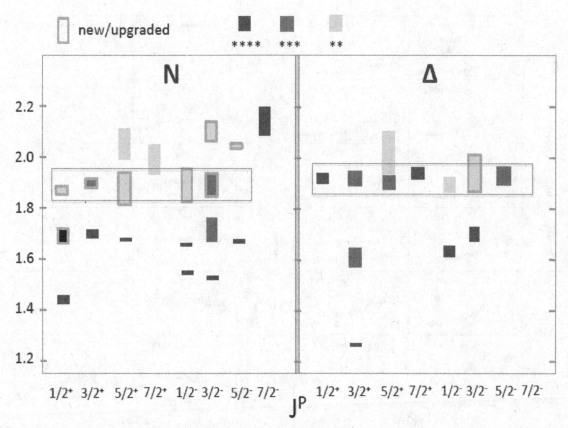

Fig. 3 Nucleon and Δ resonance spectrum below 2.2 GeV in RPP 2014 [6]. The new states and states with improved evidence observed in the recent Bonn–Gatchina multi-channel analysis are shown with the green frame. The *red frames* highlight the apparent mass degeneracy of five or six states with different spin and parity. The analysis includes all the $K^+\Lambda$ and $K^+\Sigma^\circ$ cross section and polarization data (color figure online)

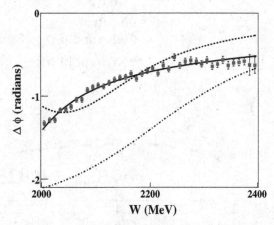

Fig. 4 Phase motion of the partial wave fit to the $\gamma p \rightarrow p\omega$ differential cross section and spin density matrix elements. Three resonant states, the subthreshold resonance $N(1680)\frac{5}{2}^+$, $N(2190)\frac{7}{2}^-$, and the missing $N(2000)\frac{5}{2}^+$ are needed to fit the data (*solid line*). Fits without $N(2000)\frac{5}{2}^+$ (*dashed-dotted line*), or without $N(1680)\frac{5}{2}^+$ (*dashed line*) cannot reproduce the data

3 Structure of Excited Nucleons

Meson photoproduction has become an essential tool in the search for new excited baryons. The exploration of the internal structure of excited states and the effective degrees of freedom contributing to s-channel resonance excitation requires use of electron beams where the virtuality (Q^2) of the exchanged photon can be varied to

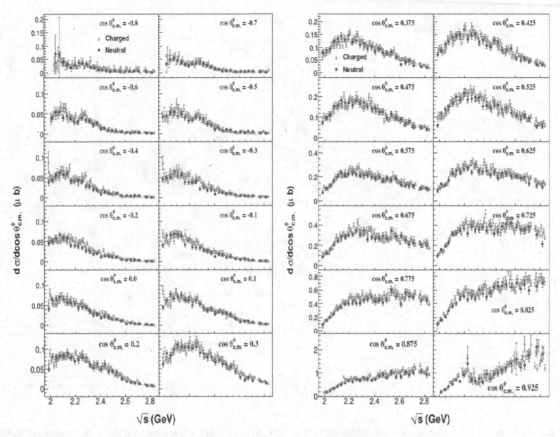

Fig. 5 Differential cross sections in a nearly full angular range for $\gamma p \rightarrow p\phi$ production

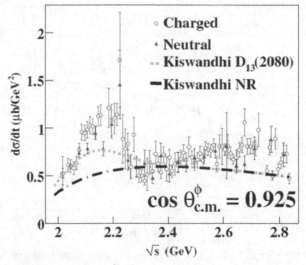

Fig. 6 Differential cross sections of $\gamma p \rightarrow p\phi$ production for the most forward angle bin. The two curves refer to fits without (*dashed*) and with (*dotted*) a known resonance at 2.08 GeV included

probe the spatial structure (Fig. 7). Electroproduction of final states with pseudoscalar mesons (e.g. $N\pi$, $p\eta$, $K\Lambda$) have been employed with CLAS, leading to new insights into the scale dependence of effective degrees of freedom, e.g. meson-baryon, constituent quark, and dressed quark contributions. Several excited states, shown in Fig. 7 assigned to their primary $SU(6) \otimes O(3)$ supermultiplets have been studied. The $N\Delta(1232)\frac{3}{2}^{+}$ transition is now well measured in a large range of Q^2 [37–39,41]. Two of the prominent higher mass states,

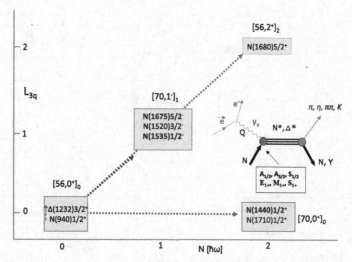

Fig. 7 Schematic of $SU(6) \otimes O(3)$ supermultiplets with excited states that have been explored in $ep \to e'\pi^+ n$, $ep \to e'p'\pi^\circ$ and $ep \to e'p'\pi^+\pi^-$. The *inset* shows the helicity amplitudes and electromagnetic multipoles extracted from the data. Only the ones highlighted in *red* are discussed here (color figure online)

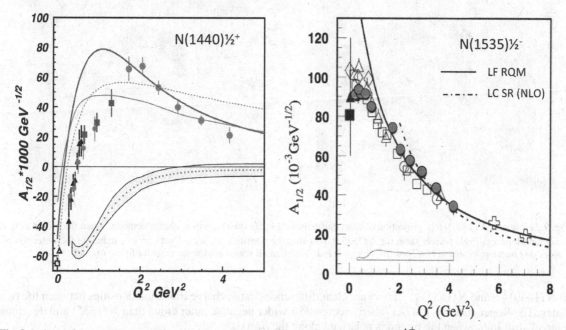

Fig. 8 *Left panel* the transverse helicity amplitudes $A_{1/2}$ for the Roper resonance $N(1440)\frac{1}{2}^+$. Data are from CLAS compared to two LF RQM with fixed quark masses (*dashed*) and with running quark mass (*solid red*), and with projections from the DSE/QCD approach. The magenta *dotted line* with error band indicates non 3-quark contributions obtained from a the difference of the DSE curve and the CLAS data. The *right panel* shows the same amplitude for the $N(1535)\frac{1}{2}^-$ compared to LF RQM calculations (*solid line*) and QCD computation within the LC Sum Rule approach (color figure online)

the Roper resonance $N(1440)\frac{1}{2}^+$ and $N(1535)\frac{1}{2}^-$ are shown in Fig. 8 as representative examples [40,41] from a wide program at JLab [42–47]. For these two states advanced relativistic quark model calculations [48] and QCD calculations from Dyson–Schwinger equation [49] and Light Cone sum rule [50] have recently become available, for the first time employing QCD-based modeling of the excitation of the quark core. There is agreement with the data at $Q^2 > 1.5\,\text{GeV}^2$. The calculations deviate significantly from the data at lower Q^2, which indicates significant non quark core effects. For the Roper resonance such contributions have been described successfully in dynamical meson-baryon models [51] and in effective field theory [52].

Knowledge of the helicity amplitudes in a large Q^2 allows for the determination of the transition charge densities on the light cone in transverse impact parameter space (b_x, b_y) [53]. Figure 9 shows the comparison

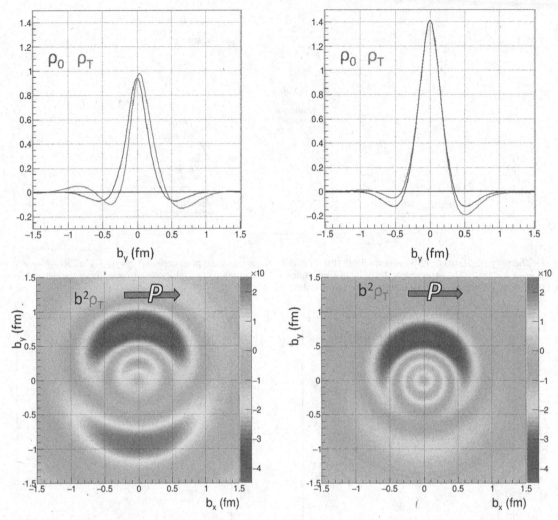

Fig. 9 *Left panels N(1440), top:* projection of charge densities on b_y, *bottom* transition charge densities when the proton is spin polarized along b_x. *Right panels* same for $N(1535)$. Note that the densities are scaled with b^2 to emphasize the outer wings. *Color code* negative charge is *blue*, positive charge is *red*. Note that all scales are the same (color figure online)

of $N(1440)\frac{1}{2}^+$ and $N(1535)\frac{1}{2}^-$. There are clear differences in the charge transition densities between the two states. The Roper state has a softer positive core and a wider negative outer cloud than $N(1535)$ and develops a larger shift in b_y when the proton is polarized along the b_x axis.

4 Conclusions and Outlook

Over the past five years eight baryon states in the mass range from 1.85 to 2.15 GeV have been either discovered or evidence for the existence of states has been significantly strengthened. To a large degree this is the result of adding very precise photoproduction data in open strangeness channels to the data base that is included in multi-channel partial wave analyses, especially the Bonn–Gatchina PWA. The possibility to measure polarization observables in these processes has been critical. In the mass range above 2 GeV more complex processes such as vector mesons or $\Delta\pi$ may have sensitivity to states with higher masses but require more complex analyses techniques to be brought to bear. Precision data in such channels have been available for a few years but remain to be fully incorporated in multi-channel partial wave analyses processes. The light-quark baryon spectrum is likely also populated with hybrid excitations [7] where the gluonic admixtures to the wave function are dominating the excitation. These states appear with the same quantum numbers as ordinary quark excitations, and can only be isolated from ordinary states due to the Q^2 dependence of their helicity amplitudes [54], which is expected to be quite different from ordinary quark excitations. This requires new electroproduction data especially at low Q^2 [55] with different final states and at masses above 2 GeV.

Despite the very significant progress made in recent years to further establish the light-quark baryon spectrum and explore the internal structure of excited states, much remains to be done. A vast amount of precision data already collected needs to be included in the multi-channel analysis frameworks, and polarization data are still to be analyzed. There are approved proposals to study resonance excitations at much higher Q^2 and with higher precision at Jefferson Lab with CLAS12 [56] that may reveal the transition to the bare quark core contributions at short distances.

Acknowledgments I like to thank Inna Aznauryan and Viktor Mokeev for numerous discussions on the subjects discussed in this presentation. This work was supported by the US Department of Energy under Contract No. DE-AC05-06OR23177.

References

1. Anderson, H.L., Fermi, E., Long, E.A., Nagle, D.E.: Total cross-sections of positive pions in hydrogen. Phys. Rev. **85**, 936 (1952)
2. Gell-Mann, M.: A schematic model of baryons and mesons. Phys. Lett. **8**, 214 (1964)
3. Zweig, G.: CERN Reports, TH 401 and 412 (1964)
4. Greenberg, O.W.: Spin and unitary spin independence in a paraquark model of baryons and mesons. Phys. Rev. Lett. **13**, 598 (1964). arXiv:0803.0992 [physics.hist-ph]
5. Burkert, V.D., Lee, T.S.H.: Electromagnetic meson production in the nucleon resonance region. Int. J. Mod. Phys. E **13**, 1035 (2004)
6. Olive, K.A., et al., Particle Data Group: Chin. Phys. C **38**, 090001 (2014)
7. Dudek, J.J., Edwards, R.G.: Phys. Rev. D **85**, 054016 (2012)
8. Suzuki, N., et al.: Disentangling the dynamical origin of P-11 nucleon resonances. Phys. Rev. Lett. **104**, 042302 (2010). arXiv:0909.1356 [nucl-th]
9. Klempt, E., Richard, J.M.: Baryon spectroscopy. Rev. Mod. Phys. **82**, 1095 (2010)
10. Tiator, L., Drechsel, D., Kamalov, S.S., Vanderhaeghen, M.: Electromagnetic excitation of nucleon resonances. Eur. Phys. J. Spec. Top. **198**, 141 (2011)
11. Aznauryan, I.G., Burkert, V.D.: Electroexcitation of nucleon resonances. Prog. Part. Nucl. Phys. **67**, 1 (2012)
12. Aznauryan, I.G., et al.: Studies of nucleon resonance structure in exclusive meson electroproduction. Int. J. Mod. Phys. E **22**, 1330015 (2013)
13. Crede, V., Roberts, W.: Progress towards understanding baryon resonances. Rep. Prog. Phys. **76**, 076301 (2013)
14. Bazavov, A., et al.: Additional strange hadrons from QCD thermodynamics and strangeness freezeout in heavy ion collisions. Phys. Rev. Lett. **113**(7), 072001 (2014). arXiv:1404.6511 [hep-lat]
15. Bazavov, A., et al.: The melting and abundance of open charm hadrons. Phys. Lett. B **737**, 210 (2014). arXiv:1404.4043 [hep-lat]
16. Capstick, S., Roberts, W.: Quasi two-body decays of nonstrange baryons. Phys. Rev. D **49**, 4570 (1994)
17. Dugger, M., et al.: Eta-prime photoproduction on the proton for photon energies from 1.527-GeV to 2.227-GeV. Phys. Rev. Lett. **96**, 062001 (2006)
18. Dugger, M., et al.: Eta-prime photoproduction on the proton for photon energies from 1.527-GeV to 2.227-GeV. Phys. Rev. Lett. **96**, 169905 (2006)
19. Dugger, M., et al., CLAS Collaboration: pi+ photoproduction on the proton for photon energies from 0.725 to 2.875-GeV. Phys. Rev. C **79**, 065206 (2009). doi:10.1103/PhysRevC.79.065206
20. Williams, M., et al., CLAS Collaboration: Differential cross sections for the reactions gamma p → p eta and gamma p → p eta-prime. Phys. Rev. C **80**, 045213 (2009)
21. Williams, M., et al., CLAS Collaboration: Partial wave analysis of the reaction $\gamma p \to$ p omega and the search for nucleon resonances. Phys. Rev. C **80**, 065209 (2009)
22. Williams, M., et al., CLAS Collaboration: Differential cross sections and spin density matrix elements for the reaction gamma p → p omega. Phys. Rev. C **80**, 065208 (2009)
23. Bradford, R.K., et al., CLAS Collaboration: First measurement of beam-recoil observables C(x) and C(z) in hyperon photoproduction. Phys. Rev. C **75**, 035205 (2007)
24. Bradford, R., et al., CLAS Collaboration: Differential cross sections for $\gamma + p \to$ K + +Y for Lambda and Sigma0 hyperons. Phys. Rev. C **73**, 035202 (2006)
25. McCracken, M.E., et al., CLAS Collaboration: Differential cross section and recoil polarization measurements for the gamma p to K+ Lambda reaction using CLAS at Jefferson Lab. Phys. Rev. C **81**, 025201 (2010)
26. Dey, B., et al., CLAS Collaboration: Differential cross sections and recoil polarizations for the reaction $\gamma p \to K^+ \Sigma^0$. Phys. Rev. C **82**, 025202 (2010)
27. McNabb, J.W.C., et al., CLAS Collaboration: Hyperon photoproduction in the nucleon resonance region. Phys. Rev. C **69**, 042201 (2004)
28. Anisovich, A., Beck, R., Klempt, E., Nikonov, V., Sarantsev, A., Thoma, U.: Properties of baryon resonances from a multi-channel partial wave analysis. Eur. Phys. J. A **48**, 15 (2012)
29. Julia-Diaz, B., Lee, T.-S.H., Matsuyama, A., Sato, T.: Dynamical coupled-channel model of pi N scattering in the W 2-GeV nucleon resonance region. Phys. Rev. C **76**, 065201 (2007)
30. Rönchen, D., et al.: Photocouplings at the pole from pion photoproduction. Eur. Phys. J. A **50**(6), 101 (2014)

31. Edwards, R.G., Dudek, J.J., Richards, D.G., Wallace, S.J.: Excited state baryon spectroscopy from lattice QCD. Phys. Rev. D **84**, 074508 (2011)
32. Seraydaryan, H., et al., CLAS Collaboration: ϕ-meson photoproduction on Hydrogen in the neutral decay mode. Phys. Rev. C **89**(5), 055206 (2014)
33. Dey, B., et al., CLAS Collaboration: Data analysis techniques, differential cross sections, and spin density matrix elements for the reaction $\gamma p \to \phi p$. Phys. Rev. C **89**(5), 055208 (2014)
34. Sarantsev, A.: The BoGa amplitude analysis methods and its extensions to higher W and to electroproduction. Talk Presented at the Workshop Nucleon Resonances: From Photoproduction to High Photon Virtualities. ECT, Trento (2015)
35. Liu, B.C., Zou, B.S.: Mass and K Lambda coupling of N*(1535). Phys. Rev. Lett. **96**, 042002 (2006)
36. Lebed, R .F.: Do the P_c^+ pentaquarks have strange siblings? Phys. Rev. D **92**(11), 114030 (2015)
37. Joo, K., et al., CLAS Collaboration: Q**2 dependence of quadrupole strength in the $\gamma^* p \to \Delta + (1232) \to p\pi^0$ transition. Phys. Rev. Lett. **88**, 122001 (2002)
38. Ungaro, M., et al., CLAS Collaboration: Measurement of the N $\to \Delta + (1232)$ transition at high momentum transfer by pi0 electroproduction. Phys. Rev. Lett. **97**, 112003 (2006)
39. Frolov, V.V., et al.: Electroproduction of the $\Delta(1232)$ resonance at high momentum transfer. Phys. Rev. Lett. **82**, 45 (1999)
40. Aznauryan, I.G., et al., CLAS Collaboration: Electroexcitation of the Roper resonance for $1.7 < Q^{**}2 < 4.5$-GeV2 in vec-ep $\to$ en pi+. Phys. Rev. C **78**, 045209 (2008)
41. Aznauryan, I.G., et al., CLAS Collaboration: Electroexcitation of nucleon resonances from CLAS data on single pion electroproduction. Phys. Rev. C **80**, 055203 (2009)
42. Mokeev, V.I., et al., CLAS Collaboration: Experimental Study of the $P_{11}(1440)$ and $D_{13}(1520)$ resonances from CLAS data on $ep \to e'\pi^+\pi^- p'$. Phys. Rev. C **86**, 035203 (2012)
43. Denizli, H., et al., CLAS Collaboration: Q*2 dependence of the S(11)(1535) photocoupling and evidence for a P-wave resonance in eta electroproduction. Phys. Rev. C **76**, 015204 (2007)
44. Armstrong, C.S., et al., Jefferson Lab E94014 Collaboration: Electroproduction of the S(11)(1535) resonance at high momentum transfer. Phys. Rev. D **60**, 052004 (1999)
45. Egiyan, H., et al., CLAS Collaboration: Single pi+ electroproduction on the proton in the first and second resonance regions at 0.25-GeV**2 $< Q^{**}2 > 0.65$-GeV**2 using CLAS. Phys. Rev. C **73**, 025204 (2006)
46. Park, K., et al., CLAS Collaboration: Cross sections and beam asymmetries for vec(e) p $\to$ en pi+ in the nucleon resonance region for $1.7 \Leftarrow Q^{**}2 \Leftarrow 4.5$-(GeV)**2. Phys. Rev. C **77**, 015208 (2008)
47. Park, K., et al., CLAS Collaboration: Measurements of $ep \to e'\pi^+ n$ at W = $1.6 - 2.0$ GeV and extraction of nucleon resonance electrocouplings at CLAS. Phys. Rev. C **91**, 045203 (2015)
48. Aznauryan, I.G., Burkert, V.D.: Electroexcitation of the $\Delta(1232)\frac{3}{2}^+$ and $\Delta(1600)\frac{3}{2}^+$ in a light-front relativistic quark model. Phys. Rev. C **92**(3), 035211 (2015)
49. Segovia, J., et al.: Completing the picture of the Roper resonance. Phys. Rev. Lett. **115**(17), 171801 (2015)
50. Anikin, I.V., Braun, V.M., Offen, N.: Electroproduction of the $N^*(1535)$ nucleon resonance in QCD. Phys. Rev. D **92**(1), 014018 (2015)
51. Obukhovsky, I.T., et al.: Electroproduction of the Roper resonance on the proton: the role of the three-quark core and the molecular $N\sigma$ component. Phys. Rev. D **84**, 014004 (2011)
52. Bauer, T., Scherer, S., Tiator, L.: Electromagnetic transition form factors of the Roper resonance in effective field theory. Phys. Rev. C **90**(1), 015201 (2014)
53. Tiator, L., Vanderhaeghen, M.: Empirical transverse charge densities in the nucleon-to-P(11)(1440) transition. Phys. Lett. B **672**, 344 (2009)
54. Li, Z.P., Burkert, V., Li, Z.J.: Electroproduction of the Roper resonance as a hybrid state. Phys. Rev. D **46**, 70 (1992)
55. Lanza, L.: Search for hybrid baryons. Talk Presented at the Workshop Nucleon Resonances: From Photoproduction to High Photon Virtualities. ECT, Trento (2015)
56. Gothe, R., Carman, D., Mokeev, V., Park, K.: Talk Presented at the Workshop Nucleon Resonances: From Photoproduction to High Photon Virtualities. ECT, Trento (2015)

Few-Body Syst (2016) 57:1067–1076
DOI 10.1007/s00601-016-1150-9

Craig D. Roberts · Jorge Segovia

Baryons and the Borromeo

Received: 8 March 2016 / Accepted: 6 August 2016 / Published online: 23 August 2016
© Springer-Verlag Wien (Outside the USA) 2016

Abstract The kernels in the tangible matter of our everyday experience are composed of light quarks. At least, they are light classically; but they don't remain light. Dynamical effects within the Standard Model of Particle Physics change them in remarkable ways, so that in some configurations they appear nearly massless, but in others possess masses on the scale of light nuclei. Modern experiment and theory are exposing the mechanisms responsible for these remarkable transformations. The rewards are great if we can combine the emerging sketches into an accurate picture of confinement, which is such a singular feature of the Standard Model; and looming larger amongst the emerging ideas is a perspective that leads to a Borromean picture of the proton and its excited states.

1 Introduction

Quantum chromodynamics (QCD) is a local, relativistic, non-Abelian, quantum gauge-field theory, which possesses the property of asymptotic freedom, *i.e.* QCD interactions are weaker than Coulombic at short distances. This behaviour is evident in the one-loop expression for the running coupling, $\alpha_s(Q^2)$, and verified in a host of experiments. Hence, as a necessary consequence of asymptotic freedom, $\alpha_s(Q^2)$ must increase as $Q^2/\Lambda_{QCD}^2 \to 1^+$, where $\Lambda_{QCD} \sim 200$ MeV is the natural mass-scale of QCD, whose dynamical generation through quantisation spoils the conformal invariance of the classical massless theory [1–3]. In fact, at $Q^2 \approx 4\,\text{GeV}^2 =: \zeta_2^2$, which corresponds to a length-scale on the order of 10 % of the proton's radius, it is empirically known that $\alpha_s(\zeta_2^2) \gtrsim 0.3$. These observations describe a peculiar circumstance, *viz.* an interaction that becomes stronger as the participants try to separate. It leads one to explore some curious possibilities: If the coupling grows so strongly with separation, then perhaps it is unbounded; and perhaps it would require an infinite amount of energy in order to extract a quark or gluon from the interior of a hadron? Such thinking has led to the

> *Confinement Hypothesis*: Colour-charged particles cannot be isolated and therefore cannot be directly observed. They clump together in colour-neutral bound-states.

Confinement seems to be an empirical fact; but a mathematical proof is lacking. Partly as a consequence, the Clay Mathematics Institute offered a "Millennium Problem" prize of $1-million for a proof that $SU_c(3)$

This article belongs to the special issue "Nucleon Resonances".

C. D. Roberts (✉)
Physics Division, Argonne National Laboratory, Argonne, IL 60439, USA
E-mail: cdroberts@anl.gov

J. Segovia
Physik-Department, TU-München, D-85748 Garching, Germany
E-mail: jorge.segovia@tum.de

gauge theory is mathematically well-defined [4], one consequence of which will be an answer to the question of whether or not the confinement conjecture is correct in pure-gauge QCD.

There is a problem with that, however: no reader of this article can be described within pure-gauge QCD. The presence of quarks is essential to understanding all known visible matter, so a proof of confinement which deals only with pure-gauge QCD is chiefly irrelevant to our Universe. We exist because Nature has supplied two light quarks; and those quarks combine to form the pion, which is unnaturally light ($m_\pi < \Lambda_{QCD}$) and hence very easily produced.

One may bring this arcanum into sharper focus by noting that one aspect of the Yang–Mills millennium problem [4] is to prove that pure-gauge QCD possesses a mass-gap $\Delta > 0$. There is strong evidence supporting this conjecture, found especially in the fact that numerical simulations of lattice-regularised QCD (lQCD) predict $\Delta \gtrsim 1.5\,\text{GeV}$ [5]. However, with $\Delta^2/m_\pi^2 \gtrsim 100$, can the mass-gap in pure Yang–Mills really play any role in understanding confinement when dynamical chiral symmetry breaking (DCSB), very likely driven by the same dynamics, ensures the existence of an almost-massless strongly-interacting excitation in our Universe? If the answer is not *no*, then it should at least be that one cannot claim to provide a pertinent understanding of confinement without simultaneously explaining its connection with DCSB. The pion must play a critical role in any explanation of confinement in the Standard Model; and any discussion that omits reference to the pion's role is *practically irrelevant*.

This perspective is canvassed elsewhere [6] and can be used to argue that the potential between infinitely-heavy quarks measured in numerical simulations of quenched lQCD – the so-called static potential [7] – is disconnected from the question of confinement in our Universe. This is because light-particle creation and annihilation effects are essentially nonperturbative in QCD, so it is impossible in principle to compute a quantum mechanical potential between two light quarks [8–10]. Consequently, there is no measurable flux tube in a Universe with light quarks and hence the classical flux tube cannot be the correct paradigm for confinement.

As highlighted already, DCSB is the key here. It ensures the existence of pseudo-Nambu-Goldstone modes; and in the presence of these modes, no flux tube between a static colour source and sink can have a measurable existence. To verify this, consider such a tube being stretched between a source and sink. The potential energy accumulated within the tube may increase only until it reaches that required to produce a particle-antiparticle pair of the theory's pseudo-Nambu-Goldstone modes. Simulations of lQCD show [8,9] that the flux tube then disappears instantaneously along its entire length, leaving two isolated colour-singlet systems. The length-scale associated with this effect in QCD is $r_{c\bar{c}} \simeq (1/3)$ fm and hence if any such string forms, it would dissolve well within a hadron's interior.

An alternative realisation associates confinement with dramatic, dynamically-driven changes in the analytic structure of QCD's propagators and vertices. That leads coloured n-point functions to violate the axiom of reflection positivity and hence forces elimination of the associated excitations from the Hilbert space associated with asymptotic states [11]. This is certainly a sufficient condition for confinement [12–15]. It should be noted, however, that the appearance of such alterations when analysing some truncation of a given theory does not mean that the theory itself is truly confining: unusual spectral properties can be introduced by approximations, leading to a truncated version of a theory which is confining even though the complete theory is not, *e.g.* Refs. [16, 17]. Notwithstanding such exceptions, a computed violation of reflection positivity by coloured functions in a veracious treatment of QCD does express confinement. Moreover, via this mechanism, it is achieved as the result of an essentially dynamical process. It is known that both quarks and gluons acquire a running mass in QCD [18–26]; and the generation of these masses leads to the emergence of a length-scale $\varsigma \approx 0.5$ fm, whose existence and magnitude is evident in all existing studies of dressed-gluon and -quark propagators, and which characterises the striking change in their analytic structure that has just been described. In models based on such features [27], once a gluon or quark is produced, it begins to propagate in spacetime; but after each "step" of length ς, on average, an interaction occurs so that the parton loses its identity, sharing it with others. Finally a cloud of partons is produced, which coalesces into colour-singlet final states. This picture of parton propagation, hadronisation and confinement can be tested in experiments at modern and planned facilities [28–30].

2 Enigma of Mass

DCSB is a crucial emergent phenomenon in QCD. It is expressed in hadron wave functions, not in vacuum condensates [6, 31–34]; and contemporary theory argues that DCSB is responsible for more than 98 % of the

Springer

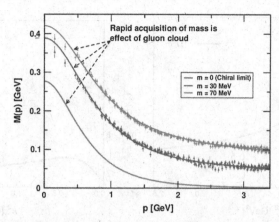

Fig. 1 Mass function, $M(p)$, a characterising feature of the dressed-quark propagator in Eq. (1). *Solid curves* DSE results, explained in Refs. [18,20], "data" numerical simulations of lattice-regularised QCD (lQCD) [19]. (NB. $m = 70\,\text{MeV}$ is the uppermost curve and current-quark mass decreases from top to bottom.) The current-quark of perturbative QCD evolves into a constituent-quark as its momentum becomes smaller. The constituent-quark mass arises from a cloud of low-momentum gluons attaching themselves to the current-quark. This is DCSB: an essentially nonperturbative effect that generates a quark *mass from nothing*; namely, it occurs even in the chiral limit

visible mass in the Universe. Given that classical massless-QCD is a conformally invariant theory, this means that DCSB is the origin of *mass from nothing*. This effect is evident in the dressed-quark propagator:

$$S(p) = 1/[i\gamma \cdot pA(p^2) + B(p^2)] = Z(p^2)/[i\gamma \cdot p + M(p^2)], \tag{1}$$

where $M(p^2)$ is the dressed-quark mass-function, the behaviour of which is depicted and explained in Fig. 1. It is important to insist on the term "dynamical," as distinct from spontaneous, because nothing is added to QCD in order to effect this remarkable outcome and there is no simple change of variables in the QCD action that will make it apparent. Instead, through the act of quantising the classical chromodynamics of massless gluons and quarks, a large mass-scale is generated.

DCSB is very clearly revealed in properties of the pion, whose structure is described by a Bethe-Salpeter amplitude:

$$\Gamma_\pi(k; P) = \gamma_5 \left[iE_\pi(k; P) + \gamma \cdot PF_\pi(k; P) + \gamma \cdot k\, G_\pi(k; P) + \sigma_{\mu\nu}k_\mu P_\nu H_\pi(k; P) \right], \tag{2}$$

where k is the relative momentum between the valence-quark and -antiquark constituents (defined here such that the scalar functions in Eq. (2) are even under $k \cdot P \to -k \cdot P$) and P is their total momentum. $\Gamma_\pi(k; P)$ is simply related to an object that would be the pion's Schrödinger wave function if a nonrelativistic limit were appropriate. In QCD if, and only if, chiral symmetry is dynamically broken, then one has in the chiral limit [35,36]:

$$f_\pi E_\pi(k; 0) = B(k^2). \tag{3}$$

This identity is miraculous.[1] It is true in any covariant gauge, independent of the renormalisation scheme; and it means that the two-body problem is solved, nearly completely, once the solution to the one body problem is known. (Details are provided in Refs. [35,36].) Eq. (3) is a quark-level Goldberger-Treiman relation. It is also the most basic expression of Goldstone's theorem in QCD, *viz.*

> *Goldstone's theorem is fundamentally an expression of equivalence between the one-body problem and the two-body problem in QCD's colour-singlet pseudoscalar channel.*

Consequently, pion properties are an almost direct measure of the dressed-quark mass function depicted in Fig. 1; and the reason a pion is massless in the chiral limit is simultaneously the explanation for a proton mass of around 1 GeV. Thus, enigmatically, properties of the nearly-massless pion are the cleanest expression of the mechanism that is responsible for almost all the visible mass in the Universe.

---

[1] Equation. (3) has many corollaries, *e.g.* it ensures that chiral-QCD generates a massless pion in the absence of a Higgs mechanism; predicts $m_\pi^2 \propto m$ on $m \simeq 0$, where m is the current-quark mass; and entails that the chiral-limit leptonic decay constant vanishes for all excited-state 0^- mesons with nonzero isospin [37,38].

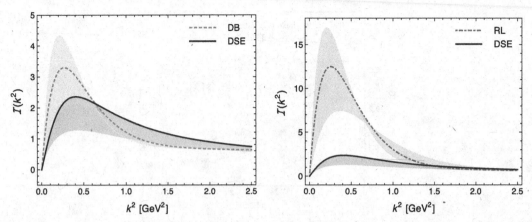

Fig. 2 Comparison between top–down results for the gauge-sector interaction, $I(k^2)$, and those obtained using the bottom–up approach based on hadron observables. *Left panel solid curve* within *grey band*, top–down result for the RGI interaction; and *dashed curve* within *pale-green band*, advanced bottom–up result obtained in the DCSB-improved (DB) truncation, detailed in Refs. [39–41]. *Right panel – solid curve* within *grey band*, top–down result for the RGI interaction, as in the *left panel*; and *dot-dashed curve* within *pale-red band*, bottom–up result obtained in the rainbow-ladder (RL) truncation, which is leading-order in the most widely-used DSE truncation scheme [42]. In all cases, the bands denote the existing level of theoretical uncertainty in extraction of the interaction. All curves are identical on the perturbative domain: $k^2 > 2.5\,\text{GeV}^2$. (Figures provided by D. Binosi, modelled after those in Ref. [43].)

3 Continuum-QCD and Ab Initio Predictions of Hadron Observables

Confidence in the insights drawn from continuum analyses of nonperturbative QCD has recently received a major boost owing to a unification of two common methods for determining the momentum-dependence of the interaction between quarks [26]: namely, the top–down approach, which works toward an *ab initio* computation of the interaction via direct analysis of the gauge-sector gap equations; and the bottom–up scheme, which aims to infer the interaction by fitting data within a well-defined truncation of those equations in the matter sector that are relevant to bound-state properties. The unification is illustrated in Fig. 2: the left panel presents a comparison between the top–down RGI interaction (solid-black curve) and the sophisticated DB-truncation bottom–up interaction (green band containing dashed curve). Plainly, the interaction predicted by modern analyses of QCD's gauge sector is in near precise agreement with that required for a veracious description of measurable hadron properties using the most sophisticated matter-sector gap and Bethe-Salpeter kernels available today. This is remarkable, given that there had previously been no serious attempt at communication between practitioners from the top–down and bottom–up hemispheres of continuum-QCD. It bridges a gap that had lain between nonperturbative continuum-QCD and the *ab initio* prediction of bound-state properties.

A comparison between the top–down prediction and that inferred using the simple DSE-RL kernel (red band containing dot-dashed curve in the right panel) is also important. One observes that the DSE-RL result has the correct shape but is too large in the infrared. This is readily explained [26]; and it follows that whilst the RL truncation supplies a useful computational link between QCD's gauge sector and measurable hadron properties, the model interaction it delivers is *not a pointwise-accurate* representation of ghost-gluon dynamics. Notwithstanding this, it remains true that the judicious use of RL truncation can yield reliable predictions for a known range of hadron observables, with an error that may be estimated and whose origin is understood.

4 Structure of Baryons

This workshop focused on the electroproduction of nucleon resonances; and highlighted just how crucial it has become to address the three valence-quark bound-state problem in QCD with the same level of sophistication that is now available for mesons [44,45]. A principal modern goal must be to correlate the properties of meson and baryon ground- and excited-states within a single, *symmetry-preserving* framework. Here, symmetry-preserving means that the analysis respects Poincaré covariance and satisfies the relevant Ward-Green-Takahashi identities. Constituent-quark models have hitherto been the most widely applied spectroscopic tools; and despite their imperfections, they are of continuing value because there is nothing yet that is providing a bigger picture. Nevertheless, they possess no connection with quantum field theory; and they are not symmetry-preserving and hence cannot veraciously connect meson and baryon properties.

A comprehensive approach to QCD will provide a unified explanation of both mesons and baryons. We have emphasised that DCSB is a keystone of the Standard Model, evident in the momentum-dependence of the dressed-quark mass function – Fig. 1; and it is just as important to baryons as it is to mesons. Crucially, the DSEs furnish the only extant framework that can simultaneously and transparently connect meson and baryon observables with this basic feature of QCD, having provided, *e.g.* a direct correlation of meson and baryon properties via a single interaction kernel, which preserves QCD's one-loop renormalisation group behaviour and can systematically be improved. This is evident in Refs. [46–52] and in many contributions to these proceedings, *e.g.* Refs. [53–55].

In order to illustrate the insights that have been enabled by DSE analyses, consider the proton, which is a composite object whose properties and interactions are determined by its valence-quark content: $u + u + d$, *i.e.* two up (u) quarks and one down (d) quark. So far as is now known, bound-states seeded by two valence-quarks do not exist; and the only two-body composites are those associated with a valence-quark and -antiquark, *i.e.* mesons. These features are supposed to derive from colour confinement, whose complexities are discussed in Sect. 1.

Such observations have led to a position from which the proton may be viewed as a Borromean bound-state [53,56], *viz.* a system constituted from three bodies, no two of which can combine to produce an independent, asymptotic two-body bound-state. In QCD the complete picture of the proton is more complicated, owing, in large part, to the loss of particle number conservation in quantum field theory and the concomitant frame- and scale-dependence of any Fock space expansion of the proton's wave function. Notwithstanding that, the Borromean analogy provides an instructive perspective from which to consider both quantum mechanical models and continuum treatments of the nucleon bound-state problem in QCD. It poses a crucial question: *Whence binding between the valence quarks in the proton,* i.e. *what holds the proton together*?

In numerical simulations of lQCD that use static sources to represent the proton's valence-quarks, a "Y-junction" flux-tube picture of nucleon structure is produced [57,58]. This might be viewed as originating in the three-gluon vertex, which signals the non-Abelian character of QCD and is the source of asymptotic freedom. Such results and notions would suggest a key role for the three-gluon vertex in nucleon structure *if* they were equally valid in real-world QCD wherein light dynamical quarks are ubiquitous. However, as explained in Sect. 1, they are not; and so a different explanation of binding within the nucleon must be found.

DCSB has numerous corollaries that are crucial in determining the observable features of the Standard Model; but one particularly important consequence is often overlooked. Namely, any interaction capable of creating pseudo-Nambu-Goldstone modes as bound-states of a light dressed-quark and -antiquark, and reproducing the measured value of their leptonic decay constants, will necessarily also generate strong colour-antitriplet correlations between any two dressed quarks contained within a baryon. This assertion is based upon evidence gathered in 20 years of studying two- and three-body bound-state problems in hadron physics. No counter examples are known; and the existence of such diquark correlations is also supported by lQCD [59,60].

The properties of diquark correlations have been charted. Most importantly, diquarks are confined. Additionally, owing to properties of charge-conjugation, a diquark with spin-parity J^P may be viewed as a partner to the analogous J^{-P} meson [61]. It follows that scalar, isospin-zero and pseudovector, isospin-one diquark correlations are the strongest in ground-state J^+-baryons; and whilst no pole-mass exists, the following mass-scales, which express the strength and range of the correlation and are each bounded below by the partnered meson's mass, may be associated with these diquarks [59–62]: $m_{[ud]_{0^+}} \approx 0.7 - 0.8\,\text{GeV}$, $m_{\{uu\}_{1^+}} \approx 0.9 - 1.1\,\text{GeV}$, with $m_{\{dd\}_{1^+}} = m_{\{ud\}_{1^+}} = m_{\{uu\}_{1^+}}$ in the isospin symmetric limit. Realistic diquark correlations are also soft. They possess an electromagnetic size that is bounded below by that of the analogous mesonic system, *viz.* [63,64]: $r_{[ud]_{0^+}} \gtrsim r_\pi$, $r_{\{uu\}_{1^+}} \gtrsim r_\rho$, with $r_{\{uu\}_{1^+}} > r_{[ud]_{0^+}}$. As with mesons, these scales are set by that associated with DCSB.

The RGI interactions depicted in Fig. 2 characterise a realistic class that generates strong attraction between two quarks and thereby produces tight diquark correlations in analyses of the three valence-quark scattering problem. The existence of such correlations considerably simplifies analyses of baryon bound states because it reduces that task to solving a Poincaré covariant Faddeev equation [65], depicted in Fig. 1, left panel, of Ref. [53]. The three gluon vertex is not explicitly part of the bound-state kernel in this picture of the nucleon. Instead, one capitalises on the fact that phase-space factors materially enhance two-body interactions over $n \geq 3$-body interactions and exploits the dominant role played by diquark correlations in the two-body subsystems. Then, whilst an explicit three-body term might affect fine details of baryon structure, the dominant effect of non-Abelian multi-gluon vertices is expressed in the formation of diquark correlations. Such a nucleon is then a compound system whose properties and interactions are primarily determined by its quark+diquark structure.

A nucleon (and any kindred baryon) with these features is a Borromean bound-state, the binding within which has two contributions. One part is expressed in the formation of tight diquark correlations; but that is augmented by attraction generated by quark exchange (depicted in the shaded area of Fig. 1, left panel, in Ref. [53]). This exchange ensures that diquark correlations within the nucleon are fully dynamical: no quark holds a special place because each one participates in all diquarks to the fullest extent allowed by its quantum numbers. The continual rearrangement of the quarks guarantees, *inter alia*, that the nucleon's dressed-quark wave function complies with Pauli statistics.

One cannot overstate the importance of appreciating that these fully dynamical diquark correlations are vastly different from the static, pointlike "diquarks" which featured in early attempts [66] to understand the baryon spectrum and to explain the so-called missing resonance problem [67–69]. Modern diquarks are soft and enforce certain distinct interaction patterns for the singly- and doubly-represented valence-quarks within the proton, as reviewed in Refs. [53,70]. On the other hand, the number of states in the spectrum of baryons obtained from the Faddeev equation [71] is similar to that found in the three constituent-quark model, just as it is in today's lQCD spectrum calculations [72].

5 Roper Resonance

There was much discussion of the Roper resonance at this workshop. That is unsurprising, given that the Roper has long resisted understanding. Recently, however, JLab experiments [73–78] have yielded precise nucleon-Roper ($N \to R$) transition form factors and thereby exposed the first zero seen in any hadron form factor or transition amplitude. Additionally, Ref. [52] has provided the first continuum treatment of this problem using the power of relativistic quantum field theory. A summary of that study is presented in these proceedings [53,55]; and in this contribution, therefore, just a few points will be highlighted.

The analysis in Ref. [52] is distinguished by using dressed-quark propagators that express a running quark mass which connects textbook knowledge about QCD's ultraviolet behaviour with continuum- and lattice-QCD predictions for the mass function's infrared behaviour. In addition, it capitalises on the existence of diquark correlations generated dynamically by the same mechanism that produces the dressed-quark mass function, *viz.* DCSB. These diquarks are soft, as predicted by QCD, and are active participants in all scattering processes. They are also intrinsically active, *i.e.* in a three-quark system, all diquarks participate in all correlations to the fullest extent allowed by their quantum numbers. Finally, all these quantum field theoretical phenomena are combined via a Poincaré-covariant Faddeev equation and electromagnetic current in order to produce the nucleon and Roper masses and the $N \to R$ transition form factors, with no parameters varied, so that the results are true predictions upon which the framework must stand or fall.

A particular feature of Ref. [52] is that the computation yields only the contribution to the form factors from a rigorously defined dressed-quark core. The mismatch between that result and data is then defined therein to be the effect of meson-baryon final-state interactions (MB-FSIs). One may improve upon the estimate of MB-FSIs by recognising that the dressed-quark core component of the baryon Faddeev amplitudes should be renormalised by inclusion of meson-baryon "Fock-space" components, with a maximum strength of 20 % [79–81]. Naturally, since wave functions in quantum field theory evolve with resolving scale, the magnitude of this effect is not fixed. Instead $I_{MB} = I_{MB}(Q^2)$, where Q^2 measures the resolving scale of any probe and $I_{MB}(Q^2) \to 0^+$ monotonically with increasing Q^2. Now, form factors in QCD possess power-law behaviour, so it is reasonable to renormalise the dressed-quark core contributions via

$$F_{\text{core}}(Q^2) \to [1 - I_{MB}(Q^2)]F_{\text{core}}(Q^2), \quad I_{MB}(Q^2) = [1 - 0.8^2]/[1 + Q^2/\Lambda_{MB}^2], \quad (4)$$

with $\Lambda_{MB} = 1\,\text{GeV}$ marking the midpoint of the transition between the nonperturbative and perturbative domains of QCD as measured by the behaviour of the dressed-quark mass-function in Fig. 1.

Following the procedure just described, one obtains the results depicted in Fig. 3, which highlight a number of important facts. (*i*) Incorporating a meson-baryon Fock-space component in the baryon Faddeev amplitudes does not materially affect the nature of the inferred meson-cloud contribution. (*ii*) Regarding $A_{1/2}$, the contribution of MB-FSIs inferred herein and that determined by EBAC are quantitatively in agreement on $x > 1.5$. However, our result disputes the EBAC suggestion that MB-FSIs are solely responsible for the $x = 0$ value of the helicity amplitude: the quark-core contributes at least two-thirds of the result. (*iii*) Regarding $S_{1/2}$, our

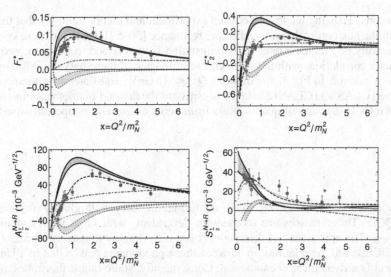

Fig. 3 $N \to R$ transition form factors and helicity amplitudes obtained therefrom. *Legend. Grey band* within *black curves* – dressed-quark core contribution with up-to 20 % Faddeev amplitude renormalisation from MB-FSIs, implemented according to Eq. (4). The transition form factor curve with smallest magnitude at $x = 1$ has the maximum renormalisation. *Green band* within *green dotted* curves – inferred MB-FSI contribution. The band demarcates the range of uncertainty arising from 0 to 20 % renormalisation of the dressed-quark core. *Blue dashed* curve – least-squares fit to the data on $x \in (0, 5)$. *Red dot-dashed* curve – contact interaction result [82]. *Pink dot-dashed* curve, *lower panels* – meson-cloud contribution to the helicity amplitude as determined by the excited baryon analysis center (EBAC). Data: circles (blue) [74]; triangle (gold) [73]; squares (purple) [76,77]; and star (green) [83]

inference for the MB-FSIs raises serious questions about the EBAC result. Indeed, the EBAC curve cannot realistically be connected with this helicity amplitude because there is necessarily a large quark-core contribution on $x < 1$. Moreover, the core and MB contributions are commensurate on $1 < x < 4$, and the core is dominant on $x > 4$. In our view, therefore, the green bands in these panels represent the best inference available today for the strength of MB-FSIs on the $N \to R$ transition form factors and helicity amplitudes.

One may summarise the main results of Ref. [52], complemented by those presented herein, as follows. A range of properties of the dressed-quark core of the proton's first radial excitation were computed. They provide an excellent understanding and description of data on the $N \to R$ transition and related quantities derived using dynamical coupled channels models. The analysis is based on a sophisticated continuum framework for the three-quark bound-state problem; all elements employed possess an unambiguous link with analogous quantities in QCD; and no parameters were varied in order to achieve success. Moreover, no material improvement in the results can be envisaged before either the novel spectral function methods introduced in Ref. [84] are extended and applied to the entire complex of nucleon, Δ-baryon and Roper-resonance properties that are unified by Refs. [50–52] or numerical simulations of lQCD become capable of reaching the same breadth of application and accuracy. On the strength of these results and remarks one may confidently conclude that the observed Roper resonance is at heart the nucleon's first radial excitation and consists of a well-defined dressed-quark core augmented by a meson cloud that reduces its (Breit-Wigner) mass by approximately 20 %. Concerning the transition form factors, a meson-cloud obscures the dressed-quark core from long-wavelength probes; but that core is revealed to probes with $Q^2 \gtrsim 3m_N^2$. This feature is typical of nucleon-resonance transitions; and hence measurements of resonance electroproduction on this domain can serve as an incisive probe of quark-gluon dynamics within the Standard Model, assisting greatly in mapping the evolution between QCD's nonperturbative and perturbative domains.

In connection with the last point and given the theme of this workshop, it is worth highlighting that, following completion of the JLab 12 GeV upgrade, CLAS12 will be capable of determining the electrocouplings, $g_{vNN^*}(Q^2)$, of most prominent N^* states at unprecedented photon virtualities: $Q^2 \in [6, 12]$ GeV^2 [85,86]. On this domain, Fig. 3 suggests that these electrocouplings are primarily determined by the dressed-quark cores within baryons. Consequently, the experimental programme employing CLAS12 will be unique in providing access to the dressed-quark cores of a diverse array of baryons. It will therefore deliver empirical information that is necessary in order to address a wide range of critical issues, *e.g.*: is there an environment sensitivity of DCSB and the dressed-quark mass function; and are quark-quark correlations an essential element in the structure of all baryons? Existing feedback between experiment and theory indicates that there is no environment sensitivity for the nucleon, Δ-baryon and Roper resonance [50–52]: DCSB in these systems is expressed in ways that can readily be predicted once its manifestation is understood in the pion, and this includes the generation of diquark correlations with the same character in each of these baryons. Moreover, regarding the dressed-quark mass-function in Fig. 1, the domain $Q^2 \leq 12\,\text{GeV}^2$ translates into momenta $k \lesssim 1.2\,\text{GeV}$. Therefore, combined CLAS and CLAS12 data are sensitive to the dressed-quark mass function throughout the domain upon which QCD dressing of quarks shifts from being essentially nonperturbative to perturbative in character.

6 Conclusion

It is worth reiterating a few points. Owing to the conformal anomaly, both gluons and quarks acquire mass dynamically in QCD. Those masses are momentum dependent, with large values at infrared momenta: $m(k^2 \simeq 0) > \Lambda_{\text{QCD}}$. The appearance of these nonperturbative running masses is intimately connected with confinement and DCSB; and the relationship between those phenomena entails that in a Universe with light-quarks, confinement is a dynamical phenomenon. Consequently, static-quark flux tubes are not the correct paradigm for confinement and it is practically meaningless to speak of linear potentials and Regge trajectories in connection with observable properties of light-quark hadrons. In exploring the connection between QCD's gauge and matter sectors, top–down and bottom–up DSE analyses have converged on the form of the renormalisation-group-invariant interaction in QCD. This outcome paves the way to parameter-free predictions of hadron properties. Decades of studying the three valence-body problem in QCD have provided the evidence necessary to conclude that diquark correlations are a reality; but diquarks are complex objects, so their existence does not restrict the number of baryon states in any obvious way. This effort has led to a sophisticated understanding of the nucleon, Δ-baryon and Roper resonance: all may be viewed as Borromean bound-states, and the Roper is at heart the nucleon's first radial excitation.

The progress summarised herein highlights the capacity of DSEs in QCD to connect the quark-quark interaction, expressed, for instance, in the dressed-quark mass function, $M(p^2)$, with predictions for a wide range of hadron observables; and therefore serves as strong motivation for new experimental studies of, *inter alia*, nucleon elastic and transition form factors, which exploit the full capacity of JLab 12 in order to chart $M(p^2)$ and thereby explain the origin of more than 98 % of the visible mass in the Universe. This must shed light on confinement, which is one of the most fundamental problems in modern physics, whose solution is unlikely to be found in a timely fashion through theoretical analysis alone. A multipronged approach is required, involving constructive feedback between experiment and theory of the type illustrated herein.

Acknowledgments Both the results described and the insights drawn herein are fruits from collaborations we have joined with many colleagues and friends throughout the world; and we are very grateful to them all. We would also like to thank Ralf Gothe, Victor Mokeev and Elena Santopinto for enabling our participation in the ECT* Workshop: "Nucleon Resonances: From Photoproduction to High Photon Virtualities", 12–16 October 2015, which proved very rewarding. This work was supported by the U.S. Department of Energy, Office of Science, Office of Nuclear Physics, under contract no. DE-AC02-06CH11357; and the Alexander von Humboldt Foundation.

References

1. Collins, J.C., Duncan, A., Joglekar, S.D.: Trace and dilatation anomalies in gauge theories. Phys. Rev. D **16**, 438–449 (1977)
2. Nielsen, N.: The energy momentum tensor in a non-Abelian quark gluon theory. Nucl. Phys. B **120**, 212–220 (1977)
3. Pascual, P., Tarrach, R.: QCD: Renormalization for the Practitioner. Springer-Verlag, Berlin, Lecture Notes in Physics 194 (1984)
4. Jaffe, A.M.: The millennium grand challenge in mathematics. Not. Am. Math. Soc. **53**, 652–660 (2006)
5. McNeile, C.: Lattice status of gluonia/glueballs. Nucl. Phys. Proc. Suppl. **186**, 264–267 (2009)
6. Cloët, I.C., Roberts, C.D.: Explanation and prediction of observables using continuum strong QCD. Prog. Part. Nucl. Phys. **77**, 1–69 (2014)
7. Wilson, K.G.: Confinement of quarks. Phys. Rev. D **10**, 2445–2459 (1974)
8. Bali, G.S., et al.: Observation of string breaking in QCD. Phys. Rev. D **71**, 114513 (2005)
9. Prkacin, Z., et al.: Anatomy of string breaking in QCD. PoS **LAT2005**, 308 (2006)
10. Chang, L., Cloët, I.C., El-Bennich, B., Klähn, T., Roberts, C.D.: Exploring the light-quark interaction. Chin. Phys. C **33**, 1189–1196 (2009)
11. Glimm, J., Jaffee, A.: Quantum Physics. A Functional Point of View. Springer-Verlag, New York (1981)

12. Stingl, M.: Propagation properties and condensate formation of the confined Yang-Mills field. Phys. Rev. D **34**, 3863–3881 (1986). [Erratum: Phys. Rev.D36,651(1987)]
13. Roberts, C.D., Williams, A.G., Krein, G.: On the implications of confinement. Int. J. Mod. Phys. A **7**, 5607–5624 (1992)
14. Hawes, F.T., Roberts, C.D., Williams, A.G.: Dynamical chiral symmetry breaking and confinement with an infrared vanishing gluon propagator. Phys. Rev. D **49**, 4683–4693 (1994)
15. Roberts, C.D., Williams, A.G.: Dyson-Schwinger equations and their application to hadronic physics. Prog. Part. Nucl. Phys. **33**, 477–575 (1994)
16. Krein, G., Nielsen, M., Puff, R.D., Wilets, L.: Ghost poles in the nucleon propagator: vertex corrections and form-factors. Phys. Rev. C **47**, 2485–2491 (1993)
17. Bracco, M.E., Eiras, A., Krein, G., Wilets, L.: Selfconsistent solution of the Schwinger–Dyson equations for the nucleon and meson propagators. Phys. Rev. C **49**, 1299–1308 (1994)
18. Bhagwat, M., Pichowsky, M., Roberts, C., Tandy, P.: Analysis of a quenched lattice QCD dressed quark propagator. Phys. Rev. C **68**, 015203 (2003)
19. Bowman, P.O., et al.: Unquenched quark propagator in Landau gauge. Phys. Rev. D **71**, 054507 (2005)
20. Bhagwat, M.S., Tandy, P.C.: Analysis of full-QCD and quenched-QCD lattice propagators. AIP Conf. Proc. **842**, 225–227 (2006)
21. Aguilar, A., Binosi, D., Papavassiliou, J.: Gluon and ghost propagators in the Landau gauge: deriving lattice results from Schwinger–Dyson equations. Phys. Rev. D **78**, 025010 (2008)
22. Aguilar, A., Binosi, D., Papavassiliou, J., Rodríguez-Quintero, J.: Non-perturbative comparison of QCD effective charges. Phys. Rev. D **80**, 085018 (2009)
23. Boucaud, P., et al.: The infrared behaviour of the pure Yang-Mills green functions. Few Body Syst. **53**, 387–436 (2012)
24. Pennington, M.R., Wilson, D.J.: Are the dressed gluon and ghost propagators in the Landau Gauge presently determined in the confinement regime of QCD? Phys. Rev. D **84**, 119901 (2011)
25. Ayala, A., Bashir, A., Binosi, D., Cristoforetti, M., Rodriguez-Quintero, J.: Quark flavour effects on gluon and ghost propagators. Phys. Rev. D **86**, 074512 (2012)
26. Binosi, D., Chang, L., Papavassiliou, J., Roberts, C.D.: Bridging a gap between continuum-QCD and ab initio predictions of hadron observables. Phys. Lett. B **742**, 183–188 (2015)
27. Stingl, M.: A systematic extended iterative solution for quantum chromodynamics. Z. Phys. A **353**, 423–445 (1996)
28. Accardi, A., Arleo, F., Brooks, W.K., D'Enterria, D., Muccifora, V.: Parton propagation and fragmentation in QCD matter. Riv. Nuovo Cim. **32**, 439–553 (2010)
29. Dudek, J., et al.: Physics opportunities with the 12 GeV upgrade at Jefferson lab. Eur. Phys. J. A **48**, 187 (2012)
30. Accardi, A. et al. Electron ion collider: the next QCD frontier – understanding the glue that binds us all. arXiv:1212.1701 [nucl-ex]
31. Brodsky, S.J., Shrock, R.: Condensates in quantum chromodynamics and the cosmological constant. Proc. Nat. Acad. Sci. **108**, 45–50 (2011)
32. Brodsky, S.J., Roberts, C.D., Shrock, R., Tandy, P.C.: New perspectives on the quark condensate. Phys. Rev. C **82**, 022201(R) (2010)
33. Chang, L., Roberts, C.D., Tandy, P.C.: Expanding the concept of in-hadron condensates. Phys. Rev. C **85**, 012201(R) (2012)
34. Brodsky, S.J., Roberts, C.D., Shrock, R., Tandy, P.C.: Confinement contains condensates. Phys. Rev. C **85**, 065202 (2012)
35. Maris, P., Roberts, C.D., Tandy, P.C.: Pion mass and decay constant. Phys. Lett. B **420**, 267–273 (1998)
36. Qin, S.-X., Roberts, C.D., Schmidt, S.M.: Ward-Green-Takahashi identities and the axial-vector vertex. Phys. Lett. B **733**, 202–208 (2014)
37. Höll, A., Krassnigg, A., Roberts, C.D.: Pseudoscalar meson radial excitations. Phys. Rev. C **70**, 042203(R) (2004)
38. Ballon-Bayona, A., Krein, G., Miller, C.: Decay constants of the pion and its excitations in holographic QCD. Phys. Rev. D **91**, 065024 (2015)
39. Chang, L., Roberts, C.D.: Sketching the Bethe-Salpeter kernel. Phys. Rev. Lett. **103**, 081601 (2009)
40. Chang, L., Liu, Y.-X., Roberts, C.D.: Dressed-quark anomalous magnetic moments. Phys. Rev. Lett. **106**, 072001 (2011)
41. Chang, L., Roberts, C.D.: Tracing masses of ground-state light-quark mesons. Phys. Rev. C **85**, 052201(R) (2012)
42. Binosi, D., Chang, L., Papavassiliou, J., Qin, S.-X., Roberts, C.D.: Symmetry preserving truncations of the gap and Bethe–Salpeter equations. Phys. Rev. D **93**, 096010 (2016)
43. Binosi, D.: From continuum QCD to hadron observables. EPJ Web Conf. **113**, 05002 (2016)
44. Chang, L., Roberts, C.D., Tandy, P.C.: Selected highlights from the study of mesons. Chin. J. Phys. **49**, 955–1004 (2011)
45. Horn, T., Roberts, C.D.: The pion: an enigma within the standard model. J. Phys. G **43**, 073001/1–46 (2016)
46. Eichmann, G., Alkofer, R., Cloët, I.C., Krassnigg, A., Roberts, C.D.: Perspective on rainbow-ladder truncation. Phys. Rev. C **77**, 042202(R) (2008)
47. Eichmann, G., Cloët, I.C., Alkofer, R., Krassnigg, A., Roberts, C.D.: Toward unifying the description of meson and baryon properties. Phys. Rev. C **79**, 012202(R) (2009)
48. Eichmann, G.: Baryon form factors from Dyson-Schwinger equations. PoS **QCD–TNT–II**, 017 (2011)
49. Chang, L., Roberts, C.D., Schmidt, S.M.: Dressed-quarks and the nucleon's axial charge. Phys. Rev. C **87**, 015203 (2013)
50. Segovia, J., Cloët, I.C., Roberts, C.D., Schmidt, S.M.: Nucleon and Δ elastic and transition form factors. Few Body Syst. **55**, 1185–1222 (2014)
51. Roberts, C.D.: Hadron physics and QCD: just the basic facts. J. Phys. Conf. Ser. **630**, 012051 (2015)
52. Segovia, J., et al.: Completing the picture of the Roper resonance. Phys. Rev. Lett. **115**, 171801 (2015)
53. Segovia, J.: $\gamma_v NN^*$ Electrocouplings in Dyson-Schwinger Equations, (2016), arXiv:1602.02768 [nucl-th]
54. Eichmann, G.: Progress in the calculation of nucleon transition form factors, (2016), arXiv:1602.03462 [hep-ph]
55. El-Bennich, B., Krein, G., Rojas, E., and Serna, F.E.: Excited hadrons and the analytical structure of bound-state interaction kernels, (2016) arXiv:1602.06761 [nucl-th]
56. Segovia, J., Roberts, C.D., Schmidt, S.M.: Understanding the nucleon as a Borromean bound-state. Phys. Lett. B **750**, 100–106 (2015)

57. Bissey, F., et al.: Gluon flux-tube distribution and linear confinement in baryons. Phys. Rev. D **76**, 114512 (2007)
58. Bissey, F., Signal, A., Leinweber, D.: Comparison of gluon flux-tube distributions for quark-diquark and quark-antiquark hadrons. Phys. Rev. D **80**, 114506 (2009)
59. Alexandrou, C., de Forcrand, Ph, Lucini, B.: Evidence for diquarks in lattice QCD. Phys. Rev. Lett. **97**, 222002 (2006)
60. Babich, R., et al.: Diquark correlations in baryons on the lattice with overlap quarks. Phys. Rev. D **76**, 074021 (2007)
61. Cahill, R.T., Roberts, C.D., Praschifka, J.: Calculation of diquark masses in QCD. Phys. Rev. D **36**, 2804 (1987)
62. Maris, P.: Effective masses of diquarks. Few Body Syst. **32**, 41–52 (2002)
63. Maris, P.: Electromagnetic properties of diquarks. Few Body Syst. **35**, 117–127 (2004)
64. Roberts, H.L.L., et al.: π- and ρ-mesons, and their diquark partners, from a contact interaction. Phys. Rev. C **83**, 065206 (2011)
65. Cahill, R.T., Roberts, C.D., Praschifka, J.: Baryon structure and QCD. Austral. J. Phys. **42**, 129–145 (1989)
66. Lichtenberg, D.B., Tassie, L.J.: Baryon mass splitting in a Boson–Fermion model. Phys. Rev. **155**, 1601–1606 (1967)
67. Ripani, M., et al.: Measurement of $ep \to e'p\pi^+\pi^-$ and baryon resonance analysis. Phys. Rev. Lett. **91**, 022002 (2003)
68. Burkert, V.D.: Evidence of new nucleon resonances from electromagnetic meson production. EPJ Web Conf. **37**, 01017 (2012)
69. Kamano, H., Nakamura, S.X., Lee, T.S.H., Sato, T.: Nucleon resonances within a dynamical coupled-channels model of πN and γN reactions. Phys. Rev. C **88**, 035209 (2013)
70. Roberts, C.D.: Three lectures on hadron physics. J. Phys. Conf. Ser. **706**, 022003 (2016)
71. Chen, C., et al.: Spectrum of hadrons with strangeness. Few Body Syst. **53**, 293–326 (2012)
72. Edwards, R.G., Dudek, J.J., Richards, D.G., Wallace, S.J.: Excited state baryon spectroscopy from lattice QCD. Phys. Rev. D **84**, 074508 (2011)
73. Dugger, M., et al.: π^+ photoproduction on the proton for photon energies from 0.725 to 2.875 GeV. Phys. Rev. C **79**, 065206 (2009)
74. Aznauryan, I., et al.: Electroexcitation of nucleon resonances from CLAS data on single pion electroproduction. Phys. Rev. C **80**, 055203 (2009)
75. Aznauryan, I., Burkert, V.: Electroexcitation of nucleon resonances. Prog. Part. Nucl. Phys. **67**, 1–54 (2012)
76. Mokeev, V.I., et al.: Experimental study of the $P_{11}(1440)$ and $D_{13}(1520)$ resonances from CLAS data on $ep \to e'\pi^+\pi^-p'$. Phys. Rev. C **86**, 035203 (2012)
77. Mokeev, V.I., et al.: New results from the studies of the $N(1440)1/2^+$, $N(1520)3/2^-$, and $\Delta(1620)1/2^-$ resonances in exclusive $ep \to e'p'\pi^+\pi^-$ electroproduction with the CLAS detector. Phys. Rev. C **93**, 025206 (2016)
78. Burkert, V.D.: Nucleon Resonance Physics, arXiv:1603.00919 [nucl-ex]
79. Cloët, I.C., Roberts, C.D.: Form Factors and Dyson-Schwinger Equations. PoS, **LC2008**, 047 (2008)
80. Bijker, R., Santopinto, E.: Unquenched quark model for baryons: magnetic moments, spins and orbital angular momenta. Phys. Rev. C **80**, 065210 (2009)
81. Cloët, I.C., Bentz, W., Thomas, A.W.: Role of diquark correlations and the pion cloud in nucleon elastic form factors. Phys. Rev. C **90**, 045202 (2014)
82. Wilson, D.J., Cloët, I.C., Chang, L., Roberts, C.D.: Nucleon and Roper electromagnetic elastic and transition form factors. Phys. Rev. C **85**, 025205 (2012)
83. Olive, K.A., et al.: Review of particle physics. Chin. Phys. C **38**, 090001 (2014)
84. Chang, L., et al.: Imaging dynamical chiral symmetry breaking: pion wave function on the light front. Phys. Rev. Lett. **110**, 132001 (2013)
85. Gothe, R.W. et al.: Nucleon Resonance Studies with CLAS12. JLab 12 Experiment: E12-09-003. (2009)
86. Carman, D.S. et al.: Exclusive $N^* \to KY$ Studies with CLAS12. JLab 12 Experiment: E12-06-108A. (2014)

Few-Body Syst (2016) 57:917–924
DOI 10.1007/s00601-016-1128-7

Ralf W. Gothe · Ye Tian

Impact of the γ_νNN* Electrocoupling Parameters at High Photon Virtualities and Preliminary Cross Sections off the Neutron

Received: 12 April 2016 / Accepted: 1 June 2016 / Published online: 28 June 2016
© Springer-Verlag Wien 2016

Abstract Meson-photoproduction measurements and their reaction-amplitude analyses can establish more sensitively, and in some cases in an almost model-independent way, nucleon excitations and non-resonant reaction amplitudes. However, to investigate the strong interaction from already explored—where meson-cloud degrees of freedom contribute substantially to the baryon structure—to still unexplored distance scales—where quark degrees of freedom dominate and the transition from dressed to current quarks occurs—we depend on experiments that allow us to measure observables that are probing this evolving non-perturbative QCD regime over its full range. Elastic *and* transition form factors are uniquely suited to trace this evolution by measuring elastic electron scattering *and* exclusive single-meson and double-pion electroproduction cross sections off the nucleon. These exclusive measurements will be extended to higher momentum transfers with the energy-upgraded CEBAF beam at JLab to study the quark degrees of freedom, where their strong interaction is responsible for the ground and excited nucleon state formations. After establishing unprecedented high-precision data, the imminent next challenge is a high-quality analysis to extract these relevant electrocoupling parameters for various resonances that can then be compared to state-of-the-art models and QCD-based calculations. The vast majority of the available exclusive electroproduction cross sections are off the proton. Hence flavor-dependent analyses of excited light-quark baryons are lacking experimental data off the neutron. The goal is to close this gap by providing exclusive $\gamma_\nu(n) \to p^+\pi^-$ reaction cross section off deuterium and to establish a kinematical final-state-interaction (FSI) correction factor (R) map that can be determined from the data set itself. The "e1e" Jefferson Lab CLAS data set, that is analyzed, includes both a hydrogen and deuterium target run period, which allows a combined analysis of the pion electroproduction off the free proton, the bound proton, and the bound neutron under the same experimental conditions. Hence it will provide the experimentally best possible information on the off-shell and FSI effects in deuterium, which must be considered in order to extract the information off the neutron. The cross section analysis of this data set, that is currently underway, will considerably improve our knowledge of the Q^2 evolution of resonance states off bound protons and neutrons. Recent results presented here and in these proceedings are demonstrating the status and continuous progress of data analyses and their theoretical descriptions, as well as highlighting the experimental and theoretical outlook of what shall and may be achieved in the new era of the 12-GeV upgraded transition form factor program.

This article belongs to the special issue "Nucleon Resonances".

R. W. Gothe (✉)
University of South Carolina, Columbia, SC 29208, USA
E-mail: gothe@sc.edu
Tel.: +1-803-777-9025

1 Introduction

Already in the early inclusive high-energy deep inelastic scattering (DIS) experiments at SLAC [1], scaling and quasi-free scattering off still dressed quarks was observed at the then highest beam energies of up to $E = 20$ GeV but yet moderate four-momentum transfers of $Q^2 < 2 \, (\text{GeV/c})^2$. In these early inclusive measurements, as shown and discussed previously [2], the quasi-free peak becomes visible at high beam energies and high center-of-mass energies W, where the electrons seem to scatter off constituent quarks. Although the absolute strength of quasi-free scattering starts to dominate the elastic and resonance contributions with increasing W and Q^2, it is pushed out so far in W [3] that its relative contribution in the resonance regions becomes even smaller with increasing Q^2. The resonance-to-background ratio further increases due to the decreasing meson-baryon contributions to the baryon structure at increasing Q^2. Both effects are less evident if observed relative to the $\Delta(1232)$ resonance yields, since these drop faster with Q^2 than the resonance contributions in the second and third resonance region [3], where the individual resonance contributions can only be separated in exclusive electroproduction measurements like those carried out with the large-acceptance spectrometer CLAS at Jefferson Lab.

Mapping out the transition from exclusive resonance production to quasi-free scattering over W and Q^2 in detail, lays the experimental foundation to investigate quark-hadron duality [4,5], scaling [6–8], the bound-quark structure, confinement, dynamical mass generation, and the structure of baryons [9]. The accepted research proposal E12-09-003 at Jefferson Lab, Nucleon Resonance Studies with CLAS12 [10], will establish this experimental foundation to address in a unique way these most pressing questions in QCD. Properly extracting and interpreting the results from the measured electron scattering data, particularly for transition form factors to high-lying excited nucleon states, might even pose a greater challenge than the measurement itself. A steadily growing collaboration of experimentalists and theorists is working together to enable the measurements, the analysis of the data, and the QCD-based interpretation of the results. The progress in this field presented here and in this proceedings has also been summarized in recent review articles [11,12] and most comprehensively in, Studies of Nucleon Resonance Structure in Exclusive Meson Electroproduction [9].

The nucleon resonance (N^*) studies are crucial to our understanding of the structure and interaction of hadrons, and are poised to push the development of quark models and QCD-based calculations forward. In the perturbative regime at large Q^2, QCD describes the strong interaction successfully based on current quarks and gauge gluons as the fundamental degree of freedom, however when Q^2 drops into the non-perturbative regime, a transition to completely different degrees of freedom, the dressed quarks and gluons as well as the mesons and baryons, happens. This transition is neither experimentally mapped out nor theoretically understood from first principles. Therefore a sufficiently complete electroexcitation data base has to be established to pin down the distance-dependent baryon structure and to aid the development of a QCD-based strong interaction theory. On the experimental side, most of the low-lying excited states of the proton have already been studied in the low-Q^2 range, but there is still very little data available on the neutron excitations [13]. Because of the inherent difficulty in obtaining a free neutron target, a deuterium target is the next best alternative to investigate the isospin-dependent structure of the nucleon and its excited states.

2 Hadronic Structure Analysis

The general analogy to the hydrogen atom—which is the simplest atom bound by electromagnetic fields of well-known dynamics—that the ground state can be unambiguously described by the spectrum of its excited states, does not hold for the nucleon—which is the lightest three-quark system bound by strong fields—since the evolution of the strong interaction from small to large distance scales is not known. Hence even the spectrum of *all* excited states is by itself not sufficient to pin down the baryonic structure, but it is the best possible approach to disentangle the individual interfering resonance and background amplitudes in an almost model-independent way by so-called complete experiments. Here in the simplest case of pseudo-scalar meson photoproduction, the cross section can be decomposed into four gauge- and Lorentz-invariant complex amplitudes. In a combination of unpolarized, beam-, target-, and recoil-polarization experiments, a total of up to 16 observables can be measured with a large solid angle detector, where only eight (or seven, taking into account an overall undetermined phase) are linearly independent. With the caveat that most baryon resonances, except the lowest lying ones, decay dominantly into vector-meson or multi-meson channels, complete experiments in single pseudo-scalar meson photoproduction and corresponding partial wave analyses will allow for the highest quality extraction of resonance parameters under minimal model assumptions. New complete sets

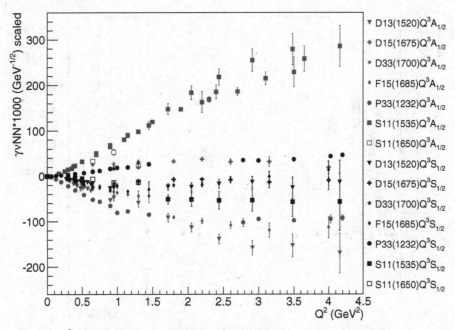

Fig. 1 (Color online) The Q^2 dependence of the $\gamma_v NN^*$ helicity amplitudes $A_{1/2}$ and $S_{1/2}$ off the proton that have been extracted from CLAS data [11,20,21] is shown. These helicity conserving amplitudes for resonances with masses up to 1700 MeV are scaled by Q^3 as predicted for baryons with three effective constituents only

of observables [14,15] that exploit the high analyzing power of some hyperon decays led to several new or PDG-upgraded [16] excited states seen in the recently updated Bonn-Gatchina coupled-channel analysis [17].

Beyond baryon spectroscopy at the real photon point $Q^2 = 0 \,(\text{GeV/c})^2$, electron scattering experiments are essential to investigate the strong interaction and thereby the internal hadronic structure at various distance scales by tuning the four-momentum transfer from $Q^2 \approx 0 \,(\text{GeV/c})^2$, where the meson cloud contributes significantly to the baryon structure, over intermediate Q^2, where the three constituent-quark core starts to dominate, to Q^2 up to $12 \,(\text{GeV/c})^2$, attainable after the 12 GeV upgrade at JLab [10], see Fig. 3, where the constituent quark gets more and more undressed towards the bare current quark [18–20].

Although originally derived in the high Q^2 limit [7,8], constituent counting rules describe in more general terms how the transition form factors and the corresponding helicity amplitudes scale with Q^2, dependent on the number of effective constituents [6]. The available CLAS results for $Q^2 < 4.5 \,(\text{GeV/c})^2$ [11,20,21] may already indicate, particularly for the helicity conserving transition amplitudes $A_{1/2}$ and $S_{1/2}$, the onset of the predicted scaling with $1/Q^3$ for effective three-quark systems, as shown in Fig. 1. Whether this constituent counting rule scaling truly sets in at such low momentum transfers, will be one of the first results based on the experimental data gathered in the first run period with the 12-GeV upgraded CLAS detector (CLAS12) in Hall B [10]. If verified at higher Q^2, this further indicates that in these cases the meson-baryon contributions become negligible in comparison to those of the three constituent-quark core, which coincides with the Argonne-Osaka dynamical coupled channel calculation [9,23].

Along the same line of reasoning perturbative QCD (pQCD) predicts in the high-Q^2 limit, by neglecting higher twist contributions, that helicity is conserved. The fact that this predicted behavior seems to set in already at much lower Q^2 values than expected [11,20,21] challenges our current understanding of baryons even further. For $N(1520)D_{13}$ the helicity conserving amplitude $A_{1/2}$ starts to dominate the helicity non-conserving amplitude $A_{3/2}$ at $Q^2 \approx 0.5 \,(\text{GeV/c})^2$, see Fig. 2, as typically documented by the zero crossing of the corresponding helicity asymmetry $A_{hel} = (A_{1/2}^2 - A_{3/2}^2)/(A_{1/2}^2 + A_{3/2}^2)$. The $N(1675)D_{15}$ and $N(1680)F_{15}$ resonances show a similar behavior with zero crossings below $Q^2 \approx 1 \,(\text{GeV/c})^2$, see Fig. 2, whereas the $\Delta(1232)P_{33}$ helicity asymmetry stays negative with no indication of an upcoming zero crossing; and even more surprising are the results for the $N(1720)P_{13}$ $A_{1/2}$ amplitude, which decreases so rapidly with Q^2 that the helicity asymmetry shows an inverted behavior with a zero crossing from positive to negative around $Q^2 \approx 0.5 \,(\text{GeV/c})^2$. Figure 2 also illustrates the statistical limitations of the currently available data, because

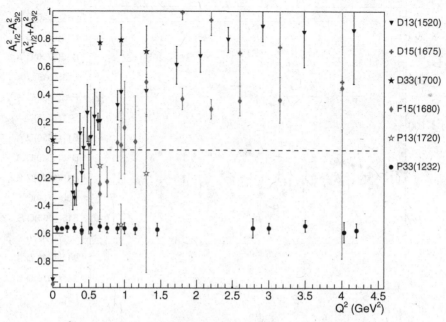

Fig. 2 (Color online) The Q^2 dependence of the helicity amplitude asymmetries off the proton that have been extracted from CLAS data [11,20,21] is shown for resonances with masses up to 1720 MeV

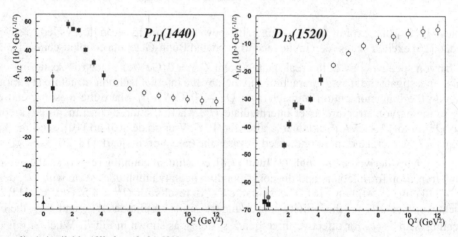

Fig. 3 (Color online) Available (*filled symbols*) [21] and projected CLAS12 [10] (*open symbols*) $A_{1/2}$ electrocouplings of the $P_{11}(1440)$ (*left*) and the $D_{13}(1520)$ (*right*) excited states

the helicity asymmetry is due to the difference of large numbers more error sensitive than the helicity amplitudes themselves, and higher statistics and thus higher accuracy would be desirable.

This essentially different behavior of transition form factors to various excited states with different quantum numbers underlines that it is necessary but not sufficient to extend the measurements of the elastic form factors to higher momentum transfers. To comprehend the strong interaction at intermediate distance scales, where dressed quarks degrees of freedom are responsible for the formation and diverse behavior of baryons in distinctively different quantum states, the Q^2 evolution of transition form factors to multiple resonance states up to $12\,(\text{GeV}/c)^2$ is absolutely crucial [9,10]. Figure 3 shows two examples of projected results for the $A_{1/2}$ helicity amplitudes of the p to $P_{11}(1440)$ and $D_{13}(1520)$ transitions. Attempts to extract N to N^* transition form factors in vector-meson electroproduction and off the neutron in deuterium are currently pursued to further complement the data base.

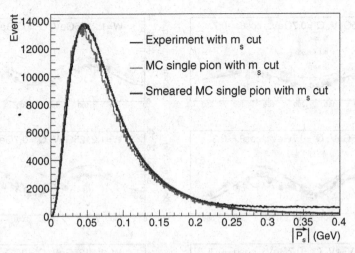

Fig. 4 (Color online) The *black line* represents the missing momentum distribution of the unmeasured proton from data. The Monte-Carlo simulated proton momentum distribution (*red line*) is based on the Bonn potential [24] and has been smeared according to the detector resolution (*blue line*)

3 Preliminary Results off the Neutron

In order to extract from electron scattering off deuterium the exclusive $\gamma_v(n) \rightarrow p^+\pi^-$ reaction cross section, the FSI corrections and the off-shell effects need to be studied thoroughly. The kinematical FSI correction factor can be extracted directly from the measured data by using

$$R_{FSI} = \frac{\left(\frac{d\sigma^{quasi-free}}{d\Omega^*_{\pi^-}}\right)}{\left(\frac{d\sigma^{full}}{d\Omega^*_{\pi^-}}\right)}, \tag{1}$$

where the exclusive quasi-free process can be isolated by applying two cuts, one on the missing mass square of the spectator proton (m_s^2) and one on the magnitude of its missing momentum ($|\mathbf{P_s}|$), whereas the exclusive full process is obtained by only cutting on the missing mass square of the spectator proton. The comparison of the measured spectator momentum distribution (black line) with the generated proton momentum distribution from the Bonn potential [24] that has been smeared according to the detector resolution (blue line) for $|\mathbf{P_s}| <$ 200 MeV, seen in Fig. 4, reveals that the quasi-free process is absolutely dominant in this kinematic region, while for $|\mathbf{P_s}| > 300$ MeV the final state interactions start to dominate the process. This shows that the quasi-free process can be successfully isolated by cutting on $|\mathbf{P_s}|$ at 200 MeV. Preliminary results on the kinematical FSI correction factor R_{FSI} as a function of the pion polar angle (θ^*_π) in the Center of Mass frame (COM) indicate that the kinematical FSI corrections are largest at small angles, θ^*_π, with respect to the virtual photon direction.

After proper particle identification, fiducial cuts, and the exclusive quasi-free event selection, the cross section for the quasi-free $\gamma_v(n) \rightarrow p^+\pi^-$ reaction with unpolarized electron beam and unpolarized deuteron target is given by

$$\frac{d^4\sigma}{dW dQ^2 d\Omega^*_{\pi^-}} = \Gamma_v \frac{d\sigma}{d\Omega^*_{\pi^-}}, \tag{2}$$

where Γ_v is the virtual photon flux defined by Eq. (3), in which ϵ denotes the transverse polarization of the virtual photon, $\nu = E_{beam} - E_{scattered\ electron}$ the energy transfer, and θ_e the scattering angle of electron in the LAB frame with respect to the incoming electron direction.

$$\Gamma_v = \frac{\alpha}{4\pi} \frac{1}{E_{beam}^2 M_n^2} \frac{W(W^2 - M_n^2)}{(1-\epsilon)Q^2}, \quad \epsilon = \left(1 + 2\left(1 + \frac{\nu^2}{Q^2}\right)tan^2\frac{\theta_e}{2}\right)^{-1} \tag{3}$$

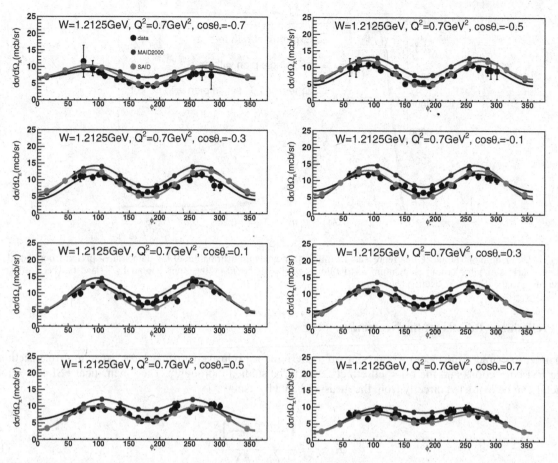

Fig. 5 A typical example of preliminary $\phi^*_{\pi^-}$ dependent cross sections in the $P_{33}(1232)$ resonance region for different $cos\theta^*_{\pi^-}$ bins at $Q^2 = 0.7\,\text{GeV}$ and $W = 1.2125\,\text{GeV}$ in comparison with MAID2000 (*blue line*) and SAID (*purple line*) predictions. *Black line* is the fit to the extracted differential cross sections (color figure online)

Finally, the quasi-free differential pion electroproduction cross section, that is corrected for FSI and off-shell effects, is given by

$$\frac{d\sigma}{d\Omega^*_{\pi^-}} = \frac{1}{\Gamma_\upsilon} \frac{1}{R_{FSI}} \frac{d^4\sigma}{dW dQ^2 d\Omega^*_{\pi^-}}. \tag{4}$$

Figure 5 shows the $\phi^*_{\pi^-}$ dependent differential cross section in the Δ resonance region as a typical subset of the available data. The preliminary cross sections are compared to two physics models MAID2000 [25] and SAID [26].

The hadronic cross section $\frac{d\sigma}{d\Omega^*_{\pi^-}}$ is fit in terms of $cos\phi^*_{\pi^-}$ (Eq. 5) to extract the underlying structure functions. The fit function has three fit parameters a, b, and c, which correspond to the structure functions $\sigma_T + \epsilon\sigma_L$, σ_{TT}, and σ_{TL}, respectively,

$$\frac{d\sigma}{d\Omega^*_{\pi^-}} = a + b\cos 2\phi^*_{\pi^-} + c\cos\phi^*_{\pi^-}, \quad a = \sigma_T + \epsilon\sigma_L, \quad b = \epsilon\sigma_{TT}, \quad c = \sqrt{2\epsilon(1+\epsilon)}\sigma_{TL}. \tag{5}$$

An example of a typical structure function separation as a function of $cos\theta^*_{\pi^-}$ for $Q^2 = 0.5$, 0.7, and $0.9\,\text{GeV}$ at $W = 1.212\,\text{GeV}$ is shown in Fig. 6 in comparison to the SAID [26] and MAID2000 [25] models. In order to gain some insight on the dominant partial wave contribution in this particular Δ-resonance energy bin, a Legendre polynomial expansion of the structure functions up to $l = 2$ has been performed, see Fig. 6.

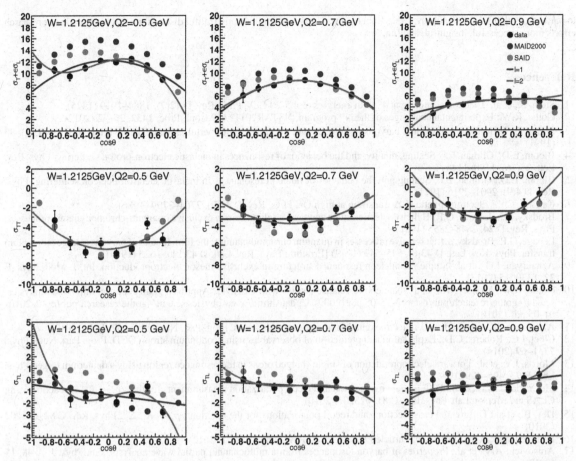

Fig. 6 Preliminary $\sigma_T + \epsilon\sigma_L$, σ_{TT}, and σ_{TL} results (*top*, *middle*, and *bottom row*) in the $P_{33}(1232)$ resonance region for different $Q^2 = 0.5, 0.7$, and 0.9 GeV (*left*, *middle*, and *right column*) are shown as *black points* and compared with the MAID2000 (*blue points*) and SAID predictions (*purple points*). The *blue* and *red lines* are Legendre polynomial fits to the *black points* for $l = 1$ and $l = 2$, respectively (color figure online)

4 Summary

All visible matter that surrounds us is made of atoms, which are made of electrons and nuclei; the latter are made of nucleons, which are finally made of quarks and gluons. Contrary to the more publicized discussions, the Higgs, or also frequently called the God Particle, is not responsible for the generation of all mass. It is already known that 98 % of all visible mass is non-perturbatively generated by strong fields. Establishing an experimental and theoretical program that provides access to

- the flavor-dependent dynamics of the non-perturbative strong interaction among dressed quarks, their emergence from QCD, and their confinement in baryons,
- the dependence of the light quark mass on the momentum transfer and thereby how the constituent quark mass arises from dynamical chiral-symmetry breaking, and
- the behavior of the universal QCD β-function in the infrared regime,

is indeed most challenging on all levels, but recent progress and future commitments [9] bring a solution of these most fundamental remaining QCD problems into reach. Single- and double-polarization experiments are essential to establish the baryon spectrum, branching ratios, and a detailed separation of individual resonance and background contributions. Elastic and particularly transition form factors are on the other hand needed to uniquely track non-perturbative QCD from long to short distance scales.

Acknowledgments This work was supported in part by the National Science Foundation, particularly under Grant No. 0856010, the U.S. Department of Energy, and other international funding agencies supporting research groups at Jefferson Lab. The authors are very grateful to the many collaborators and friends, who made the measurements and the presented results possible, and want to particularly thank Viktor Mokeev, Craig Roberts, and Harry Lee for the many fruitful discussions that have made our research efforts more successful, insightful, and fun.

References

1. Stein, S., et al.: Electron scattering at 4° with energies of 4.5-20 GeV. Phys. Rev. D **12**(7), 1884–1919 (1975)
2. Gothe, R.W.: Experimental challenges of the N* program. NSTAR2011. AIP Conf. Proc. **1432**, 26–32 (2012)
3. Stoler, P.: Form factors of excited baryons at high Q^2 and the transition to perturbative QCD. II. Phys. Rev. D **44**, 73–80 (1991)
4. Bloom, E.D., Gilman, F.J.: Scaling, duality, and the behavior of resonances in inelastic electron-proton scattering. Phys. Rev. Lett. **25**(16), 1140–1144 (1970)
5. Bloom, E.D., Gilman, F.J.: Scaling and the behavior of nucleon resonances in inelastic electron-nucleon scattering. Phys. Rev. D **4**(9), 2901–2916 (1971)
6. Carlson, C.E.: Electromagnetic N-Δ transition at high Q^2. Phys. Rev. D **34**(9), 2704–2709 (1986)
7. Brodsky, S.J., Lepage, G.P.: Helicity selection rules and tests of gluon spin in exclusive quantum-chromodynamic processes. Phys. Rev. D **24**, 2848–2855 (1981)
8. Lepage, G.P., Brodsky, S.J.: Exclusive processes in quantum chromodynamics: the form factors of baryons at large momentum transfer. Phys. Rev. Lett. D **43**(8), 545–549 (1979) [Erratum: Phys. Rev. Lett. D **43**(21), 1625 (1979)]
9. Aznauryan, I.G., et al.: Studies of nucleon resonance structure in exclusive meson electroproduction. Int. J. Mod. Phys. E **22**, 1330015 (2013)
10. Gothe, R.W., Mokeev, V.I., et al.: Nucleon resonance studies with CLAS12, Approved JLab Proposal http://www.physics.sc.edu/~gothe/research/pub/nstar12-12-08.pdf (2009), Update http://www.physics.sc.edu/~gothe/research/pub/ns12-2010-01-05.pdf (2010)
11. Aznauryan, I.G., Burkert, V.D.: Electroexcitation of nucleon resonances. Prog. Part. Nucl. Phys. **67**, 1–54 (2012)
12. Clöet, I.C., Roberts, C.D.: Explanation and prediction of observables using continuum strong QCD. Prog. Part. Nucl. Phys. **77**, 1–69 (2014)
13. Morris, J.V., et al.: Forward electroproduction of single charged pions in the resonance region using a deuterium target. Phys. Lett. B **73**, 495–499 (1978)
14. McCracken, M.E., et al.: Differential cross section and recoil polarization measurements for the $\gamma p \rightarrow K^+ \Lambda$ reaction using CLAS at Jefferson Lab. Phys. Rev. C **81**, 025201 (2010)
15. Dey, B., et al.: Differential cross sections and recoil polarizations for the reaction $\gamma p \rightarrow K^+ \Sigma^0$. Phys. Rev. C **82**, 025202 (2010)
16. Particle Data Group.: Review of Particle Physics. Phys. Rev. D **86**, 010001 (2012)
17. Anisovich, A.V., et al.: Properties of baryon resonances from a multichannel partial wave analysis. Eur. Phys. J. A **48**, 15 (2012)
18. Bhagwat, M.S., et al.: Analysis of a quenched lattice-QCD dressed-quark propagator. Phys. Rev. C **68**, 015203 (2003)
19. Bhagwat, M.S., et al.: Analysis of full-QCD and quenched-QCD lattice propagators. AIP C. P. **842**, 225 (2006)
20. Bowman, P.O., et al.: Unquenched quark propagator in Landau gauge. Phys. Rev. D **71**, 015203 (2005)
21. Mokeev, V.I.: Nucleon Resonance Photo- and Electrocouplings. https://userweb.jlab.org/~mokeev/resonance_electrocouplings/ (2016)
22. Aznauryan, I.G., et al.: Electroexcitation of nucleon resonances from CLAS data on single pion electroproduction. Phys. Rev. C **80**, 055203 (2009)
23. Julia-Diaz, B., et al.: Dynamical coupled-channels effects on pion photoproduction. Phys. Rev. C **77**, 045205 (2008)
24. Machleidt, R.: High-precision, charge-dependent Bonn nucleon-nucleon potential. Phys. Rev. C **63**, 024001 (2000)
25. Drechsel, D., et al.: A Unitary isobar model for pion photoproduction and electroproduction on the proton up to 1-GeV. Nucl. Phys. A. **645**, 145–174 (1999), The full database can be accessed via website http://portal.kph.uni-mainz.de/MAID/
26. Arndt, R.A., et al.:Baryon resonance analysis from SAID. Chin. Phys. C. **33**(12), 1063–1068. The full database can be accessed via website http://gwdac.phys.gwu.edu (2009)

Few-Body Syst (2016) 57:909–916
DOI 10.1007/s00601-016-1127-8

V. I. Mokeev

Updates on the Studies of N^* Structure with CLAS and the Prospects with CLAS12

Received: 14 February 2016 / Accepted: 1 June 2016 / Published online: 16 June 2016
© Springer-Verlag Wien 2016

Abstract The recent results on $\gamma_v p N^*$ electrocouplings from analyses of the data on exclusive meson electro-production off protons measured with the CLAS detector at Jefferson Lab are presented. The impact of these results on the exploration of the excited nucleon state structure and non-perturbative strong interaction dynamics behind its formation is outlined. The future extension of these studies in the experiments with the CLAS12 detector in the upgraded Hall-B at JLab will provide for the first time $\gamma_v p N^*$ electrocouplings of all prominent resonances at the still unexplored distance scales that correspond to extremely low (0.05 GeV^2 < Q^2 < 0.5 GeV^2) and the highest photon virtualities (5.0 GeV^2 < Q^2 < 12.0 GeV^2) ever achieved in the exclusive electroproduction measurements. The expected results will address the most important open problems of the Standard Model: on the nature of more than 98 % of hadron mass, quark–gluon confinement and emergence of the excited nucleon state structure from the QCD Lagrangian, as well as allowing a search for the new states of hadron matter predicted from the first principles of QCD, the so-called hybrid baryons.

1 Introduction

The studies of helicity amplitudes that describe the transitions between the initial real/virtual photon–ground state proton and the final N^*/Δ^* states, the so-called $\gamma_{r,v} p N^*$ photo-/electrocouplings, represent the important and absolutely needed part of efforts in exploration of non-perturbative strong interaction dynamics behind the generation of the ground and excited nucleon states from quarks and gluons. These studies, carried out in a wide range of photon virtualities Q^2 and over all prominent excited nucleon states, are the only source of information on different manifestations of the non-perturbative strong interaction in the generation of excited nucleons of different quantum numbers [1–3]. Furthermore, the recent studies of nucleon structure within the framework of the Dyson–Schwinger equations of QCD (DSEQCD) [4–7] conclusively demonstrated the critical importance of combined analysis of the data on elastic and transition $p \to N^*$ form factors in order to provide credible access to the momentum dependence of the running dynamical mass of dressed quarks. This fundamental ingredient of the non-perturbative strong interaction elucidates the generation of more than 98 % of hadron mass and the emergence of quark–gluon confinement. It makes the studies of the elastic and transition $p \to N^*$ form factors, which are directly related to $\gamma_{r,v} p N^*$ photo-/electrocouplings, one of the central focuses of contemporary hadron physics.

In this proceedings we outline the current status in the studies of the $\gamma_v p N^*$ electrocouplings from the data on exclusive meson electroproduction off protons measured with the CLAS detector at JLab [2,3,8–11], as well as the insight to the structure of excited nucleon states offered by these results. We discuss also the prospects for the future extension of the resonance electrocoupling studies in the experiments foreseen with

---

This article belongs to the special issue "Nucleon Resonances".

V. I. Mokeev (✉)
Thomas Jefferson National Accelerator Facility, Newport News, VA 23606, USA
E-mail: mokeev@jlab.org

the CLAS12 detector. The results expected from these experiments will address the key open problems of the Standard Model on the nature of more than 98 % of the ground and excited nucleon masses, the emergence of quark–gluon confinement and the nucleon spectrum and structure from QCD [12]. Furthermore, they will allow us to search for the new states of baryon matter predicted in the Lattice QCD studies of the N^* spectrum [13], the so-called the hybrid baryons [10,14].

2 Evaluation of $\gamma_v p N^*$ Electrocouplings from the CLAS Data on Exclusive Meson Electroproduction Off Protons

The CLAS detector has contributed the lion's share of the world data on all essential exclusive meson electroproduction channels in the resonance excitation region including πN, ηp, KY, and $\pi^+\pi^- p$ electroproduction off protons with nearly complete coverage of the final hadron phase space [8]. The observables measured with the CLAS detector are stored in the CLAS Physics Data Base [15]. The available observables in the resonance excitation region are listed in Table 1. In the near future the data in the resonance region will be extended by new results on π^+n, $\pi^0 p$, and $\pi^+\pi^- p$ exclusive electroproduction at $W < 1.9$ GeV and 0.3 GeV$^2 < Q^2 < 1.0$ GeV^2. The new data on $\pi^+\pi^- p$ electroproduction off protons at $W < 2.0$ GeV will become available in the Q^2 range from 2.0 to 5.0 GeV^2.

So far, most of the results on the $\gamma_v p N^*$ electrocouplings have been extracted from independent analyses of π^+n, $\pi^0 p$, and $\pi^+\pi^- p$ exclusive electroproduction data off the proton. A total of nearly 160,000 data points (d.p.) on unpolarized differential cross sections, longitudinally polarized beam asymmetries, and longitudinal target and beam-target asymmetries for πN electroproduction off protons were obtained with the CLAS detector at $W < 2.0$ GeV and 0.2 GeV$^2 < Q^2 < 6.0$ GeV^2. The data have been analyzed within the framework of two conceptually different approaches: a unitary isobar model (UIM) and dispersion relations (DR) [16,17]. The UIM describes the πN electroproduction amplitudes as a superposition of N^* electroexcitations in the s-channel, non-resonant Born terms, and ρ- and ω- t-channel contributions. The latter are reggeized, which allows for a better description of the data in the second- and third-resonance regions. The final-state interactions are treated as πN rescattering in the K-matrix approximation [16]. In the DR approach, dispersion relations relate the real to the imaginary parts of the invariant amplitudes that describe the πN electroproduction. Both approaches provide a good and consistent description of the πN data in the range of $W < 1.7$ GeV and $Q^2 < 5.0$ GeV^2, resulting in $\chi^2/d.p. < 2.9$.

The $\pi^+\pi^- p$ electroproduction data from CLAS [18,19] provide information for the first time on nine independent single-differential and fully-integrated cross sections binned in W and Q^2 in the mass range $W < 2.0$ GeV and at photon virtualities of 0.25 GeV$^2 < Q^2 < 1.5$ GeV^2. The analysis of the data have allowed us to develop the JM reaction model [3,20,21] with the goal of extracting resonance electrocouplings, as well as $\pi\Delta$ and ρp hadronic decay widths. This model incorporates all relevant reaction mechanisms in

Table 1 Observables for exclusive meson electroproduction off protons that have been measured with the CLAS detector in the resonance excitation region and stored in the CLAS Physics Data Base [15]: CM-angular distributions for the final mesons ($\frac{d\sigma}{d\Omega}$); beam, target, and beam-target asymmetries ($A_{LT'}$, A_t, A_{et}); and recoil hyperon polarizations (P', P^0)

Hadronic final state	W-range GeV	Q^2-range GeV^2	Measured observables
π^+n	1.10–1.38	0.16–0.36	$\frac{d\sigma}{d\Omega}$
	1.10–1.55	0.30–0.60	$\frac{d\sigma}{d\Omega}$
	1.10–1.70	1.70–4.50	$\frac{d\sigma}{d\Omega}$, $A_{LT'}$
	1.60–2.00	1.80–4.50	$\frac{d\sigma}{d\Omega}$
$\pi^0 p$	1.10–1.38	0.16–0.36	$\frac{d\sigma}{d\Omega}$
	1.10–1.68	0.40–1.15	$\frac{d\sigma}{d\Omega}$, $A_{LT'}$, A_t, A_{et}
	1.10–1.39	3.00–6.00	$\frac{d\sigma}{d\Omega}$
ηp	1.50–2.30	0.20–3.10	$\frac{d\sigma}{d\Omega}$
$K^+\Lambda$	1.62–2.60	1.40–3.90	$\frac{d\sigma}{d\Omega}$
	1.62–2.60	0.70–5.40	P', P^0
$K^+\Sigma^0$	1.62–2.60	1.40–3.90	$\frac{d\sigma}{d\Omega}$
	1.62–2.60	0.70–5.40	P'
$\pi^+\pi^- p$	1.30–1.60	0.20–0.60	Nine single-differential cross sections
	1.40–2.10	0.50–1.50	

the $\pi^+\pi^-p$ final-state channel that contribute significantly to the measured electroproduction cross sections off protons in the resonance region, including the $\pi^-\Delta^{++}$, $\pi^+\Delta^0$, $\rho^0 p$, $\pi^+N(1520)\frac{3}{2}^-$, $\pi^+N(1685)\frac{5}{2}^+$, and $\pi^-\Delta(1620)\frac{3}{2}^+$ meson–baryon channels, as well as the direct production of the $\pi^+\pi^-p$ final state without formation of intermediate unstable hadrons. In collaboration with JPAC [22] a special approach has been developed allowing us to remove the contributions from the s-channel resonances to the reggeized t-channel non-resonant terms in the $\pi^-\Delta^{++}$, $\pi^+\Delta^0$, $\rho^0 p$ electroproduction amplitudes. The contributions from well established N^* states in the mass range up to 2.0 GeV were included into the amplitudes of the $\pi\Delta$ and ρp meson–baryon channels by employing a unitarized version of the Breit–Wigner ansatz [20]. The JM model provides a good description of the $\pi^+\pi^-p$ differential cross sections at $W < 1.8$ GeV and 0.2 GeV$^2 < Q^2 < 1.5$ GeV^2 with $\chi^2/d.p. < 3.0$. The quality for the description of the CLAS data suggests the unambiguous and credible separation between the resonant/non-resonant contributions achieved fitting the CLAS data [3]. The credible isolation of the resonant contributions makes it possible to determine the resonance electrocouplings, along with the $\pi\Delta$, and ρN decay widths from the resonant contributions employing for their description the amplitudes of the unitarized Breit–Wigner ansatz [20] that fully accounts for the unitarity restrictions on the resonant amplitudes.

Electrocouplings of nucleon resonances and their KY hadronic decay widths can also be determined from analyses of the CLAS data on exclusive KY electroproduction off protons [23]. Future analyses of the large body of these data (Table 1) will improve the knowledge on electrocouplings and KY hadronic decay widths of high-lying N^* states with masses above 1.6 GeV. The decay to the πN final state for many of these resonances are suppressed. Currently, the preliminary results on the electrocouplings of these states are available from the studies of $\pi^+\pi^-p$ electroproduction off protons only [11,24]. Consistent results on $\gamma_v pN^*$ electrocouplings of the aforementioned resonances obtained in independent analyses of KY electroproduction will validate credible extraction of these fundamental quantities. The development of the analysis tools for extraction of the resonance parameters from KY electroproduction data measured with CLAS is urgently needed.

3 Selected Results on Resonance Electrocouplings and Their Impact on the Insight into N^* Structure

Resonance electrocouplings have been obtained from various CLAS data in the exclusive channels: π^+n and $\pi^0 p$ at $Q^2 < 5.0$ GeV^2 in the mass range up to 1.7 GeV, ηp at $Q^2 < 4.0$ GeV^2 in the mass range up to 1.6 GeV, and $\pi^+\pi^-p$ at $Q^2 < 1.5$ GeV^2 in the mass range up to 1.8 GeV [3,8,16,17,20,24]. The studies of the $N(1440)1/2^+$ and $N(1520)3/2^-$ resonances with the CLAS detector [3,16,20] have provided the dominant part of the information available worldwide on their electrocouplings in a wide range of photon virtualities 0.25 GeV$^2 < Q^2 < 5.0$ GeV^2. Currently the $N(1440)1/2^+$ and $N(1520)3/2^-$ states, together with the $\Delta(1232)3/2^+$ and $N(1535)1/2^-$ resonances [8], represent the most explored excited nucleon states. Furthermore, results on the $\gamma_v pN^*$ electrocouplings for the high-lying $N(1675)5/2^-$, $N(1680)5/2^+$, and $N(1710)1/2^+$ resonances have been determined for the first time from the CLAS πN data at 1.5 GeV$^2 < Q^2 < 4.5$ GeV^2 [17]. The first results on the electrocouplings of $\Delta(1620)1/2^-$ resonance that decays preferentially to the $N\pi\pi$ final states have recently become available from analysis of the CLAS data on $\pi^+\pi^-p$ electroproduction off protons. The up to date numerical results on electrocouplings of most nucleon resonances in the mass range up to 1.8 GeV available from the exclusive electroproduction data from CLAS and elsewhere are maintained on our web page [24].

Consistent results for the $\gamma_v pN^*$ electrocouplings of the $N(1440)1/2^+$ and $N(1520)3/2^-$ resonances, which have been determined in independent analyses of the dominant meson electroproduction channels, πN and $\pi^+\pi^-p$ shown in Fig. 1 (left) and in Fig. 2 (left), demonstrate that the extraction of these fundamental quantities is reliable, since a good data description is achieved in the major electroproduction channels having quite different background contributions. We have developed special procedures to test the reliability of the resonance $\gamma_v pN^*$ electrocouplings extracted from the charged double pion electroproduction data only. In this case, we carried out the extraction of the resonance parameters independently, fitting the CLAS $\pi^+\pi^-p$ electroproduction data [18] in overlapping intervals of W. The non-resonant amplitudes in each of the presented in Fig. 1 (center) W-intervals are different, while the resonance parameters should remain the same as they are determined from the data fit in different W-intervals. The electrocouplings of the $\Delta(1620)1/2^-$ state determined in this procedure are shown in the center of Fig. 1. The consistent results on these electrocouplings from the independent analyses of different W-intervals strongly support their reliable extraction. The tests described above demonstrated the capability of the models outlined in Sect. 2 to provide reliable information

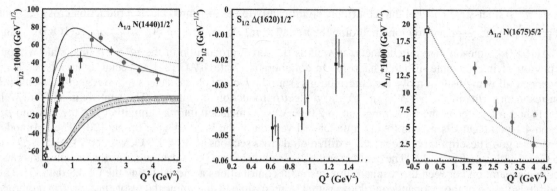

Fig. 1 $A_{1/2}\gamma_v pN^*$ electrocouplings of the $N(1440)1/2^+$ (*left*), $N(1675)5/2^-$ (*right*), and $S_{1/2}\gamma_v pN^*$ electrocouplings of the $\Delta(1620)1/2^-$ (*center*) resonances from analyses of the CLAS electroproduction data off protons in the πN—[16,17] (*red circles in the left and right panels*) and $\pi^+\pi^- p$ channels [20] (*black triangles in the left panel*), with new results from the $\pi^+\pi^- p$ channel [3] (*blue squares in the left panel*). The central panel shows the $\Delta(1620)1/2^-$ electrocouplings obtained from analyses of $\pi^+\pi^- p$ electroproduction data off protons [3] carried out independently in three intervals of W: 1.51 GeV → 1.61 GeV (*black squares*), 1.56 GeV → 1.66 GeV (*red circles*), and 1.61 GeV → 1.71 GeV (*blue triangles*). Photocouplings are taken from the RPP [25] (*open squares*) and the CLAS data analysis [26] of πN photoproduction (*open triangles*). The electrocoupling results in the *left panel* are shown in comparison with the DSEQCD—[4] (*blue thick solid*) and constituent quark model calculations [27] (*thin red solid*), [28] (*thin red dashed*). The meson–baryon cloud contributions, determined as described in Sect. 3, are shown by the magenta area. For the case of the $N(1675)5/2^-$ resonance (*right*), the absolute values of the meson–baryon cloud amplitudes at the resonance poles taken from Argonne–Osaka coupled channel analysis [29] are shown by *dashed magenta line* together with the estimate for the quark core contribution from a quark model [30] (*black solid line*) (color figure online)

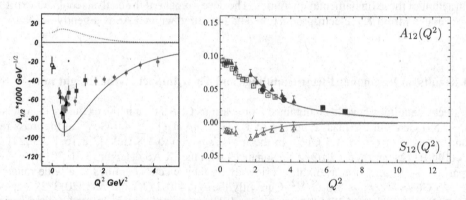

Fig. 2 $A_{1/2}$ photo-/electrocouplings of the $N(1520)3/2^-$ (*left*) and $N(1535)1/2^-$ (*right*) resonances from analyses of the CLAS electroproduction data off protons [3,16,20,32]. The *symbol* meaning in the *left panel* is the same as in Fig. 1 (*left*). The $N(1535)1/2^-$ electrocouplings in the *right panel* from πN [16] and ηN [32] electroproduction are shown by the *triangles* and *rectangles*, respectively. The analysis [32] of the ηN data was carried out assuming $S_{1/2} = 0$. The electrocoupling results for the $N(1520)3/2^-$ are shown in comparison with the model estimates for the quark core [30] (*black solid*). The meson–baryon cloud contributions taken from the Argonne–Osaka coupled channel analysis [29] are shown by the *magenta dashed line*. In case of the $N(1535)1/2^-$ resonance (*right*) the data are compared with the results of the LCSR model [1] with the quark DA normalization parameters from the LQCD results obtained starting from the QCD Lagrangian [33] (color figure online)

on the $\gamma_v pN^*$ resonance electrocouplings from independent analyses of the data on exclusive πN and $\pi^+\pi^- p$ electroproduction.

Experimental results on $\gamma_v pN^*$ electrocouplings offer insight into the non-perturbative strong interaction dynamics behind the emergence of the N^* structure from the QCD Lagrangian. Two conceptually different approaches have been developed that allow us to relate the information on the Q^2 evolution of the resonance electrocouplings to the first principles of QCD.

Due to the rapid progress in the field of DSEQCD studies of excited nucleon states [4–6,31], the first evaluations of the transition $p \rightarrow \Delta(1232)3/2^+$ form factors and the $N(1440)1/2^+$ resonance electrocouplings starting from the QCD Lagrangian have recently become available. The $A_{1/2}$ electrocouplings of the $N(1440)1/2^+$ resonance computed in [4] are shown in Fig. 1 (left) by the solid blue lines. The evaluation [4] is applicable at photon virtualities $Q^2 > 2.0\,\text{GeV}^2$ where the contributions of the inner quark core to the resonance electrocouplings are much larger than from the external meson baryon cloud. In this range of photon virtualities, the evaluation [4] offers a good description of the experimental results on the transition $p \rightarrow \Delta(1232)3/2^+$ form factors and the $N(1440)1/2^+$ resonance electrocouplings achieved with a momentum dependence of the dressed quark mass that is *exactly the same* as the one employed in the previous evaluations of the elastic electromagnetic nucleon form factors [5]. This success strongly supports: (a) the relevance of dynamical dressed quarks, with properties predicted by the DSEQCD approach [12], as constituents of the quark core in the structure of the ground and excited nucleon states; and (b) the capability of the DSEQCD approach [4,5] to map out the dressed quark mass function from the experimental results on the Q^2-evolution of the nucleon elastic—and $p \rightarrow N^*$ electromagnetic transition form factors or rather $\gamma_v p N^*$ electrocouplings.

The model [1] employs the light cone some rules (LCSR) in order to relate the quark distribution amplitudes (DA) of the excited nucleon states to the Q^2 evolution of their electrocouplings. Analysis of the $N(1535)1/2^- \gamma_v p N^*$ electrocouplings from CLAS within the framework of this model has provided access to the quark DA's of excited nucleons for the first time. The model [1] describes successfully the CLAS results shown in Fig. 2 (right) at $Q^2 > 2.0$ GeV^2 where LCSR's are applicable, with the values of the normalization parameters for the leading twist $N(1535)1/2^-$ quark DA obtained from the QCD Lagrangian within the framework of lattice QCD [33]. The results on the $N(1535)1/2^-$ electrocouplings in the much broader range of photon virtualities up to 12 GeV^2 expected from the future N^* studies with the CLAS12 detector in Hall B [23,34,35] will considerably improve knowledge of the $N(1535)1/2^-$ quark DA, allowing us to determine from electrocoupling data not only the normalization parameters, but also most of the shape parameters of the quark DA's. These studies will be extended by analyses of other excited nucleon states. Confronting the quark DA's of excited nucleon states determined from the experimental results on $\gamma_v p N^*$ electrocouplings to the lattice QCD expectations, offers an alternative way with respect to DSEQCD approaches to study the emergence of the resonance structure from the first principles of the QCD.

The quark DA's of excited nucleon states are also expected from the future DSEQCD studies [7]. Consistency between the expectations for the N^* quark DA parameters obtained within the framework of LQCD/DSEQCD and their values from the fit to the data on resonance electrocouplings within the framework of LCSR will validate credible access to the fundamental ingredients of the non-perturbative strong interaction behind the emergence of the N^* structure from the QCD Lagrangian.

Analysis of the CLAS results on the $\gamma_v p N^*$ electrocouplings over most excited nucleon states in the mass range up to 1.8 GeV has revealed the N^* structure for $Q^2 < 5.0$ GeV^2 as a complex interplay between an inner core of three dressed quarks and an external meson–baryon cloud. The two extended quark models [27,28], that account for the meson–baryon cloud and the quark core contributions combined, provided a better description of the $N(1440)1/2^+$ electrocouplings shown in Fig. 1 (left) at $Q^2 < 2.0$ GeV^2, demonstrating the increasing importance of the meson–baryon contributions as Q^2 decreases. The credible DSEQCD evaluation of the quark core contributions to the electrocouplings of the $N(1440)1/2^+$ state has allowed us to derive the meson–baryon cloud contributions to this resonance as the difference between the experimental data on resonance electrocouplings and the quark core electroexcitation amplitudes from DSEQCD [4] shown by the blue line in Fig. 1 (left). The obtained meson–baryon cloud contributions are presented in Fig. 1 (left) by magenta dashed area. The relative contributions of the quark core and the meson–baryon cloud depend strongly on the quantum numbers of the excited nucleon state. The quark core becomes the dominant contributor to the $A_{1/2}$ electrocouplings of the $N(1440)1/2^-$ and $N(1520)3/2^-$ resonances at $Q^2 > 2.0$ GeV^2, as can be seen in Figs. 1 (left) and 2 (left), respectively. The results on these state electrocouplings offer almost direct access to the dressed quark contributions for $Q^2 > 2.0$ GeV^2. Instead, electrocouplings of the $N(1675)5/2^-$ state, shown in Fig. 1 (right), are dominated by meson–baryon cloud, allowing us to explore this component from the electrocoupling data. The relative contributions of the meson–baryon cloud to the electrocouplings of all resonances studied with the CLAS decrease with Q^2 in gradual transition towards quark core dominance at photon virtualities above 5.0 GeV^2.

4 Prospects for the Future Studies of Excited Nucleon States with the CLAS12

After completion of the Jefferson Lab 12 GeV Upgrade Project, the CLAS12 detector in the upgraded Hall B will be the only foreseen facility worldwide capable of studying nucleon resonances at the still unexplored ranges of the smallest photon virtualities 0.05 GeV$^2 < Q^2 < 0.5$ GeV^2 and the highest photon virtualities ever achieved in exclusive reaction measurements up to 12 GeV^2 [9,10,34].

The studies of nucleon resonances at small photon virtualities are driven by the search for new states of baryon matter, the so-called hybrid-baryons [36]. Small Q^2 is preferential for the observation of these

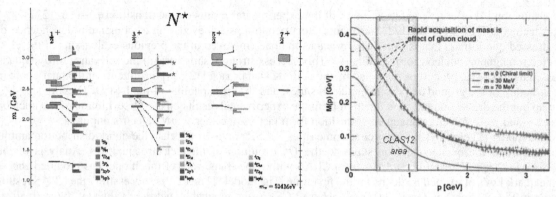

Fig. 3 *Left* Spectrum of N^* states computed in LQCD starting from the QCD Lagrangian [13] will be mapped out with CLAS12 with the primary objective of the hybrid baryon search [10,36]. The contributions from the relevant quark and glue configurations to the resonance structure are shown by the *horizontal bars* connected by *dashed lines to the boxes* showing the N^* masses and full widths with a pion mass of 524 MeV. The expected hybrid baryons with dominant contributions from glue are shown by the *blue boxes*. *Right* Momentum dependencies of the running dressed quark mass from LQCD at two values of the bare quark masses, 30 and 70 MeV, are shown by *points with error bars* [37] in comparison with the DSEQCD results (*green and magenta lines*) [12] for the same bare quark masses as in the LQCD studies. Evaluations in the chiral limit of massless bare quarks, which are close to the bare masses of the light u and d quarks, are currently available from DSEQCD only and shown by the *red line* [12]. The dressed quark mass function in the momentum range that will be accessible for the first time from the studies of the Q^2 evolution of the resonance electrocouplings from exclusive meson electroproduction data with CLAS12 [23,34] is shown in the *shadowed area*, allowing us to explore the nature of >98 % of hadron mass, quark–gluon confinement, and the emergence of the nucleon resonance structure from QCD [9] (color figure online)

new states that contain three dressed quarks and, in addition, glue as the structural component. The LQCD studies of the N^* spectrum starting from the QCD Lagrangian [13] predict several such states of the positive parities shown in Fig. 3 (left). The evaluations [13] were carried with much higher quark mass than needed to reproduce the pion mass. In order to estimate the physics masses of hybrid states we corrected the results of [13] reducing the predicted hybrid mass values by the differences between the experimental results on the masses of the known lightest N^* of the same spin-parities as for the expected hybrid baryons and their values from LQCD. In the experiment with CLAS12 we will search for the hybrid signal as the presence of extra states in the conventional resonance spectrum of $J^P = 1/2^+, 3/2^+$ in the mass range from 2.0 to 2.5 GeV from the data of exclusive KY and $\pi^+\pi^- p$ electroproduction off protons [36]. The hybrid nature of the new baryon states will be identified looking for the specific Q^2 evolution of their electrocouplings. We expect the specific behavior of the hybrid state electrocouplings with Q^2 because one might imagine that the three quarks in a hybrid baryon should be in a color-octet state in order to create a colorless hadron in combination with the glue constituent in a color-octet state. Instead, in regular baryons, constituent quarks should be in color-singlet state. So pronounced differences for quark configurations in the structure of conventional and hybrid baryons should results in a peculiar Q^2 evolution of hybrid baryon electrocouplings. The studies on the N^* structure at low Q^2 over spectrum of all prominent resonances will also extend our knowledge on the meson–baryon cloud in the resonance structure and will offer unique information on the N^* excitations by longitudinal photons as $Q^2 \to 0$. The extension of amplitude analysis [38–40] and coupled channel analysis [41,42] methods, which were employed successfully in the studies of exclusive meson photoproduction, for electroproduction off protons at small photon virtualities is of particular importance for the success of the aforementioned efforts.

The experiments on the studies of excited nucleon state structure in exclusive πN, KY, and $\pi^+\pi^- p$ electroproduction off protons at 5.0 GeV^2 < Q^2 < 12.0 GeV^2 [23,35] are scheduled in the first year of running with the CLAS12 detector. For the first time electrocouplings of all prominent nucleon resonances will become available at the highest photon virtualities ever achieved in the exclusive electroproduction studies. These distance scales correspond to the still unexplored regime for the N^* electroexcitation where the resonance structure is dominated by the quark core with almost negligible meson–baryon cloud contributions. The foreseen experiments offer almost direct access to the properties of dressed quarks inside N^* states of different quantum numbers. Consistent results on the dressed quark mass function derived from independent analyses of the data on the $\gamma_v p N^*$ electrocouplings of the resonances of distinctively different structure, such as radial excitations, spin-flavor flip, orbital excitations, will validate credible access to this fundamental ingredient of the non-perturbative strong interaction supported by the experimental data. The expected data on the $\gamma_v p N^*$ electrocouplings will provide for the first time access to the dressed quark mass function in the range of momenta/distance scales where the transition from the quark–gluon confinement to the pQCD regimes of strong interaction takes full effect, as is shown in Fig. 3 (right). Exploring the dressed quark mass function at these distances will allow us to address the most challenging open problems of the Standard Model: on the nature of >98 % of hadron mass, quark–gluon confinement, and the emergence of the N^* structure from the QCD Lagrangian [6,9,43,44].

In order to provide reliable evaluation of resonance electrocouplings at high photon virtualities, the reaction models for resonance electrocoupling extraction should be further developed, implementing explicitly quark degrees of freedom in the description of the non-resonant amplitudes. The efforts on implementation of the hand-bag diagrams for the part of non-resonant amplitudes in πN electroproduction off protons in the N^* region are underway [45].

Acknowledgments This material is based upon work supported by the U.S. Department of Energy, Office of Science, Office of Nuclear Physics under contract DE-AC05-06OR23177.

References

1. Anikin, I.V., Braun, V.M., Offen, N.: Electroproduction of the N*(1535) nucleon resonance in QCD. Phys. Rev. D **92**, 074044 (2015)
2. Aznauryan, I.G., Burkert, V.D.: Extracting meson–baryon contributions to the electroexcitation of the N(1675)$\frac{5}{2}^-$ nucleon resonance. Phys. Rev. C **92**, 015203 (2015)
3. Mokeev, V.I., Burkert, V.D., Carman, D.S., et al.: New results from the studies of the $N(1440)\frac{1}{2}^-$, $N(1520)\frac{3}{2}^-$, and $\Delta(1620)\frac{1}{2}^-$ resonances in exclusive $ep \to e'p'\pi^+\pi^-$ electroproduction with the CLAS detector. arXiv:1505.05460 [nucl-ex], accepted by Phys. Rev. C
4. Segovia, J., El-Bennich, B., Rojas, E., et al.: Completing the picture of the Roper resonance. Phys. Rev. Lett. **115**, 015203 (2015)
5. Segovia, J., Clöet, I.C., Roberts, C.D., et al.: Nucleon and Δ elastic and transition form factors. Few Body Syst. **55**, 1185 (2015)
6. Roberts, C.D.: Hadron physics and QCD: just the basic facts. J. Phys. Conf. Ser. **630**, 012051 (2015)
7. Roberts, C.D.: These Proceedings
8. Aznauryan, I.G., Burkert, V.D.: Electroexcitation of nucleon resonances. Prog. Part. Nucl. Phys. **67**, 1 (2012)
9. Aznauryan, I.G., et al.: Studies of nucleon resonance structure in exclusive meson electroproduction. Int. J. Mod. Phys. E **22**, 1330015 (2013)
10. Burkert, V.D.: These Proceedings
11. Mokeev, V.I., Aznauryan, I.G., Burkert, V.D., Gothe, R.W.: Recent results on the nucleon resonance spectrum and structure from the CLAS detector. arXiv:1508.04088 [nucl-ex]
12. Clöet, I.C., Roberts, C.D.: Explanation and prediction of observables using continuum strong QCD. Prog. Part. Nucl. Phys. **77**, 1 (2014)
13. Dudek, J.J., Edwards, R.G.: Hybrid Baryons in QCD. Phys. Rev. D **85**, 054016 (2012)
14. Lanza, L.: These Proceedings
15. CLAS Physics Data Base. http://clas.sinp.msu.ru/cgi-bin/jlab/db.cgi
16. Aznauryan, I.G., CLAS Collaboration, et al.: Electroexcitation of nucleon resonances from CLAS data on single pion electroproduction. Phys. Rev. C **80**, 055203 (2009)
17. Park, K., CLAS Collaboration, et al.: Measurements of $ep \to e'\pi^+n$ at $W = 1.6 - 2.0$ GeV and extraction of nucleon resonance electrocouplings at CLAS. Phys. Rev. C **91**, 045203 (2015)
18. Ripani, M., et al.: Measurement of $ep \to e'p\pi^+\pi^-$ and baryon resonance analysis. Phys. Rev. Lett. **91**, 022002 (2003)
19. Fedotov, G.V., CLAS Collaboration, et al.: Electroproduction of $p\pi^+\pi^-$ off protons at $0.2 < Q^2 < 0.6$-GeV^2 and 1.3 $< W < 1.57$ GeV with CLAS. Phys. Rev. C **79**, 015204 (2009)
20. Mokeev, V.I., et al.: Experimental Study of the $P_{11}(1440)$ and $D_{13}(1520)$ resonances from CLAS data on $ep \to e\pi^+\pi^-$. Phys. Rev. C **86**, 055203 (2012)
21. Mokeev, V.I., et al.: Model analysis of the $p\pi^+\pi^-$ electroproduction reaction on the proton. Phys. Rev. C **80**, 045212 (2009)
22. http://www.indiana.edu/~jpac/index.html
23. Carman, D.S., et al.: Exclusive $N^* \to KY$ studies with CLAS12.JLab Experiment E12-06-108A
24. Nucleon Resonance Photo-/Electrocouplings Determined from Analyses of Experimental Data on Exclusive Meson Electroproduction off Protons. https://userweb.jlab.org/~mokeev/resonance_electrocouplings
25. Beringer, J., et al.: Review of particle physics. Phys. Rev. D **86**, 1 (2012)
26. Dugger, M., et al.: CLAS Collaboration: π^+ photoproduction on the proton for photon energies from 0.725 to 2.875 GeV. Phys. Rev. C **79**, 065206 (2009)
27. Aznauryan, I.G., Burkert, V.D.: Nucleon electromagnetic form factors and electroexcitation of low lying nucleon resonances in a light-front relativistic quark model. Phys. Rev. C **85**, 055202 (2012)
28. Obukhovsky, I.T., et al.: Electromagnetic structure of the nucleon and the Roper resonance in a light-front quark approach. Phys. Rev. D **89**, 0142032 (2014)
29. Julia-Diaz, B., et al.: Dynamical coupled-channels effects on pion photoproduction. Phys. Rev. C **77**, 045205 (2008)

30. Santopinto, E., Giannini, M.M.: Systematic study of longitudinal and transverse helicity amplitudes in the hypercentral constituent quark model. Phys. Rev. C **86**, 065202 (2012)
31. Roberts, C.D.: Running masses in the nucleon and its resonances. arXiv:1509.08952 [nucl-th]
32. Denizli, H., et al.: Q^2 dependence of the $S_{11}(1535)$ photocoupling and evidence for a P-wave resonance in eta electroproduction. Phys. Rev. C **76**, 015204 (2007)
33. Braun, V.M., et al.: Light-cone distribution amplitudes of the nucleon and negative parity nucleon resonances from lattice QCD. Phys. Rev. D **89**, 094511 (2014)
34. Gothe, R.W.: These Proceedings
35. Gothe, R.W. et al.: Nucleon resonance studies with CLAS12. JLab Experiment E12-09-003
36. D'Angelo, A., et al.: Search for hybrid Baryons with CLAS12 in Hall B, Jefferson Lab Letter of Intent LOI12-15-004
37. Bowman, P.O., et al.: Quark propagator from LQCD and its physical implications. Lecture Notes in Physics, vol. **663**, p. 17 (2005)
38. Mathieu, V., Fox, G., Szczepaniak, A.P.: Neutral pion photoproduction in a Regge model. Phys. Rev. D **92**, 074013 (2015)
39. Sarantsev, A.V.: These Proceedings
40. Strakovsky, I., et al.: SAID analysis of meson photoproduction: determination of neutron and proton EM couplings EPJ Web Conf. **73**, 04003 (2014)
41. Lee, T.-S.H.: These Proceedings
42. Kamano, H.: These Proceedings
43. Brodsky, S.J., et al.: QCD and Hadron Physics. arXiv:1502.05728 [hep-ph]
44. Dudek, J., et al.: Physics opportunities with the 12 GeV upgrade at Jefferson Lab. Eur. Phys. J. A **48**, 187 (2012)
45. Kroll, P.: These Proceedings

Few-Body Syst (2016) 57:941–948
DOI 10.1007/s00601-016-1131-z

Daniel S. Carman · CLAS Collaboration

Nucleon Resonance Structure Studies via Exclusive *KY* Electroproduction

Received: 9 January 2016 / Accepted: 1 June 2016 / Published online: 16 June 2016
© Springer-Verlag Wien 2016

Abstract Studying the structure of excited nucleon states employing the electroproduction of exclusive reactions is an important avenue for exploring the nature of the non-perturbative strong interaction. The electrocouplings of N^* states in the mass range below 1.8 GeV have been determined from analyses of CLAS πN, ηN, and $\pi \pi N$ data. This work has made it clear that consistent results from independent analyses of several exclusive channels with different couplings and non-resonant backgrounds but the same N^* electro-excitation amplitudes, is essential to have confidence in the extracted results. In terms of hadronic coupling, many high-lying N^* states preferentially decay through the $\pi \pi N$ channel instead of πN. Data from the KY channels will therefore be critical to provide an independent analysis to compare the extracted electrocouplings for the high-lying N^* states against those determined from the πN and $\pi \pi N$ channels. A program to study excited N^* state structure in both non-strange and strange exclusive electroproduction channels using CLAS12 will measure differential cross sections and polarization observables to be used as input to extract the $\gamma_v NN^*$ electrocoupling amplitudes for the most prominent N^* states in the range of invariant energy W up 3 GeV in the virtually unexplored domain of momentum transfers Q^2 up to 12 GeV^2.

1 Introduction

Intensive spectroscopy of the nucleon excitation spectrum and detailed studies of the structure of these excited states has played a pivotal role in the development of our understanding of the strong interaction. The concept of quarks that emerged through such studies led to the development of the constituent quark model [1–3] (CQM). As a result of intense experimental and theoretical effort over the past 30 years, it is now apparent that the structure of the states in the nucleon excitation spectrum is much more complex than what can be described in terms of models based on constituent quarks alone. At the typical energy and distance scales found within the N^* states, the quark–gluon coupling is large. Therefore, we are confronted with the fact that quark–gluon confinement, hadron mass generation, and the dynamics that give rise to the N^* spectrum, cannot be understood within the framework of perturbative quantum chromodynamics (QCD). The need to understand QCD in this non-perturbative domain is a fundamental issue in nuclear physics, which the study of N^* structure can help to address. Such studies, in fact, represent a necessary step toward understanding how QCD in the regime of large quark–gluon couplings generates mass and how systems of confined quarks and gluons, i.e. mesons and baryons, are formed.

Studies of low-lying nucleon excited states using electromagnetic probes at four-momentum transfer $Q^2 <$ 5 GeV^2 have revealed that the structure of these states is a complex interplay between the internal core of three dressed quarks and an external "cloud" of quark–antiquark pairs referred to as the meson–baryon cloud.

This article belongs to the special issue "Nucleon Resonances".

D. S. Carman (✉)
Jefferson Laboratory, 12000 Jefferson Ave., Newport News, VA 23606, USA
E-mail: carman@jlab.org

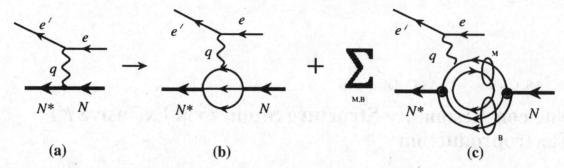

Fig. 1 Schematic representation of the $\gamma^* N \to N^*$ electroproduction process. **a** The fully dressed $\gamma_v NN^*$ electrocoupling that determines the N^* contribution to the resonant part of the meson electroproduction amplitude. **b** The contribution of the three-quark core. **c** The contribution from the meson–baryon cloud, where the sum is over all intermediate meson and baryon states. This figure is taken from Ref. [9]

N^* states of different quantum numbers have significantly different relative contributions from these two components, demonstrating distinctly different manifestations of the non-perturbative strong interaction in their generation. The relative contribution of the quark core increases with Q^2 in a gradual transition to a dominance of quark degrees of freedom for $Q^2 > 5$ GeV^2. This kinematics area still remains almost unexplored in exclusive reactions. Studies of the Q^2 evolution of N^* structure from low to high Q^2 offer access to the strong interaction between dressed quarks in the non-perturbative regime that is responsible for N^* formation.

Electroproduction reactions $\gamma^* N \to N^* \to M + B$ provide a tool to probe the inner structure of the contributing N^* resonances through the extraction of the amplitudes for the transition between the virtual photon–nucleon initial state and the excited N^* state, i.e. the $\gamma_v NN^*$ electrocoupling amplitudes, which are directly related to the N^* structure. These electrocouplings can be represented by the so-called helicity amplitudes [4], among which are $A_{1/2}(Q^2)$ and $A_{3/2}(Q^2)$, which describe the N^* resonance electroexcitation for the two different helicity configurations of a transverse photon and the nucleon, as well as $S_{1/2}(Q^2)$, which describes the N^* resonance electroexcitation by longitudinal photons of zero helicity. Detailed comparisons of the theoretical predictions for these amplitudes with their experimental measurements form the basis of progress toward gauging our understanding of non-perturbative QCD. The measurement of the $\gamma_v NN^*$ electrocouplings is needed in order to gain access to the dynamical momentum-dependent mass and structure of the dressed quark in the non-perturbative domain where the quark–gluon coupling is large [5], through mapping of the dressed quark mass function [6] and extractions of the quark distribution amplitudes for N^* states of different quantum numbers [7]. This is critical in exploring the nature of quark–gluon confinement and dynamical chiral symmetry breaking (DCSB) in baryons.

Figure 1 illustrates the two contributions to the $\gamma_v NN^*$ electrocouplings. In Fig. 1b the virtual photon interacts directly with the constituent quark, an interaction that is sensitive to the quark current and depends on the quark-mass function. However, the full meson electroproduction amplitude in Fig. 1a requires contributions to the $\gamma_v NN^*$ vertex from both non-resonant meson electroproduction and the hadronic scattering amplitudes as shown in Fig. 1c. These contributions incorporate all possible intermediate meson–baryon states and all possible meson–baryon scattering processes that eventually result in the N^* formation in the intermediate state of the reaction. These two contributions can be separated from each another using, for example, a coupled-channel reaction model [8].

Current theoretical approaches to understand N^* structure fall into two broad categories. In the first category are those that enable direct connection to the QCD Lagrangian, such as Lattice QCD (LQCD) and QCD applications of the Dyson–Schwinger equations (DSE). In the second category are those that use models inspired by or derived from our knowledge of QCD, such as quark–hadron duality, light-front holographic QCD (AdS/QCD), light-cone sum rules (LCSR), and CQMs. See Ref. [9] for an overview of these different approaches. It is important to realize that even those approaches that attempt to solve QCD directly can only do so approximately, and these approximations ultimately represent limitations that need careful consideration. As such, it is imperative that whenever possible the results of these intensive and challenging calculations be compared directly to the data on resonance electrocouplings from electroproduction experiments over a broad range of Q^2 for N^* states with different quantum numbers.

Springer

2 CLAS N^* Program

Studies of the structure of the excited nucleon states, the so-called N^* program, is one of the key cornerstones of the physics program in Hall B at Jefferson Laboratory (JLab). The large acceptance spectrometer CLAS [10], which began data taking in 1997 and was decommissioned in 2012, was designed to measure photo- and electroproduction cross sections and polarization observables for beam energies up to 6 GeV over a broad kinematic range for a host of different exclusive reaction channels. Consistent determination of N^* properties from different exclusive channels with different couplings and non-resonant backgrounds offers model-independent support for the findings.

To date photoproduction data sets from CLAS and elsewhere have been used extensively to constrain coupled-channel fits and advanced single-channel models. However, data at $Q^2 = 0$ allows us to identify N^* states and determine their quantum numbers, but tell us very little about the structure of these states. It is the Q^2 dependence of the $\gamma_v NN^*$ electrocouplings that unravel and reveal these details. In addition, electrocoupling data are promising for studies of nucleon excited states as the ratio of resonant to non-resonant amplitudes increases with increasing Q^2. An important constraint on the studies of electroproduction data is the fit parameters for the N^* states must be described by Q^2-independent resonance masses and hadronic decay widths.

The goal of the N^* program with CLAS is to study the spectrum of N^* states and their associated structure over a broad range of distance scales through studies of the Q^2 dependence of the $\gamma_v NN^*$ electrocouplings. For each final state this goal is realized through two distinct phases. The first phase consists of the measurements of the cross sections and polarization observables in as fine a binning in the relevant kinematic variables Q^2, W, $d\tau_{hadrons}$ (where $d\tau_{hadrons}$ represents the phase space of the final state hadrons) as the data support. The second phase consists of developing advanced reaction models that completely describe the data over its full phase space in order to then extract the electrocoupling amplitudes for the dominant contributing N^* states.

Electrocoupling amplitudes for most N^* states below 1.8 GeV have been extracted for the first time from analysis of CLAS data in the exclusive $\pi^+ n$ and $\pi^0 p$ channels for Q^2 up to 5 GeV^2, in ηp for Q^2 up to 4 GeV^2, and for $\pi^+\pi^- p$ for Q^2 up to 1.5 GeV^2. Figure 2 shows representative CLAS data for the $A_{1/2}$ electrocouplings for the $N(1440)\frac{1}{2}^+$, $N(1520)\frac{3}{2}^-$, and $N(1675)\frac{5}{2}^-$ [9,11–13]. Studies of the electrocouplings for N^* states of different quantum numbers at lower Q^2 have revealed a very different interplay between the inner quark core and the meson–baryon cloud as a function of Q^2. Structure studies of the low-lying N^* states, e.g. $\Delta(1232)\frac{3}{2}^+$, $N(1440)\frac{1}{2}^+$, $N(1520)\frac{3}{2}^-$, and $N(1535)\frac{1}{2}^-$, have made significant progress in recent years due to the agreement of results from independent analyses of the CLAS πN and $\pi\pi N$ final states [11]. The good agreement of the extracted electrocouplings from both the πN and $\pi\pi N$ exclusive channels is non-trivial in that these channels have very different mechanisms for the non-resonant backgrounds. The agreement thus provides compelling evidence for the reliability of the results.

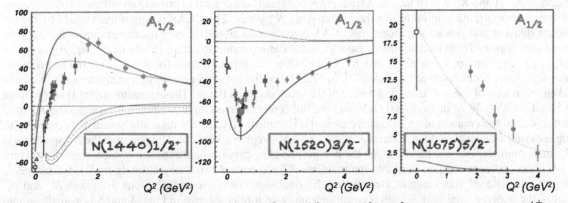

Fig. 2 The $A_{1/2}$ electrocoupling amplitudes (in units of 10^{-3} GeV$^{-1/2}$) versus Q^2 (GeV^2) for the N^* states $N(1440)\frac{1}{2}^+$ (*left*), $N(1520)\frac{3}{2}^-$ (*middle*), and $N(1675)\frac{5}{2}^-$ (*right*) from analyses of the CLAS πN (*circles*) and $\pi\pi N$ (*triangles, squares*) data [9,11–13]. (*Left*) calculation from a non-relativistic light-front quark model with a running quark mass (*red line*) [14] and calculation of the quark core from the DSE approach (*blue line*) [5]. (*Middle/right*) calculations from the hypercentral constituent quark model (*blue lines*) [15]. The magnitude of the meson–baryon cloud contributions is shown by the *magenta line* (or band) on each plot [16] (color figure online)

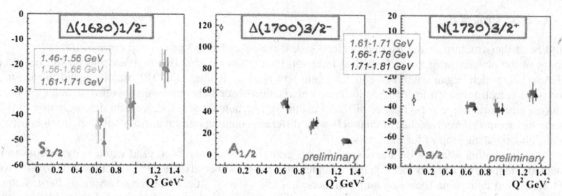

Fig. 3 CLAS results for the N^* electrocoupling amplitudes (in units of 10^{-3} GeV$^{-1/2}$) from analysis of the exclusive $\pi^+\pi^- p$ final state as a function of Q^2 (GeV^2). (*Left*) $S_{1/2}$ of the $\Delta(1620)\frac{1}{2}^-$ [12], (*middle*) preliminary extraction of $A_{1/2}$ for the $\Delta(1700)\frac{3}{2}^-$ [13], and (*right*) preliminary extraction of $A_{3/2}$ for the $N(1720)\frac{3}{2}^+$ [13]. Each electrocoupling amplitude was extracted in independent fits in different bins of W across the resonance peak width as shown for each Q^2 bin (points in each Q^2 bin offset for clarity)

The size of the meson–baryon dressing amplitudes are maximal for $Q^2 < 1$ GeV^2 (see Fig. 2). For increasing Q^2, there is a gradual transition to the domain where the quark degrees of freedom begin to dominate, as seen by the improved description of the N^* electrocouplings obtained within the DSE approach, which accounts only for the quark core contributions. This contribution can also be predicted within the hypercentral CQM [15]. For $Q^2 > 5$ GeV^2, the quark degrees of freedom are expected to fully dominate the N^* states [9]. Therefore, in the $\gamma_v NN^*$ electrocoupling studies for $Q^2 > 5$ GeV^2 expected with the future CLAS12 program (see Sect. 3), the quark degrees of freedom will be probed more directly with only small contributions from the meson–baryon cloud.

Analysis of CLAS data for the $\pi\pi N$ channel has provided the only detailed structural information available regarding higher-lying N^* states, e.g. $\Delta(1620)\frac{1}{2}^-$, $N(1650)\frac{1}{2}^-$, $N(1680)\frac{5}{2}^+$, $\Delta(1700)\frac{3}{2}^-$, and $N(1720)\frac{3}{2}^+$. Figure 3 shows a representative set of illustrative examples for $S_{1/2}$ for the $\Delta(1620)\frac{1}{2}^-$ [12], as well as for $A_{1/2}$ for the $\Delta(1700)\frac{3}{2}^-$ and $A_{3/2}$ for the $N(1720)\frac{3}{2}^+$ [13]. Here the analysis for each N^* state was carried out independently in different bins of W across the width of the resonance for Q^2 up to 1.5 GeV^2 with very good correspondence within each Q^2 bin. Note that most of the N^* states with masses above 1.6 GeV decay preferentially through the $\pi\pi N$ channel instead of the πN channel.

With a goal to have an independent determination of the electrocouplings for each N^* state from multiple exclusive reaction channels, a natural avenue to investigate for the higher-lying N^* states is the strangeness channels $K^+\Lambda$ and $K^+\Sigma^0$. In fact, data from the KY channels are critical to provide an independent extraction of the electrocoupling amplitudes for the higher-lying N^* states. The CLAS program has yielded by far the most extensive and precise measurements of KY electroproduction data ever measured across the nucleon resonance region. These measurements have included the separated structure functions σ_T, σ_L, $\sigma_U = \sigma_T + \epsilon\sigma_L$, σ_{LT}, σ_{TT}, and $\sigma_{LT'}$ for $K^+\Lambda$ and $K^+\Sigma^0$ [17–20], recoil polarization for $K^+\Lambda$ [21], and beam-recoil transferred polarization for $K^+\Lambda$ and $K^+\Sigma^0$ [22,23]. For the hyperon polarization measurements, we have taken advantage of the self-analyzing nature of the weak decay of the Λ. These measurements span Q^2 from 0.5 to 4.5 GeV^2, W from 1.6 to 3.0 GeV, and the full center-of-mass angular range of the K^+. The KY final states, due to the creation of an $s\bar{s}$ quark pair in the intermediate state, are naturally sensitive to coupling to higher-lying s-channel resonance states at $W > 1.6$ GeV, a region where our knowledge of the N^* spectrum is the most limited. Note also that although the two ground-state hyperons have the same valence quark structure (uds), they differ in isospin, such that intermediate N^* resonances can decay strongly to $K^+\Lambda$ final states, but intermediate Δ^* states cannot. Because $K^+\Sigma^0$ final states can have contributions from both N^* and Δ^* states, the hyperon final state selection constitutes an isospin filter. Shown in Figs. 4 and 5 is a small sample of the available data in the form of the $K^+\Lambda$ and $K^+\Sigma^0$ structure functions σ_U, σ_{LT}, σ_{TT}, and $\sigma_{LT'}$ [20,24], illustrating its broad kinematic coverage and statistical precision.

While there has been progress toward a better understanding of the low-lying N^* states in the region below 1.6 GeV, the vast majority of the predicted missing N^* and Δ^* states lie in the region from $1.6 < W < 3$ GeV. To date the PDG lists only four N^* states, $N(1650)\frac{1}{2}^-$, $N(1710)\frac{1}{2}^+$, $N(1720)\frac{3}{2}^+$, and $N(1900)\frac{3}{2}^+$, with known

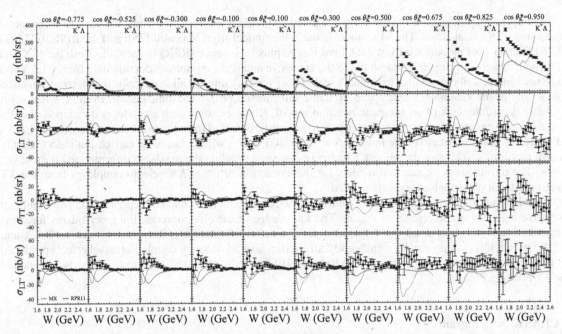

Fig. 4 Structure functions $\sigma_U = \sigma_T + \epsilon\sigma_L$, σ_{LT}, σ_{TT}, and $\sigma_{LT'}$ (nb/sr) for $K^+\Lambda$ production versus W (GeV) for $E_{beam} = 5.5$ GeV for $Q^2 = 1.80$ GeV^2 and $\cos\theta_K^*$ values as shown from CLAS data [20,24]. The *error bars* represent the statistical uncertainties only. The *red curves* are from the hadrodynamic KY model of Maxwell [25] and the *blue curves* are from the hybrid RPR-2011 KY model from Ghent [26] (color figure online)

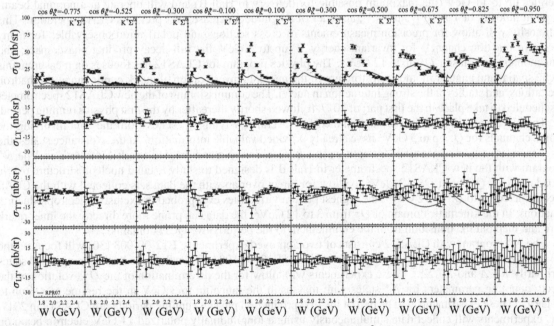

Fig. 5 Structure functions $\sigma_U = \sigma_T + \epsilon\sigma_L$, σ_{LT}, σ_{TT}, and $\sigma_{LT'}$ (nb/sr) for $K^+\Sigma^0$ production versus W (GeV) for $E_{beam} = 5.5$ GeV for $Q^2 = 1.80$ GeV^2 and $\cos\theta_K^*$ values as shown from CLAS data [20,24]. The *error bars* represent the statistical uncertainties only. The *blue curves* are from the hybrid RPR-2007 KY model from Ghent [27] (color figure online)

couplings to $K\Lambda$ and no N^* states are listed that couple to $K\Sigma$ [28]; only a single Δ^* state, $\Delta(1920)\frac{3}{2}^+$, is listed with coupling strength to $K\Sigma$. The branching ratios to KY provided for these states are typically less than 10 % with uncertainties on the order of the measured coupling. While the relevance of this core set of N^* states in the $\gamma^{(*)}p \rightarrow K^+\Lambda$ reaction has long been considered a well-established fact, this set of states falls well short of reproducing the experimental results for $W < 2$ GeV.

Figures 4 and 5 include two of the more advanced single channel reaction models for the electromagnetic production of KY final states. The MX model is the isobar model from Maxwell [25], and the RPR-2007 [27] and RPR-2011 [26] models are from the Ghent Regge plus Resonance (RPR) framework. Both the MX and RPR models were developed based on fits to the extensive and precise photoproduction data from CLAS and elsewhere and describe those data reasonably well. However, they utterly fail to describe the electroproduction data in any of the kinematic phase space. Reliable information on KY hadronic decays from N^*s is not yet available due to the lack of an adequate reaction model. However, after such a model is developed, the N^* electrocoupling amplitudes for states that couple to KY can be obtained from fits to the extensive existing CLAS KY electroproduction data over the range $0.5 < Q^2 < 4.5$ GeV^2, which should be carried out independently in different bins of Q^2 with the same KY hadronic decays, extending the available information on these N^* states. The development of reaction models for the extraction of the $\gamma_v NN^*$ electrocouplings from the KY electroproduction channels is urgently needed.

It is also important to note that the πN and $\pi\pi N$ electroproduction channels represent the two dominant exclusive channels in the resonance region. The knowledge of the electroproduction mechanisms for these channels is critically important for N^* studies in channels with smaller cross sections such as $K^+\Lambda$ and $K^+\Sigma^0$ production, as they can be significantly affected in leading order by coupled-channel effects produced by their hadronic interactions in the pionic channels. Ultimately such effects need to be properly included in the KY reaction models.

3 CLAS12 N^* Program

As part of the upgrade of the JLab accelerator from a maximum electron beam energy of 6 GeV to a maximum energy of 12 GeV, a new large acceptance spectrometer called CLAS12 was designed for experimental Hall B to replace the CLAS spectrometer. The new CLAS12 spectrometer [29] is designed for operation at beam energies up to 11 GeV (the maximum possible for delivery to Hall B) and will operate at a nominal beam-target luminosity of 1×10^{35} cm^{-2}s^{-1}, an order of magnitude increase over previous CLAS operation. This luminosity will allow for precision measurements of cross sections and polarization observables for many exclusive reaction channels for invariant energy W up to 3 GeV, the full decay product phase space, and four-momentum transfer Q^2 up to 12 GeV^2. The physics program for CLAS12 has focuses on measurements of the spatial and angular momentum structure of the nucleon, investigation of quark confinement and hadron excitations, and studies of the strong interaction in nuclei. The commissioning of the new CLAS12 spectrometer is scheduled to take place in the first part of 2017, followed shortly thereafter by the first physics running period.

The electrocoupling parameters determined for several low-lying N^* states from the data involving the pionic channels for Q^2 up to 5 GeV^2 have already provided valuable information. At these distance scales, the resonance structure is determined by both meson–baryon dressing and dressed quark contributions. The N^* program with the new CLAS12 spectrometer in Hall B is designed to study excited nucleon structure over a broad range of Q^2, from $Q^2 = 3$ GeV^2 to allow for direct overlap with the data sets collected with the CLAS spectrometer, up to $Q^2 = 12$ GeV^2, the highest photon virtualities ever probed in exclusive electroproduction reactions. In the kinematic domain of Q^2 from 3 to 12 GeV^2, the data can probe more directly the inner quark core and map out the transition from the confinement to the perturbative QCD domains.

The N^* program with CLAS12 consists of two approved experiments. E12-09-003 [30] will focus on the non-strange final states (primarily πN, ηN, $\pi\pi N$) and E12-06-108A [31] will focus on the strange final states (primarily $K^+\Lambda$ and $K^+\Sigma^0$). These experiments will allow for the determination of the Q^2 evolution of the electrocoupling parameters for N^* states with masses in the range up to 3 GeV in the regime of Q^2 up to 12 GeV^2. These experiments will be part of the first production physics running period with CLAS12 in 2017. The experiments will collect data simultaneously using a longitudinally polarized 11 GeV electron beam on an unpolarized liquid-hydrogen target.

The program of N^* studies with the CLAS12 detector has a number of important objectives. These include:

1. To map out the quark structure of the dominant N^* and Δ^* states from the acquired electroproduction data through the exclusive final states including the non-strange channels $\pi^0 p$, $\pi^+ n$, ηp, $\pi^+\pi^- p$, as well as the dominant strangeness channels $K^+\Lambda$ and $K^+\Sigma^0$. This objective is motivated by results from existing analyses such as those shown in Fig. 2, where it is seen that the meson–baryon dressing contribution to the N^* structure decreases rapidly with increasing Q^2. The data can be described approximately in terms of dressed quarks already for Q^2 up to 3 GeV^2. It is therefore expected that the data at $Q^2 > 5$ GeV^2 can be used more directly to probe the quark substructure of the N^* and Δ^* states [9]. The comparison of the

extracted resonance electrocoupling parameters from this new higher Q^2 regime to the predictions from LQCD and DSE calculations will allow for a much improved understanding of how the internal dressed quark core emerges from QCD and how the dynamics of the strong interaction are responsible for the formation of N^* and Δ^* states of different quantum numbers.

2. To investigate the dynamics of dressed quark interactions and how they emerge from QCD to generate N^* states of different quantum numbers. This work is motivated by recent advances in the DSE approach [32,33] and LQCD [34], which have provided links between the dressed quark propagator, the dressed quark scattering amplitudes, and the QCD Lagrangian. These approaches also relate the momentum dependence of the dressed quark mass function to the $\gamma_v NN^*$ electrocouplings for N^* states of different quantum numbers. DSE analyses of the extracted N^* electrocoupling parameters have the potential to allow for investigation of the origin of quark–gluon confinement in baryons and the nature of more than 98 % of the hadron mass generated non-perturbatively through DSCB, since both of these phenomena are rigorously incorporated into the DSE approach [9]. Efforts are currently underway to study the sensitivity of the proposed electromagnetic amplitude measurements to different parameterizations of the momentum dependence of the quark mass [35].

3. To offer constraints from resonance electrocoupling amplitudes on the Generalized Parton Distributions (GPDs) describing $N \rightarrow N^*$ transitions. We note that a key aspect of the CLAS12 measurement program is the characterization of exclusive reactions at high Q^2 in terms of GPDs. The elastic and $\gamma_v NN^*$ transition form factors represent the first moments of the GPDs [36,37], and they provide for unique constraints on the structure of nucleons and their excited states. Thus the N^* program at high Q^2 represents the initial step in a reliable parameterization of the transition $N \rightarrow N^*$ GPDs and is an important part of the larger overall CLAS12 program studying exclusive reactions.

4 Concluding Remarks

The study of the spectrum and structure of the excited nucleon states represents one of the key physics foundations for the measurement program in Hall B with the CLAS spectrometer. To date measurements with CLAS have provided a dominant amount of precision data (cross sections and polarization observables) for a number of different exclusive final states for Q^2 from 0 to 4.5 GeV^2. From the πN and $\pi\pi N$ data, the electrocouplings of most N^* states up to ~ 1.8 GeV have been extracted for the first time. With the development and refinement of reaction models to describe the extensive CLAS $K^+\Lambda$ and $K^+\Sigma^0$ electroproduction data, the data from the strangeness channels is expected to provide an important complement to study the electrocoupling parameters for higher-lying N^* resonances with masses above 1.6 GeV.

The N^* program with the new CLAS12 spectrometer will extend these studies up to Q^2 of 12 GeV^2, the highest photon virtualities ever probed in exclusive reactions. This program will ultimately focus on the extraction of the $\gamma_v NN^*$ electrocoupling amplitudes for the s-channel resonances that couple strongly to the non-strange final states πN, ηN, and $\pi\pi N$, as well as the strange $K^+\Lambda$ and $K^+\Sigma^0$ final states. These studies in concert with theoretical developments will allow for insight into the strong interaction dynamics of dressed quarks and their confinement in baryons over a broad Q^2 range. The data will address the most challenging and open problems of the Standard Model on the nature of hadron mass, quark–gluon confinement, and the emergence of the N^* states of different quantum numbers from QCD.

Acknowledgments This work was supported by the U.S. Department of Energy. The author is grateful for many lengthy and fruitful discussions on this topic with Victor Mokeev and Ralf Gothe. The author also thanks the organizers of the ECT* 2015 Workshop Nucleon Resonances: From Photoproduction to High Photon Virtualities for the opportunity to present this work and participate in this workshop.

References

1. Morpurgo, G.: Is a non-relativistic approximation possible for the internal dynamics of elementary particles? Physics **2**, 95 (1965)
2. Isgur, N., Karl, G.: Positive parity excited baryons in a quark model with hyperfine interactions. Phys. Rev. D **19**, 2653 (1979)
3. Capstick, S., Isgur, N.: Baryons in a relativized quark model with chromodynamics. Phys. Rev. D **34**, 2809 (1986)
4. Aznauryan, I.G., Burkert, V.D.: Electroexcitation of nucleon resonances. Prog. Part. Nucl. Phys. **67**, 1 (2012)
5. Segovia, J., et al.: Completing the picture of the Roper resonance. Phys. Rev. Lett. **115**, 171801 (2015)

6. Cloët, I.C., Roberts, C.D.: Explanation and prediction of observables using continuum strong QCD. Prog. Part. Nucl. Phys. **77**, 1 (2014)
7. Anikin, I.V., et al.: Electroproduction of the $N^*(1535)$ nucleon resonance in QCD. Phys. Rev. D **92**, 014018 (2015)
8. Kamano, H., Nakamura, S.X., Lee, T.-S.H., Sato, T.: Nucleon resonances within a dynamical coupled-channels model of πN and γN reactions. Phys. Rev. C **88**, 035209 (2013)
9. Aznauryan, I.G., et al.: Studies of nucleon resonance structure in exclusive meson electroproduction. Int. J. Mod. Phys. **E22**, 1330015 (2013)
10. Mecking, B.A., et al.: The CEBAF large acceptance spectrometer (CLAS). Nucl. Instrum. Methods A **503**, 513 (2003)
11. Mokeev, V.I., Aznauryan, I.G.: Studies of N^* structure from the CLAS meson electroproduction data. Int. J. Mod. Phys. Conf. Ser. **26**, 1460080 (2014)
12. Mokeev, V.I., et al.: arXiv:1509.05460 (submitted for publication) (2015)
13. Mokeev, V.I.: Presentation at the ECT* Workshop, Nucleon Resonances: From Photoproduction to High Photon Virtualities (2015). See these proceedings
14. Aznauryan, I.G., Burkert, V.D.: Nucleon electromagnetic form factors and electroexcitation of low lying nucleon resonances in a light-front relativistic quark model. Phys. Rev. C **85**, 055202 (2012)
15. Santopinto, E., Giannini, M.M.: Systematic study of longitudinal and transverse helicity amplitudes in the hypercentral constituent quark model. Phys. Rev. C **86**, 065202 (2012)
16. Julia-Diaz, B., et al.: Dynamical coupled-channels effects on pion photoproduction. Phys. Rev. C **77**, 045205 (2008)
17. Raue, B.A., Carman, D.S.: Ratio of σ_L/σ_T for $p(e, e'K^+)\Lambda$ extracted from polarization transfer. Phys. Rev. C **71**, 065209 (2005)
18. Ambrozewicz, P., et al. (CLAS Collaboration): Separated structure functions for the exclusive electroproduction of $K^+\Lambda$ and $K^+\Sigma^0$ final states. Phys. Rev. C **75**, 045203 (2007)
19. Nasseripour, R., et al. (CLAS Collaboration): Polarized structure function $\sigma_{LT'}$ for $^1H(\vec{e}, e'K^+)\Lambda$ in the nucleon resonance region. Phys. Rev. C **77**, 065208 (2008)
20. Carman, D.S., et al. (CLAS Collaboration): Separated structure functions for exclusive $K + \Lambda$ and $K^+\Sigma^0$ electroproduction at 5.5 GeV with CLAS. Phys. Rev. C **87**, 025204 (2013)
21. Gabrielyan, M., et al. (CLAS Collaboration): Induced polarization of Λ (1116) in Kaon electroproduction. Phys. Rev. C **90**, 035202 (2014)
22. Carman, D.S., et al. (CLAS Collaboration): First measurement of transferred polarization in the exclusive $\vec{e}p \rightarrow e'K^+\vec{\Lambda}$ reaction. Phys. Rev. Lett. **90**, 131804 (2003)
23. Carman, D.S., et al. (CLAS Collaboration): Beam-recoil polarization transfer in the nucleon resonance region in the exclusive $\vec{e}p \rightarrow e'K^+\vec{\Lambda}$ and $\vec{e}p \rightarrow e'K^+\vec{\Sigma^0}$ reactions at the CLAS spectrometer. Phys. Rev. C **79**, 065205 (2009)
24. CLAS physics database. http://clasweb.jlab.org/physicsdb
25. Maxwell, O.: Electromagnetic production of Kaons from protons and baryon electromagnetic form factors. Phys. Rev. C **85**, 034611 (2012)
26. De Cruz, L., et al.: A Bayesian analysis of Kaon photoproduction with the Regge-plus-resonance model. Phys. Rev. C **86**, 015212 (2012)
27. Corthals, T., et al.: Electroproduction of Kaons from the proton in a Regge-plus-resonance approach. Phys. Lett. B **656**, 186 (2007)
28. Beringer, J., et al. (PDG): Review of particle physics (RPP). Phys. Rev. D **86**, 010001 (2012)
29. See CLAS12 web page at http://www.jlab.org/Hall-B/clas12-web
30. Burkert, V.D., Cole, P., Gothe, R., Joo, K., Mokeev, V., Stoler, P.: JLab Experiment E12-09-003 spokespersons
31. Carman, D.S., Gothe, R., Mokeev, V.: JLab Experiment E12-06-108A, spokespersons
32. Bhagwat, M., Tandy, P.: Analysis of full-QCD and quenched-QCD lattice propagators. AIP Conf. Proc. **842**, 225 (2006)
33. Roberts, C.D.: Hadron properties and Dyson–Schwinger equations. Prog. Part. Nucl. Phys. **61**, 55 (2008)
34. Bowman, P.O., et al.: Unquenched quark propagator in Landau gauge. Phys. Rev. D **71**, 054507 (2005)
35. Cloët, I.C., Roberts, C.D., Thomas, A.W.: Revealing dressed-quarks Via the proton's charge distribution. Phys. Rev. Lett. **111**, 101803 (2013)
36. Frankfurt, L.L., et al.: Hard exclusive electroproduction of decuplet baryons in the large N(c) limit. Phys. Rev. Lett. **84**, 2589 (2000)
37. Goeke, K., Polyakov, M.V., Vanderhaeghen, M.: Hard exclusive reactions and the structure of hadrons. Prog. Part. Nucl. Phys. **47**, 401 (2001)

Few-Body Syst (2016) 57:1035–1040
DOI 10.1007/s00601-016-1145-6

Kijun Park

Exclusive Single Pion Electroproduction off the Proton: Results from CLAS

Received: 22 February 2016 / Accepted: 26 July 2016 / Published online: 13 August 2016
© Springer-Verlag Wien 2016

Abstract Exclusive meson electroproduction off protons is a powerful tool to probe the effective degrees of freedom in excited nucleon states at the varying distance scale where the transition from the contributions of both quark core and meson-baryon cloud to the quark core dominance. During the past decade, the CLAS collaboration has executed a broad experimental program to study the excited states of the proton using polarized electron beam and both polarized and unpolarized proton targets. The measurements covered a broad kinematic range in the invariant mass W and photon virtuality Q^2 with nearly full coverage in polar and azimuthal angles in the hadronic CM system. As results, several low-lying nucleon resonance states in particular from pion threshold to $W < 1.6$ GeV have been explored. These include $\Delta(1232)\frac{3}{2}^+$, $N(1440)\frac{1}{2}^+$, $N(1520)\frac{3}{2}^-$, and $N(1535)\frac{1}{2}^-$ states. In addition, we recently published the differential cross sections and helicity amplitudes of the reaction $\gamma^* p \to n\pi^+$ at higher W (1.6–2.0 GeV) which are the $N(1675)\frac{5}{2}^-$, $N(1680)\frac{5}{2}^+$, and $N(1710)\frac{1}{2}^+$ states. These excited states with isospin 1/2 and with masses near 1.7 GeV can be accessed in single $n\pi^+$ production as there are no isospin 3/2 states present in this mass range with the same spin-parity assignments. I will briefly discuss these states from CLAS results of the single charged pion electroproduction data.

1 Introduction

The structure of nucleon and its excited states have been one of the most extensively investigated subjects in nuclear and particle physics for several decades, because it allows us to understand important aspects of the underlying theory of the strong interactions. Many different reactions can be used to study the properties of the nucleon and its excited states. The inclusive electron scattering spectrum clearly indicates four resonance regions above the elastic peak. However, it does not allow us to separate excited states with different isospin and J^P quantum numbers which make up the second and higher resonance peaks. Even in the first resonance region there is a considerable non-resonant background under the dominant $\Delta(1232)$ peak. Therefore, exclusive measurements with a full angular coverage in the hadronic center-of-mass (CM) are necessary to separate the non-resonant contributions from the resonant contributions. A fit of the angular distributions and the W dependence within reaction theories or models allows to determine relative strengths for different resonances. In particular, reonances rapidly decaying into meson nucleon final states are interesting, and because of the small mass of the pion, the single-pion-nucleon decay is the favorite decay channel for many lower mass resonances. Moreover, single-pion electroproduction has been extensively exploited to understand the structure of baryons. In order to establish a better understanding of the connection between the dressed quark regime

---

This article belongs to the special issue "Nucleon Resonances".

K. Park (✉)
Jefferson Lab, 12000 Jefferson Ave, Newport News, VA 23606, USA
E-mail: parkkj@jlab.org

and the perturbative QCD domain at high Q^2, it is important to measure fundamental observable, such as cross sections and asymmetries in the resonance region.

On the fundamental level there exists only a very limited understanding of the relationship between Quantum Chromo-Dynamics (QCD), the field theory of the strong interaction, and the constituent quark models (CQM) or alternative hadron models, although recent developments in Lattice QCD, most notably the predictions of the excited strangeness $S = 0$ baryon spectrum of N^* and Δ^* states, have shown [1] that the same symmetry of $SU(6) \otimes O(3)$ is likely at work here as is underlying the spectrum in the CQM. The various current resonance models predict not only different excitation spectra but also different Q^2-dependence of transition form factors [2]. The mapping of the transition form factors of resonances in the full invariant W range, will help us to better understand the underlying quark or hadronic structures [3]. Experimentally, sufficient and complete data will help to uncover unambiguously the structure of the nucleon and its excited states in the entire resonance mass range.

Recently, precise data [4–11] allowed the determination of the $\Delta(1232)\frac{3}{2}^+$ magnetic dipole and the electric and scalar quadrupole transition form factors, covering a range of $0 \text{ GeV}^2 \leq Q^2 \leq 7 \text{ GeV}^2$. One of the major results of these analyses is the clear evidence for the presence of significant meson-baryon contributions to the resonance formation, which at low Q^2 are of the same magnitude as the quark contribution, but fall off more rapidly with increasing photon virtuality Q^2. A reasonable description of the $\gamma^* p \rightarrow \Delta(1232)\frac{3}{2}^+$ transition was achieved in the models that include pion-cloud contribution [12,13] and also in the dynamical reaction models, where the missing strength has been attributed to dynamical meson-baryon interaction in the final state [14–18].

From much more data [19–21] for beyond $\Delta(1232)\frac{3}{2}^+$, similar conclusions have been drawn for the excited nucleon states $N(1440)\frac{1}{2}^+$, $N(1520)\frac{3}{2}^-$, and $N(1535)\frac{1}{2}^-$ [22–24] using a relativistic quark model with a spectator di-quark. The results of this effort has been extensively discussed in recent reviews [2,25]. Moreover, the higher mass range $W > 1.6$ GeV shows that many N^* and Δ^* resonances are populated [26]. Several of them have significant branching ratio into the $N\pi$ final state and can be investigated in the exclusive single-pion channel, while others couple more strongly to $N\pi\pi$ final states. Of course, a full exploration should be done by requiring several final states to be measured and analyzed together in a coupled channel framework. Providing essential input to full coupled-channel analyses allows us to expect for some resonances, especially $N(1675)\frac{5}{2}^-$ and $N(1680)\frac{5}{2}^+$ that a single channel analysis will yield reliable results due to the large coupling of these states to $N\pi$ and the absence of $I = \frac{3}{2}$ states with the same spin-parity in that mass range.

In this proceeding, I summarize some of the results from the analysis of differential cross sections for the process $ep \rightarrow e'\pi^+ n$ in the range of W from near pion production threshold to deep inelastic scattering regime (up to 2.4 GeV) in high photon virtuality ($1.5 \text{ GeV}^2 \leq Q^2 \leq 4 \text{ GeV}^2$) with nearly full azimuthal and polar angle coverage for the $n\pi^+$ system.

2 Summary

The CLAS detector covers a very large kinematic range in the four CM variables W, Q^2, $\cos\theta_\pi^*$, and ϕ_π. For further analysis the data binning was matched to the underlying physics to be extracted. The study of nucleon excitation requires the analysis of the azimuthal ϕ_π^* dependence of the differential cross section to determine the separated structure functions and the analysis of the polar angle dependence to identify the partial wave contributions at a given invariant mass of the hadronic final state. The binning in the hadronic mass W must accommodate variations in the cross section, taking into account the width of resonances and their threshold behavior. On the other hand the Q^2-dependence is expected to be smooth. We have collected large number of kinematic bins, the resulting over 33,000 differential cross sections, 4,000 asymmetries for 1.11 GeV < W < 1.67 GeV, 37,000 differential cross sections for 1.6 GeV < W < 2.0 GeV region, and 140 differential cross sections for the deep inelastic scattering (DIS) regime 2.0 GeV < W < 2.4 GeV through several dedicated analyses.

The extraction of the axial form factor (G_A) and the generalized to dipole form factor ratio (G_1/G_D) in near threshold $W = 1.11$ GeV is done by multipole analysis [27], which is shown in Fig. 1. The data shows a good agreement with Light Cone Sum Rule (LCSR) calculations that provide most directly a relation of the hadron form factors to nucleon distribution amplitudes (DA) that enter into pQCD calculation without double counting.

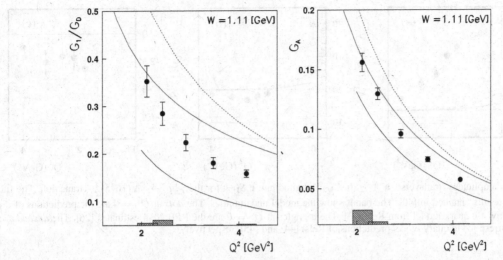

Fig. 1 (color online) Q^2 dependence for $n\pi^+$ of G_1 normalized by the dipole form factor (*left*) and axial form factor G_A (*right*). *Shaded bars* show the systematic errors. Various models are presented, *blue solid line*: MAID2007 [29] for E_{0+}/G_D and *red lines*: LCSR (*red solid* is the LCSR calculation using experimental electromagnetic form factors as input and *red dash* is pure LCSR) [28]

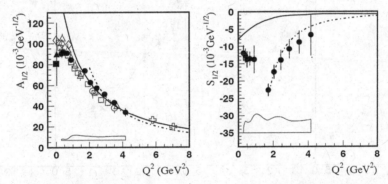

Fig. 2 Helicity amplitudes for $N(1535)1/2^-$ with $\beta_{N\pi} = 0.485$, $\beta_{N\eta} = 0.460$. The *solid* and *open boxes* are the results extracted from η photo- and electroproduction data in Ref. [31], the *open boxes* show the results from η electroproduction data [35–37]. The *solid circles* in $Q^2 = 1.5 - 4\,\text{GeV}^2$ are $n\pi^+$ and $p\pi^0$, $n\pi^+$ at low Q^2. The curves are *solid*: LFRQM [38], *dash-dot*: LCSR [39] models

A series of analyses, not only using the exclusive single-pion electroproduction, but also other meson production channels ($\pi^0, \eta, 2\pi$), allowed the extraction of several nucleon excited states in the second resonance region. These N^* results are obtained from data analyses within Unitary Isobar Models (UIM) and Dispersion Relations (DR). These data sets cover Q^2 range from 1.8 to 4.0 GeV^2. The employed approaches of UIM and DR were described in detail in Refs. [30, 34] and have been used successfully in Refs. [32–34] for the analyses of pion-electroproduction data in a wide range of Q^2 from 0.16 to 6 GeV^2.

Figure 2 shows helicity amplitudes ($A_{1/2}, S_{1/2}$) for $N(1535)1/2^-$, one of the second resonance region from various analyses including the single-pion channel. The $A_{1/2}$ confirms the Q^2-dependence of this amplitude observed in η electroproduction. The numerical comparison of the results has been carried out for the π and η photo- and electroproduction data using the branching ratios, $\beta_{\pi N}$ and $\beta_{\eta N}$ from the fit at $0\,\text{GeV}^2 \leq Q^2 \leq 4.5\,\text{GeV}^2$. It was found that $\beta_{\eta N} = 0.460 \pm 0.08 \pm 0.022$ and $\beta_{\pi N} = 0.485 \pm 0.008 \pm 0.023$. In addition, the $N\pi$ channel shows a sensitivity to the longitudinal helicity amplitude ($S_{1/2}$) as well due to strong interference between S_{11} and P_{11}.

For the high resonance region ($W > 1.6\,\text{GeV}$), in the absence of a coupled-multi-channel analysis framework for electroproduction channels, we subjected the differential cross section data to single channel energy-dependent partial wave analyses to extract the helicity amplitudes $A_{1/2}, A_{3/2}, S_{1/2}$ and their Q^2 dependence for some of the well-known isospin $\frac{1}{2}$ N^* states. Much of the model sensitivity the fits performed is due to the uncertainty in the non-resonant background amplitudes. Again, in order to have a quantitative measure of

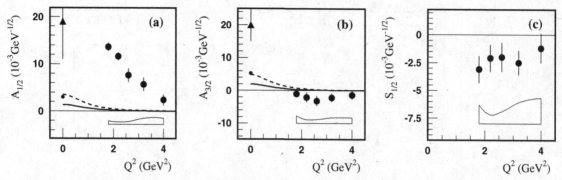

Fig. 3 Helicity amplitudes (transverse: **a** $A_{1/2}$, **b** $A_{3/2}$, longitudinal: **c** $S_{1/2}$) for the $\gamma^* p \to N(1675)\frac{5}{2}^-$ transition. The full circles are the results obtained in [40]. The bands show the model uncertainties. The *dots* at $Q^2 = 0$ are the predictions of the light-front relativistic quark model from Ref. [43]. The triangles at $Q^2 = 0$ are the RPP 2014 estimates [26]. The *dashed* and *solid curves* correspond to quark model predictions of Refs. [44] and [45], respectively

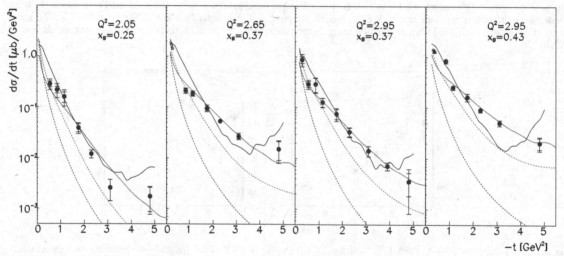

Fig. 4 (color online). Differential cross sections $d\sigma/dt$ [$\mu b/\text{GeV}^2$] integrated over ϕ_π^* for various (Q^2, x_B) bins. The *blue solid points* are the present work. The error bars (outer error) on all cross sections include both statistical (inner error) and systematic uncertainties added in quadrature. The *black open squares* ($d\sigma/dt$) [51] and *open stars* ($d\sigma_L/dt$) [52] are JLab Hall C data. The *red thick solid* ($d\sigma/dt$) and *dashed* ($d\sigma_L/dt$) *curves* are the calculations from the Laget model [48] with (Q^2, t)-dependent form factors at the photon-meson vertex. The *black thin solid* ($d\sigma/dt$) and *dashed* ($d\sigma_L/dt$) *curves* are the calculations from the Kaskulov et al. model [49]

the sensitivity to the specific modeling of the background amplitudes in the fit we employed two independent approaches, which are the unitary isobar model and the fixed-t dispersion relations.

The data [20,40] cover the mass range up to 2 GeV, and are thus sensitive to many N^* and Δ^* states. All of these states were used in the global analysis. However, the single channel analysis does not allow the separation of the different isospin contributions. We have therefore limited our analysis to the determination of those resonances that are most sensitively probed in the $n\pi^+$ channel, i.e. N^* states that do not overlap with Δ^* states of the same spin and parity. We also restricted the analysis to masses below $W = 1.8$ GeV. This leaves the three states for which we show the resulting electrocoupling amplitudes, $N(1675)\frac{5}{2}^-$, $N(1680)\frac{5}{2}^+$, and $N(1710)\frac{1}{2}^+$.

Figure 3 shows the most intriguing results of our analysis for high lying resonance region. A large $A_{1/2}$ amplitude of the transition to the $N(1675)\frac{5}{2}^-$ state has been observed at all measured Q^2. This result is in contrast to several calculations of dynamical quark models that predict an order of magnitude smaller values than what is extracted from the data due to the Moorhouse selection rule. To our knowledge this is to date the strongest and most direct evidence for dominant non-quark contribution to the electroexcitation of a nucleon resonance on the proton. The situation with quark and meson-baryon contributions will become much clearer when data on neutrons become available.

For the beyond nucleon resonance region which is shown in Fig. 4, we also have measured the cross sections $(d\sigma/dt)$ of exclusive electroproduction of π^+ mesons from protons in terms of various kinematic variables, $-t = 0.1 - 5.3\,\text{GeV}^2$, $x_B = 0.16 - 0.58$, and $Q^2 = 1.6 - 4.5\,\text{GeV}^2$ [46]. We compared our differential cross sections to four recent calculations [47–50] based on hadronic and partonic degrees of freedom. The four models give a qualitative description of the overall strength and of the t-, Q^2- and x_B-dependencies of our unseparated cross sections. There is an obvious need for L/T separated cross sections in order to distinguish between the several approaches. These separations will be possible with the upcoming JLab 12-GeV upgrade. In particular, if the handbag approach can accommodate the data, the $p(e, e'\pi^+)n$ process offers the outstanding potential to access transversity GPDs. We have also several final states data ($K^+\Lambda$, $K^+\Sigma$, $\pi^0 p$) from resonance region to this deep inelastic kinematic regime to be analyzed together in a coupled channel framework [41,42].

The selective single pion exclusive data set presented in this proceeding, for the first time has been explored in the W range from near threshold to the DIS region and a wide range of $Q^2 = 1.8 - 4.0\,\text{GeV}^2$ with nearly 4π solid CM angle. This data with future exclusive data from other reactions ($p\eta$, $p\pi^0$, $\pi\pi p$,...) will allow to determine the transition charge and current densities of individual states through a Fourier transformation of the transverse amplitudes in the light cone frame. Such data can reveal novel information of the internal structure of the excited states in the transverse impact parameter space [53].

References

1. Edwards, R.G., Dudek, J.J., Richards, D.G., Wallace, S.J.: Excited state baryon spectroscopy from lattice QCD. Phys. Rev. D **84**, 074508 (2011)
2. Aznauryan, I.G., Burkert, V.D.: Electroexcitation of nucleon resonances. Prog. Part. Nucl. Phys. **67**, 1 (2012)
3. Burkert, V.D., Lee, T.-S.H.: Electromagnetic meson production in the nucleon resonance region. Int. J. Mod. Phys. E **13**, 1035 (2004)
4. Joo, K., et al., [CLAS].: Dependence of quadrupole strength in the transition. Phys. Rev. Lett. **88**, 122001 (2002)
5. Joo, K., et al., [CLAS].: Measurement of the polarized structure function for in the resonance region. Phys. Rev. C **68**, 032201 (2003)
6. Sparveris, N.F., et al.: Measurement of the response function for electroproduction at in the transition. Phys. Rev. C **67**, 058201 (2003)
7. Biselli, A., et al., [CLAS].: Study of polarized-e polarized-p $\to e'$ p π^0 in the $\Delta(1232)$ mass region using polarization asymmetries. Phys. Rev. C **68**, 035202 (2003)
8. Kelly, J., et al.: Recoil polarization for Δ excitation in pion electroproduction. Phys. Rev. Lett. **95**, 102001 (2005)
9. Stave, S., et al.: Lowest-Q^2 measurement of the $\gamma p \to \Delta$ reaction. Eur. Phys. J. A **30**, 471–476 (2006)
10. Ungaro, M., et al., [CLAS].: Measurement of the N $\to \Delta^+(1232)$ Transition at High-Momentum Transfer by π^0 Electroproduction. Phys. Rev. Lett. **97**, 112003 (2006)
11. Sparveris, N.F. et al.: Determination of quadrupole strengths in the $\gamma^* p \to \Delta(1232)$ transition at $Q^2 = 0.20$ (GeV/c)^2. Phys. Lett. B **651**, 102–107 (2007). arXiv:nucl-ex/0611033
12. Lu, D.H., Thomas, A.W., Williams, A.G.: A chiral bag model approach to delta electroproduction. Phys. Rev. C **55**, 3108 (1997)
13. Faessler, A., Gutsche, T., Holstein, B.R., et al.: Light baryon magnetic moments and N $\to \Delta\gamma$ transition in a Lorentz covariant chiral quark approach. Phys. Rev. D **74**, 074010 (2006)
14. Kamalov, S.S., Yang, S.N.: Pion cloud and the Q 2 dependence of $\gamma^*N \leftrightarrow \Delta$ transition form factors. Phys. Rev. Lett. **83**, 4494 (1999)
15. Kamalov, S.S., et al.: A new analysis of data on $p(e, e'p)\pi^0$ at $Q^2 = 2.8$ and 4.0 GeV^2. Phys. Rev. C **64**, 032201(R) (2001)
16. Sato, T., Lee, T.-S.H.: Dynamical study of the Δ excitation in N (e, e'π) reactions. Phys. Rev. C **63**, 055201 (2001)
17. Matsuyama, A., Sato, T., Lee, T.-S.H.: Dynamical coupled-channel model of meson production reactions in the nucleon resonance region. Phys. Rep. **439**, 193 (2007)
18. Juliá-Díaz, B., Lee, T.-S.H., Matsuyama, A., Sato, T.: Dynamical coupled-channels effects on pion photoproduction. Phys. Rev. C **77**, 045205 (2008)
19. Egiyan, H., et al., [CLAS].: Single π^+ Electroproduction on the Proton in the First and Second Resonance Regions at 0.25 GeV^2 < Q^2 < 0.65 GeV^2 Using CLAS. Phys. Rev. C **73**, 025204 (2006)
20. Park, K., et al., [CLAS].: Cross sections and beam asymmetries for ep$\to$enπ^+ in the nucleon resonance region for 1.7 < Q^2 < 4.5 GeV^2. Phys. Rev. C **77**, 015208 (2008)
21. Fedotov, G.V., et al., [CLAS].: Electroproduction of $p\pi^+\pi^-$ off protons at 0.2 < Q^2 < 0.6 GeV^2 and 1.3 < W < 1.57 GeV with the CLAS detector. Phys. Rev. C **79**, 015204 (2009)
22. Ramalho, G., Tsushima, K.: Valence quark contributions for the γN $\to$ P11(1440) form factors. Phys. Rev. D **81**, 074020 (2010)
23. Ramalho, G., Pea, M.T.: Reevaluation of the parton distribution of strange quarks in the nucleon. Phys. Rev. D **89**, 094016 (2014)
24. Ramalho, G., Pena, M.T.: A covariant model for the γN $\to$ N(1535) transition at high momentum transfer. Phys. Rev. D **84**, 033007 (2011)

25. Aznauryan, I.G., et al.: Studies of nucleon resonance structure in exclusive meson electroproduction. Int. J. Mod. Phys. E **22**, 1330015 (2013)
26. Olive, K.A., et al.: Particle data group collaboration. Review of particle physics. Chin. Phys. C **38**, 090001 (2014)
27. Park, K., et al., [CLAS].: Measurement of the generalized form factors near threshold via $\gamma^* p \to n\pi^+$ at high Q^2. Phys. Rev. C **85**, 035208 (2012)
28. Braun, V.M., Ivanov, DYu., Lenz, A., Peters, A.: Deep inelastic pion electroproduction at threshold. Phys. Rev. D **75**, 014021 (2007)
29. Drechsel, D., Kamalov, S.S., Tiator, L.: Unitary isobar model-MAID2007. Eur. Phys. J. A **34**, 69 (2007)
30. Aznauryan, I.G.: Multipole amplitudes of pion photoproduction on nucleons up to 2 GeV using dispersion relations and the unitary isobar model. Phys. Rev. C **67**, 015209 (2003)
31. Aznauryan, I.G.: Resonance contributions to η photoproduction on protons found using dispersion relations and an isobar model. Phys. Rev. C **68**, 065204 (2003)
32. Aznauryan, I.G., Burkert, V.D., Egiyan, H., et al.: Electroexcitation of the P33(1232), P11(1440), D13(1520), S11(1535) at $Q^2 = 0.4$ and $0.65 (\text{GeV/c})^2$. Phys. Rev. C **71**, 015201 (2005)
33. Aznauryan, I.G., Burkert, V.D., et al.: Electroexcitation of nucleon resonances at $Q^2 = 0.65 \,\text{GeV/c}^2$ from a combined analysis of single- and double-pion electroproduction data. Phys. Rev. C **72**, 045201 (2005)
34. Aznauryan, I.G., et al., [CLAS].: Electroexcitation of nucleon resonances from CLAS data on single pion electroproduction. Phys. Rev. C **80**, 055203 (2009)
35. Armstrong, C., et al.: Electroproduction of the S11(1535) resonance at high momentum transfer. Phys. Rev. D **60**, 052004 (1999)
36. Thompson, R., et al., [CLAS].: The $ep \to e'p\eta$ reaction at and above the S11(1535) Baryon resonance. Phys. Rev. Lett. **86**, 1702 (2001)
37. Denizli, H., et al., [CLAS].: $Q2$ dependence of the S11(1535)photocoupling and evidence for a P-wave resonance in η electroproduction. Phys. Rev. C **76**, 015204 (2007)
38. Aznauryan, I.G., Burkert, V.D.: Nucleon electromagnetic form factors and electroexcitation of low-lying nucleon resonances in a light-front relativistic quark model. Phys. Rev. C **85**, 055202 (2012)
39. Anikin, I.V., Braun, V.M., Offen, N.: Electroproduction of the $N^*(1535)$ nucleon resonance in QCD. Phys. Rev. D **92**, 014018 (2015)
40. Park, K., et al., [CLAS].: Measurements of ep$\to$ e'π^+n at $1.6 < W < 2.0$ GeV and extraction of nucleon resonance electrocouplings at CLAS. Phys. Rev. C **91**, 045203 (2015)
41. Carman, D.S., Park, K., et al., [CLAS].: Separated structure functions for exclusive $K^+\Lambda$ and $K^+\Sigma 0$ electroproduction at 5.5 GeV with CLAS. Phys. Rev. C **87**, 025204 (2013)
42. Mestayer, M., Park, K., et al., [CLAS].: Flavor dependence of qq-bar creation observed in the exclusive limit. Phys. Rev. Lett. **113**(15), 152004 (2014)
43. Aznauryan, I.G., Bagdasaryan, A.S.: Radiative decays of nucleon resonances of the $[70, 1^-]$ multiplet in the relativistic quark model. Yad. Fiz. **41**, 249 (1985). (translation in Sov. J. Nucl. Phys. 41, 158 (1985))
44. Merten, D., Löring, U., Metsch, B., Petry, H.: Electromagnetic properties of baryons in a relativistic quark model. Eur. Phys. J. A **18**, 193 (2003)
45. Santopinto, E., Giannini, M.M.: Systematic study of longitudinal and transverse helicity amplitudes in the hypercentral constituent quark model. Phys. Rev. C **86**, 065202 (2012)
46. Park, K., et al., [CLAS].: Deep exclusive π^+ electroproduction off the proton at CLAS. Eur. Phy. J. A **49**, 16 (2013)
47. Goloskokov, S.V., Kroll, P.: Transversity in hard exclusive electroproduction of pseudoscalar mesons. Eur. Phys. J. A **47**, 112 (2011)
48. Laget, J.M.: Space–time structure of hard-scattering processes. Phys. Rev. D **70**, 054023 (2004)
49. Kaskulov, M.M., Gallmeister, K., Mosel, U.: Deeply inelastic pions in the exclusive reaction $p(e, e'\pi^+)n$ above the resonance region. Phys. Rev. D **78**, 114022 (2008)
50. Kaskulov, M.M., Mosel, U.: Deep exclusive electroproduction of π^+ from data measured with the HERMES detector at DESY. Phys. Rev. C **81**, 045202 (2010)
51. Qian, X., et al.: Experimental study of the $A(e, e'\pi^+)$ reaction on H1,H2,C12,Al27,Cu63, and Au197. Phys. Rev. C **81**, 055209 (2010)
52. Horn, T., et al.: Scaling study of the pion electroproduction cross sections. Phys. Rev. C **78**, 058201 (2008)
53. Tiator, L., Vanderhaeghen, M.: Empirical transverse charge densities in the nucleon-to-P11(1440) transition. Phys. Lett. B **672**, 344 (2009)

Few-Body Syst (2016) 57:901–907
DOI 10.1007/s00601-016-1125-x

Steffen Strauch · CLAS Collaboration

Baryon Spectroscopy with Polarization Observables from CLAS

Received: 23 February 2016 / Accepted: 25 April 2016 / Published online: 9 June 2016
© Springer-Verlag Wien 2016

Abstract The spectrum of nucleon excitations is dominated by broad and overlapping resonances. Polarization observables in photoproduction reactions are key in the study of these excitations. They give indispensable constraints to partial-wave analyses and help clarify the spectrum. A series of polarized photoproduction experiments have been performed at the Thomas Jefferson National Accelerator Facility with the CEBAF Large Acceptance Spectrometer. These measurements include data with linearly and circularly polarized tagged-photon beams, longitudinally and transversely polarized proton and deuterium targets, and recoil polarizations through the observation of the weak decay of hyperons. An overview of these studies and recent results will be given.

1 Introduction

The nucleon is a color-neutral object which consists of color-charged quarks and gluons. Quantum chromodynamics (QCD) is the fundamental theory of the strong interaction between quarks and gluons. Valuable information about QCD can be learned from nuclear spectroscopy; e.g. information about the internal degrees of freedom in a nucleon. Experiments, especially pion-nucleon scattering, have confirmed low-lying excited states predicted by quark models with three independent quark degrees of freedom. These models, however, predict an overabundance of higher-lying excited states compared to what has been observed until now [1]. Also, recent lattice QCD calculations [2] find a large number of not-yet-discovered nucleon resonances. To clarify the nucleon resonance spectrum is an important task in the study of QCD.

Nucleon resonances are short-lived, and the identification of resonances in partial-wave analyses of experimental data is complicated by their large width and overlap. Experimental cross sections are insufficient, and polarization observables are crucial to constrain these analyses. A complete set of certain polarization observables is necessary to unambiguously determine the amplitudes of the reaction. In the photoproduction of pseudoscalar mesons, a formally complete experiment requires at each energy and angle the measurement of at least eight carefully chosen observables [3]. In the photoproduction of two mesons, even more observables are needed [4]. It is also important to include in the analysis data from a variety of excitation and decay channels, as some of the missing states may couple only weakly to, e.g., the πN channel.

In the following, examples of recent photoproduction measurements of polarization observables from the CLAS Collaboration are presented. A similar overview has been given in [5]. These measurements include

Supported in parts by the U.S. National Science Foundation: NSF PHY-1505615.

This article belongs to the special issue "Nucleon Resonances".

S. Strauch (✉)
University of South Carolina, Columbia, SC, USA
E-mail: strauch@sc.edu

$$\gamma + p \to K^+ \Lambda$$

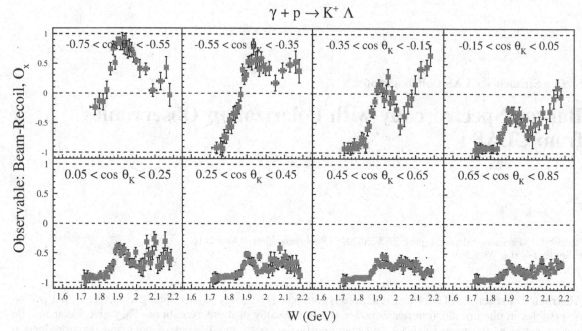

Fig. 1 Preliminary energy distributions of CLAS $\gamma p \to K^+ \Lambda$ data for the double-polarization observable O_x in various kaon angular bins. Figure from D. Ireland (University of Glasgow)

single pseudoscalar-meson, vector-meson, double-pion, and hyperon photoproduction off the proton and quasi-free off the bound neutron. The observables include single- and double-polarization observables with combinations of polarized beam, target, and the polarization of the recoiling baryon.

The experiments were performed in Experimental Hall B at the Thomas Jefferson National Accelerator Facility (JLab). The incident bremsstrahlung photon beams were energy-tagged [6] and either unpolarized, circularly, or linearly polarized. The photon beam irradiated the production target. Unpolarized liquid hydrogen and deuterium targets, as well as the newly developed polarized frozen-spin (FROST) [7] and HDice targets [8,9], have been used in these experiments. Final-state particles were detected in the CEBAF Large Acceptance Spectrometer (CLAS) [10]. Recoil polarization was accessible in hyperon-production measurements through the measurement of the decay-proton angular distribution in the parity-violating weak decay of hyperons.

2 Unpolarized Targets

The CLAS Collaboration has studied the photoproduction reactions $\gamma p \to p\pi^0$ and $\gamma p \to n\pi^+$ with linearly polarized photons in an energy range from 1.10 to 1.86 GeV [11]. The beam asymmetry observable Σ has been obtained from the pion angular distributions with respect to the polarization direction of the linearly polarized photon beam. The high statistics in these reactions allowed for precise constraints of partial-wave analyses. Resonance couplings have been extracted in fits of the SAID [12] analysis after including the new data set. The largest change from previous fits was found to occur for the 'well known' $\Delta(1700)3/2^-$ and $\Delta(1905)5/2^+$ resonances [11].

Polarization observables were also obtained in the hyperon-photoproduction reactions $\gamma p \to K^+ \Lambda$ and $\gamma p \to K^+ \Sigma^0$. Results include the Λ recoil polarization P from measurements with unpolarized photon beams [13,14] as well as the polarization-transfer observables C_x and C_z with circularly polarized beams [15]. These data were critical in a coupled-channel analysis by the Bonn–Gatchina group [16,17]. In particular, the analysis found further evidence for the, at the time, poorly known $N(1900)3/2^+$ resonance. This resonance is predicted by symmetric three-quark models, but is not expected to exist in earlier quark–diquark models. Preliminary hyperon-photoproduction data with linearly polarized photons off an unpolarized proton target have been obtained up to $W \approx 2.2$ GeV. Together with the recoil polarization of the hyperon, this gives access to five polarization observables: Σ, P, T, O_x, and O_z. Energy distributions of

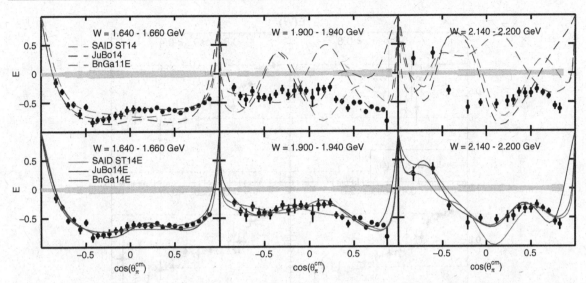

Fig. 2 CLAS data for the beam-target polarization observable E in the $\gamma p \rightarrow \pi^+ n$ reaction for selected center-of-mass energy bins. The *shaded bands* indicate systematic uncertainties. The *dashed curves in the upper panels* are results from the SAID ST14 [12], Jülich14 [25], and BnGa11E [26] analyses. The *solid curves in the lower panels* are results from updated analyses including the new CLAS E data. Figure from Ref. [23]

the preliminary results of the beam-recoil polarization observable O_x for the $K^+\Lambda$ channel are shown in Fig. 1.

3 Polarized Targets

An integral part of the experimental N* program at JLab are experiments off polarized protons [18–22]. The Jefferson Lab frozen spin target (FROST) was constructed for use inside CLAS to allow for the measurement of polarization observables with longitudinally or transversely polarized protons in the butanol target material [7].

First results from the FROST program have been published for the $\gamma p \rightarrow \pi^+ n$ reaction [23]. The double-polarization observable E has been determined in the energy range from 0.35 to 2.37 GeV from data of circularly polarized photons incident on longitudinally polarized protons. A subset of the about 900 data points is shown in the three energy bins of Fig. 2. Results from previous partial-wave analyses describe the new data at low photon energies reasonably well; at higher energies, however, significant deviations are observed. The data have been included in new analyses resulting in good descriptions of the data and in updated nucleon resonance parameters. One particularly interesting result is strengthened evidence for the poorly known $\Delta(2200)7/2^-$ resonance in improving the Bonn–Gatchina fit at the highest energies [24]. The mass of the $\Delta(2200)7/2^-$ resonance is significantly higher than the mass of its parity partner $\Delta(1950)7/2^+$, which is the lowest-mass Δ^* resonance with spin-parity $J^P = 7/2^+$.

The polarization observable E has also been measured in the $\gamma p \rightarrow \eta p$ reaction from threshold to $W = 2.15$ GeV [27]. Because η mesons have isospin zero, the reaction selects isospin-$1/2$ resonances in the nucleon resonance spectrum. Figure 3 shows the data and a fit with the Jülich–Bonn dynamical coupled-channel model. It has been shown in Ref. [27] that the observable E in η photoproduction is especially suited to disentangle electromagnetic resonance properties. Initial investigation of these results show pronounced changes in the description of this observable when these new CLAS data are included. The fit in Fig. 3 describes the data quite well without the need for an additional narrow resonance near 1.68 GeV, which was previously suggested [27].

FROST data from other single-pion photoproduction channels and for other observables are under ongoing analyses. Examples of preliminary angular distributions for the target T and beam-target F polarization observables in the reaction $\gamma p \rightarrow \pi^0 p$ are shown in Fig. 4. These observables are extracted from data with unpolarized and circularly polarized photons off transversely polarized protons, respectively. In the figure, the data are compared with present results of partial-wave analyses which do not include these preliminary data.

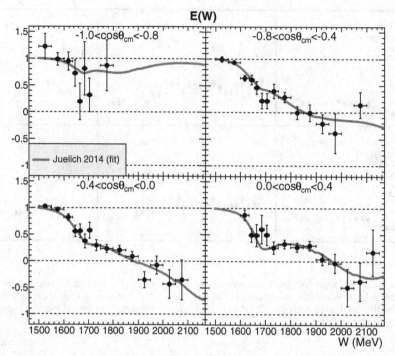

Fig. 3 Observable E in the $\gamma p \rightarrow \eta p$ reaction as a function of W for various angular bins. The *red curve* shows fits with the Jülich–Bonn dynamical coupled-channel model. Figure from Ref. [27] (color figure online)

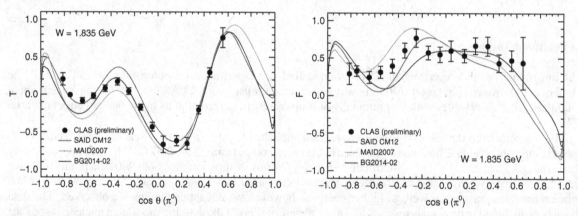

Fig. 4 Preliminary CLAS results for the observables T and F in the $\gamma p \rightarrow \pi^0 p$ reaction as a function of $\cos \theta_\pi$ for $W = 1.835$ GeV. The data are compared with various partial-wave analyses (not fitted to the data) from the SAID [12], MAID [28], and Bonn–Gatchina [26] groups. Figures from H. Jiang (University of South Carolina)

Hyperon photoproduction reactions are also studied with data from FROST. Figure 5 shows preliminary results of the beam-target observable F. The data cover energies between $W = 1.7$ GeV and $W = 2.3$ GeV. The data are compared to prior results of analyses with the RPR-Ghent [29], KAON-MAID [30], and Bonn–Gatchina [26] models. As none of the models describe the data well the FROST data will provide important new constraints to the models.

Many nucleon resonances in the mass region above 1.6 GeV decay predominantly through either $\pi \Delta$ or ρN intermediate states into $\pi \pi N$ final states. Double-pion photoproduction allows the study of resonances that have no significant coupling to the πN channel with great potential to observe previously unobserved states. Many polarization observables are accessible in two-pion photoproduction off the nucleon [4]. The CLAS Collaboration was first to study the beam-helicity asymmetry $I^\odot$ for the two-pion-photoproduction reaction. The measurements covered energies between $W = 1.35$ and 2.30 GeV [31]. The FROST group is working on the determination of twelve different polarization observables from $\gamma p \rightarrow p \pi^+ \pi^-$ data. As example, preliminary results are shown in Fig. 6 for observables that are accessible in measurements with unpolarized

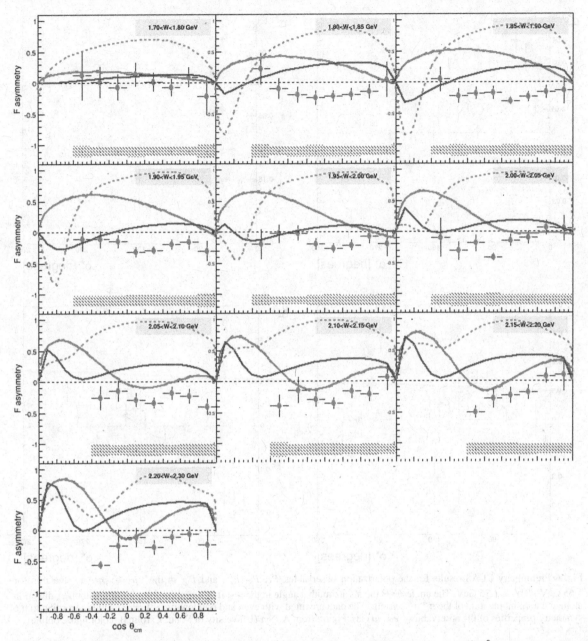

Fig. 5 Preliminary angular distributions of the polarization observable F from FROST in the $\gamma p \to K^+ \Sigma^0$ reaction for various W bins. Model curves are from the RPR-Ghent [29] (*red*), KAON-MAID [30] (*blue*), and Bonn–Gatchina [26] (*magenta*) models. Figure from N. Walford (U. Basel) (color figure online)

(P_x and P_y) and circularly polarized ($P_x^\odot$ and $P_y^\odot$) photons off transversally polarized protons. The data will strongly constrain coupled-channel analyses.

Data have predominantly been taken off proton targets, and the new data from the FROST program will expand this data base over a large range of energy with many observables for polarized proton reactions. In contrast, data off neutrons are extremely sparse. However, measurements with both proton and neutron targets are needed to completely specify the amplitude of the reaction.

The CLAS collaboration has taken production data with circularly and linearly polarized photons off a polarized solid deuterium-hydride target (HDice) [8,9] up to center-of-mass energies of $W \approx 2.3$ GeV. The run conditions were optimized for polarized neutron reactions. The ongoing analyses of this run include single- and double-pion photoproduction and hyperon photoproduction off the bound neutron.

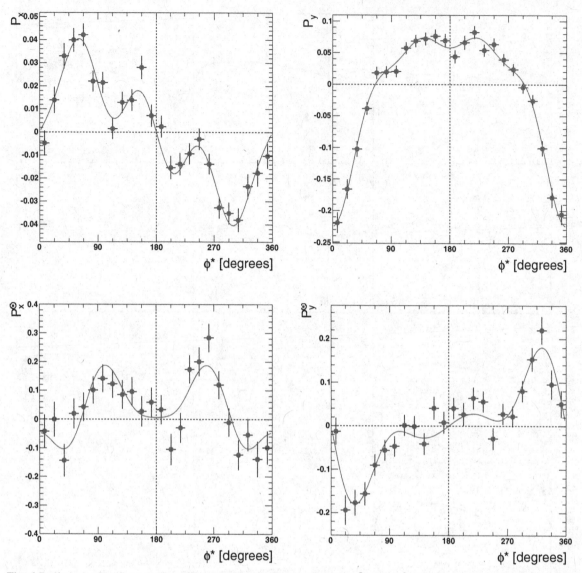

Fig. 6 Preliminary CLAS results for the polarization observables P_x, P_y, $P_x^\odot$, and $P_y^\odot$ in the $\gamma p \rightarrow p\pi^+\pi^-$ reaction for 1.65 GeV $< W <$ 1.75 GeV. The angle $\phi*$ is the π^+ azimuthal angle in the rest frame of the $\pi^+\pi^-$ system with the z direction along the total momentum of the $\pi^+\pi^-$ system. The data are fitted with even and odd low-order Fourier series according to the symmetry properties of the observables (*red curves*). Figure from A. Net (University of South Carolina) (color figure online)

4 Conclusion

During the past years, knowledge of the baryon spectrum has increased greatly, as reflected in recent N and Δ resonance updates in the Review of Particle Physics [32]. New polarized photoproduction data from CLAS off polarized and unpolarized proton and neutron targets are both under analysis and becoming available. These data will contribute to complete or nearly-complete experiments, and will challenge previously poorly constrained models. It is very likely that these data will have a tremendous impact on the understanding of baryon resonances and may provide evidence for new states found in coupled-channel analyses.

The quasi-real photon tagger for CLAS12 will allow to expand these photoproduction studies to higher beam energies with new experiments after the energy upgrade of JLab [33].

References

1. Credé, V., Roberts, W.: Progress towards understanding baryon resonances. Rept. Prog. Phys. **76**, 076301 (2013)
2. Edwards, R.G., Dudek, J.J., Richards, D.G., Wallace, S.J.: Excited state baryon spectroscopy from lattice QCD. Phys. Rev. D **84**, 074508 (2011)
3. Chiang, W.-T., Tabakin, F.: Completeness rules for spin observables in pseudoscalar meson photoproduction. Phys. Rev. C **55**, 2054 (1997)
4. Roberts, W., Oed, T.: Polarization observables for two-pion production off the nucleon. Phys. Rev. C **71**, 055201 (2005)
5. Strauch, S.: Baryon spectroscopy with polarization observables from CLAS. In: 12th Conference on the Intersections of Particle and Nuclear Physics (CIPANP 2015) Vail, Colorado, USA, May 19–24, 2015 (2015)
6. Sober, D., et al.: The bremsstrahlung tagged photon beam in Hall B at JLab. Nucl. Instrum. Method A **440**, 263 (2000)
7. Keith, C., Brock, J., Carlin, C., Comer, S., Kashy, D., McAndrew, J., Meekins, D., Pasyuk, E., Pierce, J., Seely, M.: The Jefferson Lab frozen spin target. Nucl. Instrum. Method A **684**, 27 (2012)
8. Sandorfi, A.M.: Unravelling the excitation spectrum of the nucleon. J. Phys. Conf. Ser. **424**, 012001 (2013)
9. Jefferson Lab Experiment E06-101: N* Resonances in Pseudoscalar-meson photo-production from Polarized Neutrons in $\mathbf{H} \cdot \mathbf{D}$ and a complete determination of the $\gamma n \to K^0 \Lambda$ amplitude. F. Klein and A.M. Sandorfi, spkespersons
10. Mecking, B.A., et al.: The CEBAF Large Acceptance Spectrometer (CLAS). Nucl. Instrum. Method **A503**, 513 (2003)
11. Dugger, M., et al.: Beam asymmetry for π^+ and π^0 photoproduction on the proton for photon energies from 1.102 to 1.862 GeV. Phys. Rev. C **88**, 065203 (2013)
12. Workman, R.L., Paris, M.W., Briscoe, W.J., Strakovsky, I.I.: Unified Chew–Mandelstam SAID analysis of pion photoproduction data. Phys. Rev. C **86**, 015202 (2012)
13. McNabb, J.W.C., et al.: Hyperon photoproduction in the nucleon resonance region. Phys. Rev. C **69**, 042201 (2004)
14. McCracken, M.E., et al.: Differential cross section and recoil polarization measurements for the $\gamma p \to K^+ \Lambda$ reaction using CLAS at Jefferson Lab. Phys. Rev. C **81**, 025201 (2010)
15. Bradford, R.K., et al.: First measurement of beam-recoil observables C_x and C_z in hyperon photoproduction. Phys. Rev. C **75**, 035205 (2007)
16. Sarantsev, A.V., Nikonov, V.A., Anisovich, A.V., Klempt, E., Thoma, U.: Decays of baryon resonances into ΛK^+, $\Sigma^0 K^+$ and $\Sigma^+ K^0$. Eur. Phys. J. A **25**, 441 (2005)
17. Nikonov, V.A., Anisovich, A.V., Klempt, E., Sarantsev, A.V., Thoma, U.: Further evidence for $N(1900)P_{13}$ from photoproduction of hyperons. Phys. Lett. B **662**, 245 (2008)
18. Jefferson Lab Experiment E02-112: Search for Missing Nucleon Resonances in Hyperon Photoproduction. P. Eugenio, F. Klein, and L. Todor, spokespersons
19. Jefferson Lab Experiment E03-105: Pion Photoproduction from a Polarized Target. N. Benmouna, W. Briscoe, G. O'Rielly, I. Strakovsky, S. Strauch, spokespersons
20. Jefferson Lab experiment E04-102: Helicity Structure of Pion Photoproduction. D. Crabb, M. Khandaker, and D. Sober, spokespersons
21. Jefferson Lab experiment E05-012: Measurement of polarization observables in η-photoproduction with CLAS. M. Dugger and E. Pasyuk, spokespersons
22. Jefferson Lab Experiment E06-013: Measurement of $\pi^+\pi^-$ Photoproduction in Double-Polarization Experiments using CLAS. M. Bellis, V. Credé, S. Strauch, spokespersons
23. Strauch, S., et al.: First measurement of the polarization observable E in the $\mathbf{p}(\gamma, \pi^+)n$ reaction up to 2.25 GeV. Phys. Lett. **B750**, 53 (2015)
24. Anisovich, A.V., Burkert, V., Klempt, E., Nikonov, V.A., Pasyuk, E., Sarantsev, A.V., Strauch, S., Thoma, U.: Existence of $\Delta(2200)7/2^-$ precludes chiral symmetry restoration at high mass, arXiv:1503.05774 [nucl-ex]
25. Rönchen, D., Döring, M., Huang, F., Haberzettl, H., Haidenbauer, J., Hanhart, C., Krewald, S., Meißner, U.G., Nakayama, K.: Photocouplings at the pole from pion photoproduction. Eur. Phys. J. A **50**, 101 (2014)
26. Anisovich, A.V., Beck, R., Klempt, E., Nikonov, V.A., Sarantsev, A.V., Thoma, U.: Properties of baryon resonances from a multichannel partial wave analysis. Eur. Phys. J. A **48**, 15 (2012)
27. Senderovich, I., et al.: First measurement of the helicity asymmetry E in η photoproduction on the proton. Phys. Lett. B **755**, 64 (2016)
28. Drechsel, D., Kamalov, S., Tiator, L.: Unitary Isobar Model - MAID2007. Eur. Phys. J. A **34**, 69 (2007)
29. Corthals, T., Ryckebusch, J., Van Cauteren, T.: Phys. Rev. **C73** (2006) 045207 and T. Vrancx, personal communication
30. Lee, F.X., Mart, T., Bennhold, C., Haberzettl, H., Wright, L.E.: Nucl. Phys. **A695**, 237 (2001). Kaon-MAID. http://www.kph.uni-mainz.de/MAID/kaon/kaonmaid.html
31. Strauch, S., et al.: Beam-helicity asymmetries in double-charged-pion photoproduction on the proton. Phys. Rev. Lett. **95**, 162003 (2005)
32. Olive, K.A., et al.: Review of particle physics. Chin. Phys. C **38**, 090001 (2014)
33. Glazier, D.I.: A quasi-real photon tagger for CLAS12. Nucl. Phys. Proc. Suppl. **207–208**, 204 (2010)

Few-Body Syst (2016) 57:1027–1033
DOI 10.1007/s00601-016-1144-7

Yasuhiro Yamaguchi

Exotic Baryons from a Heavy Meson and a Nucleon

Received: 15 February 2016 / Accepted: 26 July 2016 / Published online: 12 August 2016
© Springer-Verlag Wien 2016

Abstract Hadronic molecules have been considered to be one of the candidate of exotic hadron structures near the threshold. In the heavy quark sector, new symmetry, called the heavy quark symmetry, plays an important role to form the molecules. This symmetry has an essential role which is to enhance the one pion exchange potential arising through the mixing of heavy pseudoscalar and heavy vector mesons. In this study, we investigate new hadronic molecule formed by the heavy meson $P^{(*)} = \bar{D}^{(*)}$, $B^{(*)}$ and a nucleon N, being $P^{(*)}N$ and $P^{(*)}NN$ few-body states. As the interaction between P and N, the π exchange and vector meson (ρ and ω) exchanges are considered. By solving the coupled-channel schrödinger equations for $PN(N)$ and $P^{(*)}N(N)$, we predict the bound and resonant states in the charm and bottom sectors. In the bound and resonant states, the $PN - P^*N$ mixing effect is important, where the tensor force of the one pion exchange potential generates the strong attraction.

1 Introduction

Study of the exotic hadrons attracts a great deal of interest in the hadron and nuclear physics. The constituent quark model describing the baryons as qqq and mesons as $q\bar{q}$ with constituent quark $q(\bar{q})$ has been successfully applied to explain the ordinary hadron spectra [1,2]. However resent observations of new exotic states, such as XYZ states, in the charm and bottom regions, cannot be explained by the ordinary heavy quarkonium $Q\bar{Q}$ with $Q = c, b$ [3–6]. These states motivate us to investigate exotic structures of the hadron states. These exotic states are considered to be a multiquark state being $Q\bar{Q}q\bar{q}$, a hadronic molecule being a meson-antimeson state and the hybrid states being $Q\bar{Q}g$ [7,8]. In addition, a quarkonium plus a molecular state has been discussed [9–11]. In particular as candidates of the hadronic molecules, $X(3872)$ [12] and Z_b [13] which is considered to be $D\bar{D}^*$ and $B\bar{B}^*$ attract interest of us.

Hadronic molecule is a loosely bound state or a resonance of several hadrons appearing near the thresholds. This is an analogous state to the Deuteron or the atomic nuclei which is the bound state of protons and nucleons [14]. The Deuteron is a loosely bounding system with the binding energy 2.2 MeV measured from the proton-neutron threshold and the large root mean square radius 4 fm. The hadronic molecules emerging closed to the threshold is expected to have such natures.

The properties of the hadronic molecule will be characterized by the interaction between composite hadrons as seen in the nuclear physics. In the Deuteron and nuclei which are the loosely binding systems, the one pion exchange potential (OPEP) has a crucial role to form them, where it is a long range force because of the small mass of the exchanged pion, $m_\pi \sim 140$ MeV.

This article belongs to the special issue "Nucleon Resonances".

Y. Yamaguchi (✉)
Istituto Nazionale di Fisica Nucleare - Sezione di Genova, Via Dodecaneso 33, 16146 Genoa, Italy
E-mail: yasuhiro.yamaguchi@ge.infn.it
Tel: +39-010-353-6203

Recently, the strong attraction between a $\bar{D}$ (B) meson with the heavy flavor (charm and bottom) and a nucleon N has been discussed [15–21]. The attraction is generated by the OPEP which is enhanced by the mass degeneracy between heavy pseudoscalar meson $P = \bar{D}$, B and heavy vector meson $P^* = \bar{D}^*$, B^*. The small mass splittings of the P and P^* are induced by the heavy quark spin symmetry (HQS) [22–24]. It has been known that the tensor force of the OPEP plays an important role to generate the strong attraction among nucleons in the nuclei. Therefore it is expected that the appearance of the OPEP in the heavy hadron system produces the molecular structure in similar way to the nuclei.

In addition, the attraction of the hadron-nucleon system will show us interesting phenomena which are not present in normal atomic nuclei. In the strangeness sector, the strange hadrons, $\bar{K}$ mesons or hyperons, in normal nuclei cause the impurity effects such as the formation of the high density states in the $\bar{K}$ nuclei [25,26], and the shrinking of the wave functions in the hypernuclei [27,28].

The attractive $\bar{D}N$ (BN) interaction would make it possible to give the exotic nuclear systems with heavy antiquarks ($\bar{c}$ or $\bar{b}$). These states are manifestly exotic states having no lower hadronic channels coupled by a strong interaction. The $\bar{D}$ (B) mesons in nuclei with large baryon numbers have been studied by many authors as summarized in Refs. [29–36] Such states are expected to have bound states because a binding energy increases as baryon number increases. However, $\bar{D}$ (B) nuclei as a few-body system have not been investigated so far. On the other hand, the D nuclei with a quark-antiquark annihilation where a strong attraction is expected have been studied [37].

In the present work, we study the mass spectrum of the exotic nuclei with a heavy meson , being $P^{(*)}N$ for two-body systems, and $P^{(*)}NN$ for three-body systems, where $P^{(*)} = P$ and P^*. The interaction between the heavy meson and the nucleon is given by the meson exchange potential. The potential with the spin operator induces the coupled channel systems between PN and P^*N. By solving the coupled channel Schrödinger equations, we obtain the bound and resonance states of the genuinely exotic states.

This article is organized as follows. After the introduction, the effective Lagrangian and the meson exchange potential are present in Sect. 2. In Sect. 3, the numerical results of $P^{(*)}N$ and $P^{(*)}NN$ systems are given. Finally, the summary of this article is shown in Sect. 4.

2 Interactions

In the interaction between a heavy meson $P^{(*)}$ and a nucleon N, the OPEP is expected to give a strong attraction because it is enhanced by the small mass splitting between P and P^* mesons due to the HQS [15–19]. The OPEP appears through the $PP^*\pi$ and $P^*P^*\pi$ vertices, while the $PP\pi$ vertex violating the parity conservation is absent. In the experiments, small mass splitting of the spin partners in the heavy hadron sectors is found, e.g. $m_{\bar{D}^*} - m_{\bar{D}} \sim 140$ MeV for charm and $m_{B^*} - m_B \sim 45$ MeV for bottom, while $m_\rho - m_\pi \sim 600$ MeV, and $m_{K^*} - m_K \sim 400$ MeV in the up, down and strange sectors have the large mass difference. Therefore the pion coupling in the meson vertices will be important in the heavy meson sectors.

In this study, we consider the π, ρ and ω exchange potentials obtained by the meson exchange scattering amplitudes shown in Fig. 1. The amplitudes are given by the interaction Lagrangians for a heavy meson and a light meson and for a nucleon and a light meson [16–19,38]. The Lagrangian for the heavy meson and the

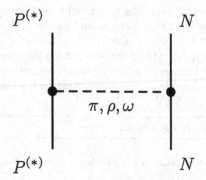

Fig. 1 Feynman diagram of the meson exchange scatering of a heavy meson $P^{(*)}$ and a nucleon N

light meson (π, V = (ρ, ω)), based on the HQS, are

$$\mathcal{L}_{\pi H H} = i g_\pi \mathrm{Tr} \left[H_b \gamma_\mu \gamma_5 A^\mu_{ba} \bar{H}_a \right], \tag{1}$$

$$\mathcal{L}_{V H H} = - i \beta \mathrm{Tr} \left[H_b v^\mu (\rho_\mu)_{ba} \bar{H}_a \right] + i \lambda \mathrm{Tr} \left[H_b \sigma^{\mu\nu} F_{\mu\nu}(\rho)_{ba} \bar{H}_a \right], \tag{2}$$

where the heavy meson field H is given by

$$H_a = \frac{1 + /v}{2} \left[P^*_{a\mu} \gamma^\mu - P_a \gamma^5 \right], \tag{3}$$

$$\bar{H}_a = \gamma^0 H_a^\dagger \gamma^0, \tag{4}$$

which are expressed by the linear combination of the heavy pseudoscalar (P) and vector (P^*) meson fields. The heavy meson fields are normalized as

$$\langle 0 | P | P(p_\mu) \rangle = \sqrt{p^0}, \tag{5}$$

$$\langle 0 | P^*_\mu | P^*(p_\mu, \alpha) \rangle = \varepsilon(\alpha)_\mu \sqrt{p^0}, \tag{6}$$

where $\varepsilon(\alpha)_\mu$ is the polarization vector of P^* with polarization $\alpha = \pm, 0$. v^μ is the four-velocity of a heavy quark where we take the static approximation $v^\mu = (1, \mathbf{0})$. The subscripts a, b are for light flavors. The axial current A^μ and the field tensor $F^{\mu\nu}$ are

$$A^\mu = \frac{i}{2 f_\pi} \partial^\mu \hat{\pi}, \quad \hat{\pi} = \sqrt{2} \begin{pmatrix} \pi^0/\sqrt{2} & \pi^+ \\ \pi^- & -\pi^0/\sqrt{2} \end{pmatrix}, \tag{7}$$

$$F^{\mu\nu} = \partial^\mu V^\nu - \partial^\nu V^\mu, \quad V^\mu = i \frac{g_V}{\sqrt{2}} \hat{V}^\mu, \tag{8}$$

$$\hat{V}^\mu = \sqrt{2} \begin{pmatrix} (\rho^0 + \omega)/\sqrt{2} & \rho^+ \\ \rho^- & (-\rho^0 + \omega)/\sqrt{2} \end{pmatrix}^\mu, \tag{9}$$

with the pion decay constant $f_\pi = 92.3$ MeV and $g_V = m_\rho/f_\pi$. The coupling constants of the Lagrangians are summarized in Table 1.

The Lagrangians for a nucleon and a light meson are given by the Bonn model [39,40] as

$$\mathcal{L}_{\pi N N} = i g_{\pi N N} \bar{N}_b \gamma^5 N_a \hat{\pi}_{ba}. \tag{10}$$

$$\mathcal{L}_{V N N} = g_{V N N} \bar{N}_b \left(\gamma^\mu (\hat{V}_\mu)_{ba} + \frac{\kappa}{2 m_N} \sigma_{\mu\nu} \partial^\nu (\hat{V}^\mu)_{ba} \right) N_a. \tag{11}$$

From those Lagrangians in Eqs. (1)–(11), we obtain the $P^{(*)}N$ potentials. The π exchange potential is given by

$$V^\pi_{PN - P^*N} = - \frac{g_\pi g_{\pi N N}}{2 m_N f_\pi} \frac{1}{3} \left[\boldsymbol{\varepsilon}^\dagger \cdot \boldsymbol{\sigma} C(r; m_\pi) + S_\varepsilon T(r; m_\pi) \right] \boldsymbol{\tau}_P \cdot \boldsymbol{\tau}_N, \tag{12}$$

$$V^\pi_{P^*N - P^*N} = \frac{g_\pi g_{\pi N N}}{2 m_N f_\pi} \frac{1}{3} \left[\mathbf{T} \cdot \boldsymbol{\sigma} C(r; m_\pi) + S_T T(r; m_\pi) \right] \boldsymbol{\tau}_P \cdot \boldsymbol{\tau}_N. \tag{13}$$

Table 1 Mass of mesons ($\alpha = \pi, \rho, \omega$) and coupling constants in Refs. [17–19]

	m_α [MeV]	g_π	β	λ [GeV^{-1}]	$g^2_{\alpha N N}$	κ
π	137.27	0.59	–	–	13.6	–
ρ	769.9	–	0.9	0.56	0.84	6.1
ω	781.94	–	0.9	0.56	20.0	0.0

and the vector meson exchange potentials are given by

$$V_{PN-PN}^{\mathrm{V}}(r) = \frac{\beta g_V\, g_{VNN}}{\sqrt{2}m_V^2} C(r; m_V)\boldsymbol{\tau}_P \cdot \boldsymbol{\tau}_N, \tag{14}$$

$$V_{PN-P^*N}^{\mathrm{V}}(r) = \frac{\beta g_V\, g_{VNN}\lambda(1+\kappa)}{\sqrt{2}m_V^2}\frac{1}{3}\left[-2\boldsymbol{\varepsilon}\cdot\boldsymbol{\sigma}\, C(r; m_V) + S_\varepsilon T(r; m_V)\right]\boldsymbol{\tau}_P \cdot \boldsymbol{\tau}_N, \tag{15}$$

$$V_{P^*N-P^*N}^{\mathrm{V}}(r) = \frac{\beta g_V\, g_{VNN}}{\sqrt{2}m_V^2} C(r; m_V)\boldsymbol{\tau}_P \cdot \boldsymbol{\tau}_N + \frac{\beta g_V\, g_{VNN}\lambda(1+\kappa)}{\sqrt{2}m_V^2}$$
$$\times \frac{1}{3}\left[2\mathbf{T}\cdot\boldsymbol{\sigma}\, C(r; m_V) - S_T T(r; m_V)\right]\boldsymbol{\tau}_P \cdot \boldsymbol{\tau}_N. \tag{16}$$

The nucleon mass is given by $m_N = 940$ MeV. $\mathbf{T}$ is the spin-one operator, and S_ε (S_T) is the tensor operator $S_{\mathcal{O}}(\hat{r}) = 3(\mathcal{O}\cdot\hat{r})(\boldsymbol{\sigma}\cdot\hat{r}) - \mathcal{O}\cdot\boldsymbol{\sigma}$ with $\hat{r} = \mathbf{r}/r$ and $r = |\mathbf{r}|$ for $\mathcal{O} = \boldsymbol{\varepsilon}$ ($\mathbf{T}$), where $\mathbf{r}$ is the relative position vector between $P^{(*)}$ and N. $\boldsymbol{\sigma}$ are Pauli matrices acting on nucleon spin. $\boldsymbol{\tau}_P$ ($\boldsymbol{\tau}_N$) are isospin operators for $P^{(*)}$ (N). For the OPEP, the $PN - PN$ term is absent because the vertex of three pseudoscalar mesons violates the parity conservation.

The functions $C(r; m_\alpha)$ and $T(r; m_\alpha)$ for $\alpha = \pi, \rho, \omega$ are expressed by

$$C(r; m_\alpha) = \int \frac{d^3\mathbf{q}}{(2\pi)^3}\frac{m_\alpha^2}{\mathbf{q}^2+m_\alpha^2}e^{i\mathbf{q}\cdot\mathbf{r}}F(\mathbf{q}, m_\alpha), \tag{17}$$

$$S_{\mathcal{O}}(\hat{r})T(r; m_\alpha) = \int \frac{d^3\mathbf{q}}{(2\pi)^3}\frac{-\mathbf{q}^2}{\mathbf{q}^2+m_\alpha^2}S_{\mathcal{O}}(\hat{q})e^{i\mathbf{q}\cdot\mathbf{r}}F(\mathbf{q}, m_\alpha), \tag{18}$$

where the dipole-type form factor

$$F(\mathbf{q}, m_\alpha) = \frac{\Lambda_N^2 - m_\alpha^2}{\Lambda_N^2 + |\mathbf{q}|^2}\frac{\Lambda_P^2 - m_\alpha^2}{\Lambda_P^2 + |\mathbf{q}|^2} \tag{19}$$

with cutoff parameters Λ_N and Λ_P is introduced for spatially extended hadrons. From a quark model estimation, we use $\Lambda_D = 1.35\Lambda_N$ for $\bar{D}^{(*)}$ meson and $\Lambda_B = 1.29\Lambda_N$ for $B^{(*)}$ meson, with $\Lambda_N = 830$ MeV, as discussed in Refs. [16–18].

For three-body PNN systems, the NN interaction is introduced, by the realistic NN potential, Argonne v_8' potential [41]. This potential is formed by a sum of 8 operators $\mathcal{O}^{p=1,8}$ which are the central forces with operators 1, $\boldsymbol{\sigma}_1 \cdot \boldsymbol{\sigma}_2$, $\boldsymbol{\tau}_1 \cdot \boldsymbol{\tau}_2$ and $\boldsymbol{\sigma}_1 \cdot \boldsymbol{\sigma}_2\,\boldsymbol{\tau}_1 \cdot \boldsymbol{\tau}_2$, the tensor forces with S_{12} and $S_{12}\,\boldsymbol{\tau}_1 \cdot \boldsymbol{\tau}_2$, and the LS forces with $\mathbf{L}\cdot\mathbf{S}$ and $\mathbf{L}\cdot\mathbf{S}\,\boldsymbol{\tau}_1 \cdot \boldsymbol{\tau}_2$.

The Hamiltonian is given by $H = K + V_{P^{(*)}N}(+V_{NN})$, where K is the kinetic term, and $V_{P^{(*)}N}$ (V_{NN}) is the $P^{(*)}N$ (NN) potential shown above. V_{NN} appears only for the three-body PNN systems. By diagonalizing the Hamiltonian, we obtain the eigenenergies. In the analysis of the two-body PN systems, the resonance is obtained by analyzing the phase shift. For the three-body PNN systems, we employ the useful numerical methods for few-body calculations. The three-body wave functions are expressed by the Gaussian expansion method [42]. The poles of resonances are calculated by the complex scaling method [43–46].

3 Numerical Results

3.1 Results of Two-Body Systems, $\bar{D}N$ and BN

In this section, the results for the bound and resonant states of the $\bar{D}N$ and BN molecules are shown. The two-body meson-nucleon systems with $J^P = 1/2^{\pm}, \ldots, 5/2^{\pm}$ are investigated. By solving the coupled channel Schrödinger equations for PN and P^*N, the bound states and resonances are obtained for the isospin $I = 0$.

The obtained energy spectra are summarized in Table 2. The binding energy is shown in a real negative value. The resonance energy is a complex value, which is given as $E_{\mathrm{re}} - i\Gamma/2$ with a resonance position E_{re} and a decay width Γ. We obtain the bound states both in $\bar{D}N$ and BN. In the bound states, we find that the tensor force of the OPEP in the $PN - P^*N$ mixing component plays an important role to generate the strong attraction. The mechanism is similar to the Deuteron where the tensor force in the $^3S_1 - {}^3D_1$ mixing component

Table 2 Energy spectra of $\bar{D}N$ and BN with $I = 0$ in Refs. [17,18]

$\bar{D}N$	$P = -$	$P = +$	BN	$P = -$	$P = +$
$J = 1/2$	-2.1	$26.8 - i65.7$	$J = 1/2$	-23.0	$5.8 - i3.0$
$J = 3/2$	$113.2 - i8.9$	$148.2 - i5.1$	$J = 3/2$	$6.9 - i0.05$	$31.8 - i14.4$
$J = 5/2$	$-$	$176.0 - i87.4$	$J = 5/2$	$-$	$58.4 - i24.8$

J and P are total angular momentum and parity, respectively. Binding energy is shown by a real negative value, while the energy of the resonance states are given by the resonance energy E_{re} and the decay width Γ as $E_{\mathrm{re}} - i\Gamma/2$. All values are measured from the PN threshold and given in units of MeV

produces the strong attraction. In the systems, the mixing of heavy pseudoscalar and vector mesons, namely $PN - P^*N$, is important rather than $PN - PN$ and $P^*N - P^*N$ components. This mixing is caused by the small mass splitting of the mesons due to the heavy quark spin symmetry. For the bottom sector, the BN binding energy is more larger than the $\bar{D}N$. The BN system obtains not only the suppression of the kinetic energy but also the enhancement of the mixing effect due to the small BB^* mass difference.

For $J^P = 1/2^+, 3/2^\pm$ and $5/2^+$, the resonance states are found above the PN threshold. In particular the states with $J^P = 3/2^-$ have the small decay width. We find the interesting structure of the resonances which is the Feshbach resonance. The states have the quasi bound states of the P^*N channels.

We obtain no bound state for $I = 1$. In this channel, the OPEP does not produce the strong attraction because of the small contribution of the isospin factor $\tau_\mathbf{P} \cdot \tau_\mathbf{N} = 1$ for $I = 1$, while $\tau_\mathbf{P} \cdot \tau_\mathbf{N} = -3$ for $I = 0$.

3.2 Results of Three-Body Systems, $\bar{D}NN$ and BNN

For the three-body systems, we investigate the $\bar{D}NN$, BNN with $J^P = 0^-$ and 1^- with isospin $I = 1/2$. As a result, we obtain the bound states with $J^P = 0^-$ and resonances with $J^P = 1^-$ in charm and bottom sectors. The obtained spectra are summarized in Table 3, where the energies are measured from the lowest thresholds.

In the bound states with $J^P = 0^-$ for the charm and bottom sectors, the binging energies are -5.2 MeV for $\bar{D}NN$ and -26.2 MeV for BNN. The BNN is more deeply bound than the $\bar{D}NN$, because the large reduced mass suppresses the kinetic term, and small mass splitting of BB^* enhances the $PNN - P^*NN$ mixing effects as seen in the two-body systems.

We find that the tensor force of the $PNN - P^*NN$ mixing provides the strong attraction in the PNN and it dominates the binding system. Although the tensor force in V_{NN} is a driving force in the Deuteron, this force is almost irrelevant, and the central force is rather dominant. This is reasonable because the component of Deuteron with angular momentum and parity 1^+ is suppressed in the main component of $\bar{D}NN$ (BNN) with $J^P = 0^-$. The NN subsystem with 0^+ is dominant in the bound state and hence the attraction of the central force in V_{NN} is stronger than that of the tensor force.

In resonant states for $(I, J^P) = (1/2, 1^-)$, the resonance energies are obtained as $E_{\mathrm{re}} - \Gamma/2 = 111.2 - i9.3$ MeV for $\bar{D}NN$, and $6.8 - i0.2$ MeV for BNN, respectively. The resonances are obtained as the quasi-bound state of $\bar{D}^*NN$ (B^*NN). Hence the resonances are also the Feshbach resonance as seen in the two-body systems.

Table 3 Energy spectra of $\bar{D}NN$, BNN with $I = 1/2$ from Ref. [19]

J^P	$\bar{D}NN$	BNN
0^-	-5.2	-26.2
1^-	$111.2 - i9.3$	$6.8 - i0.2$

All values are given in units of MeV

4 Summary

We have explored the possible existence of the genuinely exotic nuclei as PN and PNN for $P = \bar{D}, B$. The interaction is constructed by introducing the heavy quark spin symmetry which provides the mass degeneracy of the heavy pseudoscalar and vector mesons. The meson exchange potentials for π, ρ and ω with the $P - P^*$ mixing are employed as the $P^{(*)}N$ interaction. For the three-body systems, the realistic NN potential, Argonne v'_8 potential, was also introduced. By solving the coupled-channel equations for $PN(N)$ and $P^*N(N)$, we have obtained bound states and resonances for two-body and three-body systems in charm and bottom sectors. The tensor force of the OPEP mixing PN and P^*N plays an important role to produce the strong attraction. This force is enhanced by the small mass splitting of mesons due to the heavy quark spin symmetry. The mechanism is unique in the heavy hadron systems with the small mass difference of the spin partner.

The obtained exotic states can be searched in relativistic heavy ion collisions in RHIC and LHC [47,48]. Furthermore, the search for the $\bar{D}NN$ would be also carried out in J-PARC and GSI-FAIR.

Acknowledgments This work is supported in part by the INFN Fellowship Programme.

References

1. Klempt, E., Richard, J.-M.: Baryon spectroscopy. Rev. Mod. Phys. **82**, 1095 (2010)
2. Eichten, E., Godfrey, S., Mahlke, H., Rosner, J.L.: Quarkonia and their transitions. Rev. Mod. Phys. **80**, 1161 (2008)
3. Brambilla, N., Eidelman, S., Heltsley, B.K., Vogt, R., Bodwin, G.T., Eichten, E., Frawley, A.D., Meyer, A.B., et al.: Heavy quarkonium: progress, puzzles, and opportunities. Eur. Phys. J. C **71**, 1534 (2011)
4. Swanson, E.S.: The new heavy mesons: a status report. Phys. Rep. **429**, 243 (2006)
5. Voloshin, M.B.: Charmonium. Prog. Part. Nucl. Phys. **61**, 455 (2008)
6. Godfrey, S., Olsen, S.L.: The exotic XYZ charmonium-like mesons. Ann. Rev. Nucl. Part. Sci. **58**, 51 (2008)
7. Guo, P., Szczepaniak, A.P., Galata, G., Vassallo, A., Santopinto, E.: Gluelump spectrum from Coulomb gauge QCD. Phys. Rev. D **77**, 056005 (2008). doi:10.1103/PhysRevD.77.056005. arXiv:0707.3156 [hep-ph]
8. Guo, P., Szczepaniak, A.P., Galata, G., Vassallo, A., Santopinto, E.: Heavy quarkonium hybrids from Coulomb gauge QCD. Phys. Rev. D **78**, 056003 (2008). doi:10.1103/PhysRevD.78.056003. arXiv:0807.2721 [hep-ph]
9. Ferretti, J., Galat, G., Santopinto, E.: Interpretation of the $X(3872)$ as a charmonium state plus an extra component due to the coupling to the meson-meson continuum. Phys. Rev. C **88**(1), 015207 (2013). doi:10.1103/PhysRevC.88.015207. arXiv:1302.6857 [hep-ph]
10. Ferretti, J., Galat, G., Santopinto, E.: Quark structure of the $X(3872)$ and $\mathcal{X}b(3P)$ resonances. Phys. Rev. D **90**(5), 054010 (2014). doi:10.1103/PhysRevD.90.054010. arXiv:1401.4431 [nucl-th]
11. Ferretti, J., Santopinto, E.: Higher mass bottomonia. Phys. Rev. D **90**(9), 094022 (2014). doi:10.1103/PhysRevD.90.094022. arXiv:1306.2874 [hep-ph]
12. Choi, S.K., et al. [Belle Collaboration]: Observation of a narrow charmonium-like state in exclusive B+ $\longrightarrow$ K+− pi+ pi− J/psi decays. Phys. Rev. Lett. **91**, 262001 (2003)
13. Bondar, A., et al. [Belle Collaboration]: Observation of two charged bottomonium-like resonances in Y(5S) decays. Phys. Rev. Lett. **108**, 122001 (2012)
14. Tornqvist, N.A.: From the deuteron to deusons, an analysis of deuteron-like meson meson bound states. Z. Phys. C **61**, 525 (1994)
15. Cohen, T.D., Hohler, P.M., Lebed, R.F.: On the existence of heavy pentaquarks: the large N(c) and heavy quark limits and beyond. Phys. Rev. D **72**, 074010 (2005)
16. Yasui, S., Sudoh, K.: Exotic nuclei with open heavy flavor mesons. Phys. Rev. D **80**, 034008 (2009)
17. Yamaguchi, Y., Ohkoda, S., Yasui, S., Hosaka, A.: Exotic baryons from a heavy meson and a nucleon—Negative parity states. Phys. Rev. D **84**, 014032 (2011)
18. Yamaguchi, Y., Ohkoda, S., Yasui, S., Hosaka, A.: Exotic baryons from a heavy meson and a nucleon—positive parity states. Phys. Rev. D **85**, 054003 (2012)
19. Yamaguchi, Y., Yasui, S., Hosaka, A.: Exotic dibaryons with a heavy antiquark. Nucl. Phys. A **927**, 110 (2014)
20. Yasui, S., Sudoh, K., Yamaguchi, Y., Ohkoda, S., Hosaka, A., Hyodo, T.: Spin degeneracy in multi-hadron systems with a heavy quark. Phys. Lett. B **727**, 185 (2013)
21. Yamaguchi, Y., Ohkoda, S., Hosaka, A., Hyodo, T., Yasui, S.: Heavy quark symmetry in multi-hadron systems. Phys. Rev. D **91**, 034034 (2015). doi:10.1103/PhysRevD.91.034034. arXiv:1402.5222 [hep-ph]
22. Isgur, N., Wise, M.B.: Weak decays of heavy mesons in the static quark approximation. Phys. Lett. B **232**, 113 (1989)
23. Flynn, J.M., Isgur, N.: Heavy quark symmetry: ideas and applications. J. Phys. G **18**, 1627 (1992)
24. Manohar, A.V., Wise, M.B.: Heavy quark physics. Camb. Monogr. Part. Phys. Nucl. Phys. Cosmol. **10**, 1–191 (2000)
25. Yamazaki, T., Dote, A., Akaishi, Y.: Invariant mass spectroscopy for condensed single and double anti-K nuclear clusters to be formed as residues in relativistic heavy ion collisions. Phys. Lett. B **587**, 167 (2004)
26. Dote, A., Hyodo, T., Weise, W.: Variational calculation of the ppK- system based on chiral SU(3) dynamics. Phys. Rev. C **79**, 014003 (2009)
27. Hashimoto, O., Tamura, H.: Spectroscopy of Lambda hypernuclei. Prog. Part. Nucl. Phys. **57**, 564 (2006)
28. Tamura, H.: Impurity nuclear physics Hypernuclear Γ spectroscopy and future plans for neutron-rich hypernuclei. Eur. Phys. J. A **13**, 181 (2002)

29. Tsushima, K., Lu, D.-H., Thomas, A.W., Saito, K., Landau, R.H.: Charmed mesic nuclei. Phys. Rev. C **59**, 2824 (1999)
30. Mishra, A., Bratkovskaya, E.L., Schaffner-Bielich, J., Schramm, S., Stoecker, H.: Mass modification of D meson in hot hadronic matter. Phys. Rev. C **69**, 015202 (2004)
31. Lutz, M.F.M., Korpa, C.L.: Open-charm systems in cold nuclear matter. Phys. Lett. B **633**, 43 (2006)
32. Tolos, L., Ramos, A., Mizutani, T.: Open charm in nuclear matter at finite temperature. Phys. Rev. C **77**, 015207 (2008)
33. Mishra, A., Mazumdar, A.: D-mesons in asymmetric nuclear matter. Phys. Rev. C **79**, 024908 (2009)
34. Kumar, A., Mishra, A.: D mesons and charmonium states in asymmetric nuclear matter at finite temperatures. Phys. Rev. C **81**, 065204 (2010)
35. Garcia-Recio, C., Nieves, J., Salcedo, L.L., Tolos, L.: D^- mesic atoms. Phys. Rev. C **85**, 025203 (2012)
36. Yasui, S., Sudoh, K.: $\bar{D}$ and B mesons in nuclear medium. Phys. Rev. C **87**, 015202 (2013)
37. Bayar, M., Xiao, C.W., Hyodo, T., Dote, A., Oka, M., Oset, E.: Energy and width of a narrow $I = 1/2$ DNN quasibound state. Phys. Rev. C **86**, 044004 (2012)
38. Casalbuoni, R., Deandrea, A., Di Bartolomeo, N., Gatto, R., Feruglio, F., Nardulli, G.: Phenomenology of heavy meson chiral Lagrangians. Phys. Rep. **281**, 145 (1997)
39. Machleidt, R., Holinde, K., Elster, C.: The bonn meson exchange model for the nucleon nucleon interaction. Phys. Rep. **149**, 1 (1987)
40. Machleidt, R.: The high precision, charge dependent Bonn nucleon-nucleon potential (CD-Bonn). Phys. Rev. C **63**, 024001 (2001)
41. Pudliner, B.S., Pandharipande, V.R., Carlson, J., Pieper, S.C., Wiringa, R.B.: Quantum Monte Carlo calculations of nuclei with $A \leq 7$. Phys. Rev. C **56**, 1720 (1997)
42. Hiyama, E., Kino, Y., Kamimura, M.: Gaussian expansion method for few-body systems. Prog. Part. Nucl. Phys. **51**, 223 (2003)
43. Aguilar, J., Combes, J.M.: A class of analytic perturbations for one-body Schrödinger Hamiltonians. Commun. Math. Phys. **22**, 269 (1971)
44. Balslev, E., Combes, J.M.: Spectral properties of many-body Schrödinger operators with dilatation-analytic interactions. Commun. Math. Phys. **22**, 280 (1971)
45. Simon, B.: Quadratic form techniques and the Balslev–Combes theorem. Commun. Math. Phys. **27**, 1 (1972)
46. Aoyama, S., Myo, T., Katō, K., Ikeda, K.: The complex scaling method for many-body resonances and its applications to three-body resonances. Prog. Theor. Phys. **116**, 1 (2006)
47. Cho, S., et al. [ExHIC Collaboration]: Multi-quark hadrons from heavy ion collisions. Phys. Rev. Lett. **106**, 212001 (2011)
48. Cho, S., et al. [ExHIC Collaboration]: Studying exotic hadrons in heavy ion collisions. Phys. Rev. C **84**, 064910 (2011)

Part IV
**Advances and Prospects in Extraction
of Resonance Parameters from the Data**

Few-Body Syst (2016) 57:933–940
DOI 10.1007/s00601-016-1130-0

Hiroyuki Kamano

Light-Quark Baryon Spectroscopy within ANL-Osaka Dynamical Coupled-Channels Approach

Received: 8 February 2016 / Accepted: 2 June 2016 / Published online: 16 June 2016
© Springer-Verlag Wien 2016

Abstract Recent results on the study of light-quark baryons with the ANL-Osaka dynamical coupled-channels (DCC) approach are presented, which contain the N^* and Δ^* spectroscopy via the analysis of πN and γN reactions and the Λ^* and Σ^* spectroscopy via the analysis of $K^- p$ reactions. A recent application of our DCC approach to neutrino–nucleon reactions in the resonance region is also presented.

1 Introduction

The spectroscopic study of light-quark baryons (N^*, Δ^*, Λ^*, Σ^*) remains a central issue in the hadron physics. Since the excitation spectra of light-quark baryons are an embodiment of the dynamics of confinement in Quantum Chromodynamics (QCD), their precise determination through analyzing reaction data is essential to test QCD in the nonperturbative domain. In this regard, several attempts to link the (real) energy spectrum of QCD in the finite box to the complex pole masses of actual hadron resonances have been being made recently (see, e.g., Refs. [1–4]).

A particular complexity exists in extracting the light-quark baryons from reaction data. This originates from a peculiar character of the light-quark baryons: they are broad and highly overlapping resonances. The light-quark baryons are thus strongly correlated with each other and with backgrounds through reaction processes over the wide energy range, and the observed shape of cross sections and spin asymmetries represents such a complicated interference between them. As a result, the light-quark baryon resonances do not produce any clean isolated peaks in cross sections with a few exceptions. This makes identification of light-quark baryons from experiments rather difficult, and cooperative works between experiments and theoretical analyses are indispensable for the light-quark baryon spectroscopy.

To disentangle complicated correlations between the light-quark baryon resonances, one needs to perform a partial-wave analysis of various meson-production reactions comprehensively over the wide energy range. Here the use of a reaction framework satisfying multichannel unitarity is particularly important. In fact, as is well known, multichannel unitarity ensures the conservation of probabilities in the multichannel reaction processes, and this is crucial for accomplishing a comprehensive analysis of meson-production reactions with various final states consistently within a single reaction framework. Also, the multichannel unitarity properly defines the analytic structure (branch points, cuts, threshold cusps, etc.) of scattering amplitudes in the complex-energy plane, as required by the scattering theory, which is essential to extract resonance information from reaction data *correctly*.

This work was supported by the JSPS KAKENHI Grant No. 25800149, and by the HPCI Strategic Program (Field 5 "The Origin of Matter and the Universe") of MEXT of Japan.

This article belongs to the special issue "Nucleon Resonances".

H. Kamano (✉)
Research Center for Nuclear Physics (RCNP), Osaka University, Ibaraki, Osaka 567-0047, Japan
E-mail: kamano@rcnp.osaka-u.ac.jp

As an approach that enables such a multichannel analysis, a dynamical coupled-channels (DCC) model for meson-production reactions in the resonance region was developed in Ref. [5]. So far, this model has been employed in a series of works [6–15] to analyze the πN, γN, and eN reactions and to study the mass spectrum, structure, and dynamical origins of nonstrange N^* and Δ^* resonances. This model has also been applied to the study of Λ^* and Σ^* resonances with strangeness $S = -1$ through a comprehensive analysis of $K^- p$ reactions [16,17]. In addition to these studies related to the light-quark baryon spectroscopy, we have recently made an application of our DCC model to neutrino–nucleon reactions in the resonance region [18,19]. The purpose of this contribution is to give an overview of our recent efforts for these subjects.

2 ANL-Osaka DCC Model

The basic formula of our DCC model is the coupled-channels integral equation obeyed by the partial-wave amplitudes of $a \rightarrow b$ reactions that are specified by the total angular momentum (J), parity (P), and total isospin (I) (here we explain our approach by taking the N^* and Δ^* sector as an example):

$$T_{b,a}^{(JPI)}(p_b, p_a; W) = V_{b,a}^{(JPI)}(p_b, p_a; W) + \sum_c \int_C dp_c\, p_c^2 V_{b,c}^{(JPI)}(p_b, p_c; W) G_c(p_c; W) T_{c,a}^{(JPI)}(p_c, p_a; W),$$

(1)

with

$$V_{b,a}^{(JPI)}(p_b, p_a; W) = v_{b,a}(p_b, p_a) + Z_{b,a}(p_b, p_a; W) + \sum_{N_n^*} \frac{\Gamma_{b,N_n^*}(p_b)\,\Gamma_{N_n^*,a}(p_a)}{W - M_{N_n^*}^0}.$$

(2)

Here the subscripts (a, b, c) represent the considered reaction channels (the indices associated with the total spin and orbital angular momentum of the channels are suppressed); p_a is the magnitude of the relative momentum for the channel a in the center-of-mass frame; and W is the total scattering energy. For the N^* and Δ^* sector, we have taken into account the eight channels, $\gamma^{(*)}N$, πN, ηN, $K\Lambda$, $K\Sigma$, $\pi\Delta$, ρN, and σN[1], where the last three are the quasi two-body channels that subsequently decay into the three-body $\pi\pi N$ channel. The Green's functions for the meson(M)-baryon(B) channels are given by $G_c(p_c; W) = 1/[W - E_M(p_c) - E_B(p_c) + i\varepsilon]$ for $c = \pi N, \eta N, K\Lambda, K\Sigma$, while $G_c(p_c; W) = 1/[W - E_M(p_c) - E_B(p_c) + \Sigma_{MB}(p_c; W)]$ is used for $c = \pi\Delta, \sigma N, \rho N$, where $E_\alpha(p) = \sqrt{m_\alpha^2 + p^2}$ is the relativistic single-particle energy of a particle α and $\Sigma_{MB}(p_c; W)$ is the self-energy that produces the three-body $\pi\pi N$ cut. The transition potential $V_{b,a}$ [Eq. (2)] consists of three pieces. The first two, $v_{b,a}$ and $Z_{b,a}$, describe the nonresonant processes including only the ground-state mesons and baryons belonging to each flavor SU(3) multiplet, and the third one describes the propagation of the bare N^* states. In our approach, the unitary transformation method [5,20] is used to derive the potential $v_{b,a}$ from effective Lagrangians. With this method, the resulting $v_{b,a}$ becomes energy independent and its off-shell behavior is specified. On the other hand, the Z-diagram potential, $Z_{b,a}$, is derived using the projection operator method [21]. It also produces the three-body $\pi\pi N$ cut [5,14], and implementation of both the Z-diagram potential and the self-energy in the Green's functions is necessary to maintain the three-body unitarity. Furthermore, off-shell rescattering effects are also taken into account properly through the momentum integral in Eq. (1), which are usually neglected in on-shell approaches.

3 Recent Efforts for N^* and Δ^* Spectroscopy with ANL-Osaka DCC Model

Our latest published model for the N^* and Δ^* sector [14,15] was constructed by making a simultaneous analysis of $\pi p \rightarrow \pi N$, ηN, $K\Lambda$, $K\Sigma$ and $\gamma p \rightarrow \pi N$, ηN, $K\Lambda$, $K\Sigma$. This contains the data of both unpolarized differential cross sections and polarization observables up to $W = 2.1$ GeV (up to $W = 2.3$ GeV for $\pi p \rightarrow \pi N$), which results in fitting ~23, 000 data points. After the completion of this analysis, we have been mainly proceeding with two subjects for the N^* and Δ^* spectroscopy: (1) extraction of the helicity amplitudes for the $\gamma n \rightarrow N^*$ transition from the available γ 'n' $\rightarrow \pi N$ data, and (2) determination of Q^2 dependence of the p-N^* and p-Δ^* electromagnetic transition form factors by analyzing the structure function data for

---

[1] Because of the perturbative nature of the electromagnetic interactions, it is only necessary to solve the coupled-channels equations in the channel space excluding the $\gamma^{(*)}N$ channel. Thus the summation in Eq. (1) runs over only the hadronic channels.

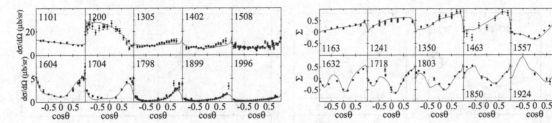

Fig. 1 Differential cross sections (*left*) and photon asymmetries (*right*) for $\gamma\,'n' \to \pi^- p$. The *numbers* shown in each panel are the corresponding total scattering energy W in MeV. The data are taken from Ref. [22]

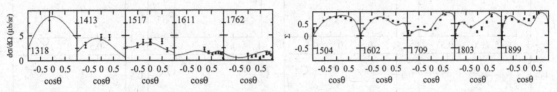

Fig. 2 Differential cross sections (*left*) and photon asymmetries (*right*) for $\gamma\,'n' \to \pi^0 n$. The *numbers* shown in each panel are the corresponding total scattering energy W in MeV. The data are taken from Ref. [22]

single-pion electroproductions off the proton target. The latter subject is discussed in Ref. [23], and in this contribution we focus on presenting the current status of our $\gamma\,'n' \to \pi N$ analysis.

There are mainly two reasons for studying meson photo- and electro-production reactions off the neutron target. One comes from the fact that the analysis of the data for both proton- and neutron-target reactions is required to decompose the matrix elements for the electromagnetic currents into the ones of isoscalar and isovector currents and uniquely determine the isospin structure of the $\gamma^{(*)}N \to N^*$ transition amplitudes. Another is because the matrix elements for such "isospin-decomposed" currents are necessary for constructing a model for neutrino-induced reactions, which will be discussed in Sect. 5.

So far, we have performed the fits to the data for $\gamma n \to \pi N$ up to $W \lesssim 2$ GeV. The data are available for $d\sigma/d\Omega$, Σ, T, and P for $\gamma n \to \pi^- p$, and $d\sigma/d\Omega$ and Σ for $\gamma n \to \pi^0 n$. This contains ~3200 data points. Some of the results of our fits are presented in Figs. 1 and 2. One can see that a reasonably good reproduction of the $\gamma n \to \pi N$ data has been accomplished for the considered energy region.

In Fig. 3, we present our calculated results for all 16 observables ($d\sigma/d\Omega$ and 15 polarization observables) of the $\gamma n \to \pi^- p$ reaction at $W = 1604$ MeV. They are predictions from our current DCC model except for $d\sigma/d\Omega$, Σ, T, and P, whose data displayed in Fig. 3 were included in our fits to determine the model parameters. The polarization data are known to provide crucial constraints on multipole amplitudes [24], and tremendous efforts are now being pursued at electron- and photon-beam facilities such as JLab, ELSA, and MAMI to extract polarization observables for the neutron-target photoproductions through the γd reaction (see, e.g., Ref. [25]).

4 Comprehensive Analysis of $K^- p$ Reactions and Extraction of Λ^* and Σ^* Resonances

Following the success of our N^* and Δ^* studies, we have recently extended our DCC approach to the strangeness $S = -1$ sector to also explore the Λ^* and Σ^* hyperon resonances. In the formulation of the DCC model in the $S = -1$ sector, we have taken into account seven channels ($\bar{K}N$, $\pi\Sigma$, $\pi\Lambda$, $\eta\Lambda$, $K\Xi$, $\pi\Sigma^*$, $\bar{K}^*N$), where the $\pi\Sigma^*$ and $\bar{K}^*N$ channels are the quasi two-body channels that subsequently decay into the three-body $\pi\pi\Lambda$ and $\pi\bar{K}N$ channels, respectively. The model parameters are then fixed by fitting to all available data of $K^- p \to \bar{K}N$, $\pi\Sigma$, $\pi\Lambda$, $\eta\Lambda$, $K\Xi$ reactions from the threshold up to $W = 2.1$ GeV. The data contain the total cross section (σ), differential cross section ($d\sigma/d\Omega$), and recoil polarization (P), and this results in fitting more than 17,000 data points.

Figure 4 shows the results of our fits to the total cross sections for all reactions we considered. Here two curves are plotted in red (Model A) and blue (Model B), respectively. This is because the existing $K^- p$ reaction data are not sufficient to determine our model parameters unambiguously, and it allowed us to have two distinct sets of the model parameters, yet both give almost the same χ^2 value. We quote Ref. [16] for all the details of our analysis including the fits to the differential cross sections and polarization observables.

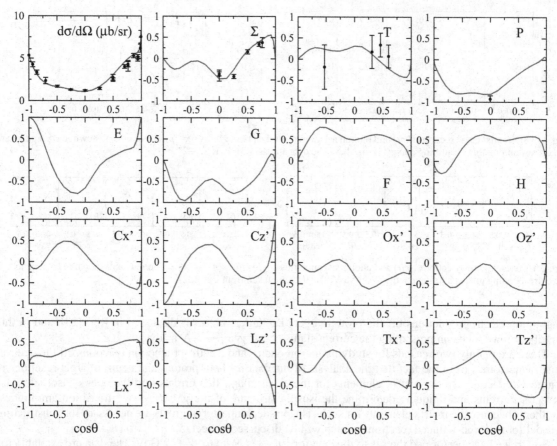

Fig. 3 Differential cross section and all polarization observables for $\gamma n \to \pi^- p$ at $W = 1604$ MeV ($E_\gamma = 900$ MeV). The results are predictions from our current DCC model except for $d\sigma/d\Omega$, Σ, T, and P. The data are taken from Ref. [22]

Once the model parameters are determined by fitting to the reaction data, we can extract various information (mass, width, pole residues, and branching ratios, etc.) on the Λ^* and Σ^* resonances by making an analytic continuation of the calculated scattering amplitudes in the complex energy plane (see Refs. [26,27] for a numerical procedure to perform the analytic continuation within our DCC model). This was done in Ref. [17], and the value of the resonance pole masses extracted obtained via such an analytic continuation is presented in Table 1. We searched pole positions of scattering amplitudes in the complex energy region with $m_{\bar{K}} + m_N \leq$ Re(W) ≤ 2.1 GeV and 0 GeV $\leq -$Re(W) ≤ 0.2 GeV, and found 18 resonances (10 Λ^* and 8 Σ^*) in Model A, while 20 resonances (10 Λ^* and 10 Σ^*) in Model B. The uncertainties assigned for the value of each extracted pole mass was deduced by following a procedure similar to the one used in Ref. [24] to determine the uncertainties in extracted multipole amplitudes for $\gamma p \to K^+\Lambda$. Here we did not assign the uncertainty for $J^P = 7/2^+$ Λ resonance in Model A because it was turned out to be too large to be meaningful.

We see in Table 1 that the pole mass of 9 resonances (6 Λ^* and 3 Σ^*) agree well between our two models within the deduced uncertainties (each resonance is aligned in the same row). These resonances would correspond to $\Lambda(1670)1/2^-$, $\Lambda(1600)1/2^+$, $\Lambda(1520)3/2^-$, $\Lambda(1690)3/2^-$, $\Lambda(1820)5/2^+$, $\Lambda(1830)5/2^-$ $\Sigma(1670)3/2^-$, $\Sigma(1775)5/2^-$, and $\Sigma(2030)7/2^+$ in the PDG notation [28]. They are assigned as four-star resonances by PDG except for $\Lambda(1600)1/2^+$ with three-star status. However, within the energy region we are currently interested in, two more resonances, $\Lambda(1890)3/2^+$ and $\Sigma(1915)5/2^+$, are also rated as four-star by PDG, but the corresponding resonances are not found in Model B. This is one example indicating that four-star resonances rated by PDG using the Breit–Wigner parameters are not confirmed by the multichannel analyses, in which the resonance parameters are extracted at pole positions [17].

A noteworthy difference between our two models is the existence of a new narrow $J^P = 3/2^+$ Λ^* resonance with $M_R = 1671^{+2}_{-8} - i(5^{+11}_{-2})$ MeV, which is found only in Model B. We refer to this resonance as $\Lambda(1671)3/2^+$. It is located close to the $\eta\Lambda$ threshold and has almost the same M_R value as the four-star $\Lambda(1670)1/2^-$ resonance [see the second $J^P(l_{I2J}) = 1/2^-(S_{01})$ resonance in Table 1]. It is known that the

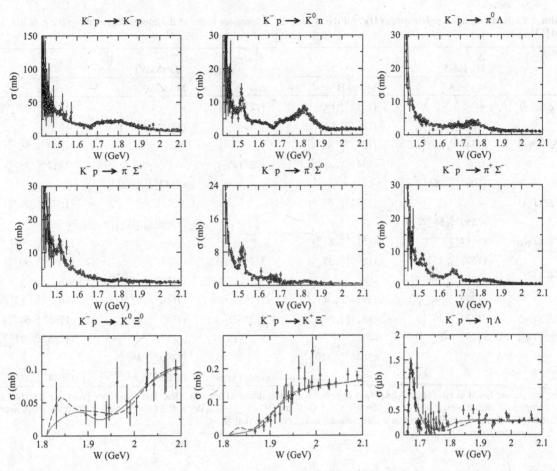

Fig. 4 Fitted results for the total cross sections of K^-p reactions [16]. *Red solid* (*blue dashed*) curves are the results of Model A (Model B)

sharp peak of the $K^-p \to \eta\Lambda$ total cross section near the threshold (see the bottom-right panel of Fig. 4) can be explained by the existence of $\Lambda(1670)1/2^-$ [29]. In addition to this, we further find that the new $\Lambda(1671)3/2^+$ resonance seems to be favored by the differential cross section data (Fig. 5). Actually, both Models A and B reproduce the $K^-p \to \eta\Lambda$ total cross section equally well, but Model A with no $\Lambda(1671)3/2^+$ does not reproduce the concave-up behavior of the differential cross section data well.

5 Application to Neutrino-Induced Reactions in Resonance Region

After the establishment of the neutrino mixing between all three flavors, major interests in the neutrino physics are shifting to the determination of the leptonic CP phase and the neutrino-mass hierarchy (see, e.g., Ref. [30]). In such neutrino-parameter searches via next-generation neutrino-oscillation experiments, the lack of knowledge of neutrino-nucleus reaction cross sections are expected to become one of the major sources of the systematic uncertainties. Thus the developments of reaction models that give the cross sections in 10 % or better accuracy are now highly desirable. Furthermore, the kinematics relevant to the neutrino-oscillation experiments extends over the quasi-elastic (QE), resonance (RES), and deep-inelastic-scattering (DIS) regions (see the left panel of Fig. 6), which are governed by rather different physics mechanisms and theoretical foundations. These facts lead experimentalists and theorists in different fields to developing a new collaboration [31,32] at the J-PARC Branch of KEK Theory Center, aiming at constructing a unified neutrino-reaction model that covers all the relevant kinematical regions mentioned above consistently. We will employ our DCC model as a base model for the RES region.

As a first step towards developing a model for neutrino-nucleus reactions in the RES region, we have recently constructed a DCC model for neutrino-nucleon (νN) reactions [19]. This is actually an important step

Table 1 Extracted complex pole masses (M_R) for the Λ^* and Σ^* resonances found in the energy region above the $\bar{K}N$ threshold [17]

$J^P(l_{I2J})$	Λ^* M_R (MeV)		$J^P(l_{I2J})$	Σ^* M_R (MeV)	
	Model A	Model B		Model A	Model B
$1/2^-(S_{01})$	–	$(1512^{+1}_{-1},185^{+1}_{-2})$	$1/2^-(S_{11})$	–	$(1551^{+2}_{-9},188^{+6}_{-1})$
	$(1669^{+3}_{-8}, 9^{+9}_{-1})$	$(1667^{+1}_{-2},12^{+3}_{-1})$		$(1704^{+3}_{-6}, 43^{+7}_{-2})$	–
$1/2^+(P_{01})$	$(1544^{+3}_{-3}, 56^{+6}_{-1})$	$(1548^{+5}_{-6}, 82^{+7}_{-7})$		–	$(1940^{+2}_{-2}, 86^{+2}_{-2})$
	–	$(1841^{+3}_{-4}, 31^{+3}_{-2})$	$1/2^+(P_{11})$	–	$(1457^{+5}_{-1}, 39^{+1}_{-4})$
	$(2097^{+40}_{-1}, 83^{+32}_{-6})$	–		$(1547^{+111}_{-59}, 92^{+43}_{-39})$	–
$3/2^+(P_{03})$	–	$(1671^{+2}_{-8}, 5^{+11}_{-2})$		–	$(1605^{+2}_{-4}, 96^{+1}_{-5})$
	$(1859^{+5}_{-7}, 56^{+10}_{-2})$	–		$(1706^{+67}_{-60}, 51^{+79}_{-42})$	–
$3/2^-(D_{03})$	$(1517^{+4}_{-4},8^{+5}_{-4})$	$(1517^{+4}_{-3}, 8^{+6}_{-6})$		–	$(2014^{+6}_{-13}, 70^{+14}_{-1})$
	$(1697^{+6}_{-6}, 33^{+7}_{-7})$	$(1697^{+6}_{-5}, 37^{+7}_{-7})$	$3/2^-(D_{13})$	–	$(1492^{+4}_{-7}, 69^{+4}_{-7})$
$5/2^-(D_{05})$	$(1766^{+37}_{-34},106^{+47}_{-31})$	–		$(1607^{+13}_{-11},126^{+15}_{-9})$	–
	$(1899^{+35}_{-37}, 40^{+50}_{-17})$	$(1924^{+52}_{-24}, 45^{+57}_{-17})$		$(1669^{+7}_{-7}, 32^{+5}_{-7})$	$(1672^{+5}_{-10}, 33^{+3}_{-3})$
$5/2^+(F_{05})$	$(1824^{+2}_{-1}, 39^{+1}_{-1})$	$(1821^{+1}_{-1}, 32^{+1}_{-1})$	$5/2^-(D_{15})$	$(1767^{+2}_{-2}, 64^{+2}_{-1})$	$(1765^{+2}_{-1}, 64^{+3}_{-1})$
$7/2^+(F_{07})$	$(1757, 73)$	–	$5/2^+(F_{15})$	–	$(1695^{+20}_{-77}, 97^{+50}_{-44})$
	–	$(2041^{+80}_{-82},119^{+57}_{-17})$		$(1890^{+3}_{-2}, 49^{+2}_{-3})$	–
			$7/2^+(F_{17})$	$(2025^{+10}_{-5}, 65^{+3}_{-12})$	$(2014^{+12}_{-1},103^{+3}_{-9})$

The masses are listed as $(\mathrm{Re}(M_R), -\mathrm{Im}(M_R))$ together with their deduced uncertainties. The resonance poles are searched in the complex W region with $m_{\bar{K}} + m_N \leq \mathrm{Re}(W) \leq 2.1$ GeV and 0 GeV $\leq -\mathrm{Im}(W) \leq 0.2$ GeV, and all of the resonances listed are located in the complex W Riemann surface nearest to the physical real W axis

because it can be used as an input to elementary processes in nuclear-target reactions. The right panel of Fig. 6 presents a schematic view of the νN reactions in the RES region. The blue rectangle represents the matrix element for the transition between the nucleon and meson–baryon (MB) states induced by the ("$V-A$"-type) weak currents, while the red circle represents the rescattering process induced by the hadronic interactions. In our DCC model, the rescattering part has been determined through the analysis of πN and γN reaction data. However, the weak-current matrix elements have to be newly constructed. The vector-current part can be determined using the data for photon- and electron-induced meson-production reactions off the proton and neutron targets (see Sect. 3 and Ref. [23]), and in Ref. [19], the vector-current matrix elements were determined in the range of $W \leq 2$ GeV and $Q^2 \leq 3$ GeV^2. The axial-current matrix elements are difficult to be determined because plenty of νN and/or νd reaction data are required for that purpose, but the existing data are far from sufficient to constrain our model parameters. Therefore, at this stage, we determined the axial-current matrix elements by making use of PCAC hypothesis, and by imposing purely phenomenological assumptions for their Q^2 dependence, for which we will not go into the details here and just quote Ref. [19]. With this setup, we can make a prediction for the neutrino–nucleon reactions in the RES region.

Figure 7 shows the results for the charged-current (CC) single- and double-pion productions predicted from our DCC model. This is the first-time fully coupled-channels calculation of νN reactions beyond the $\Delta(1232)3/2^+$ region. The contour plot of $d\sigma/dWdQ^2$ for $\nu_\mu n \to \mu^- \pi N$ (the left panel of Fig. 7) show a clear peak due to $\Delta(1232)3/2^+$ and relatively small peak at $W \sim 1.5$ GeV that comes from $N(1535)1/2^-$ and $N(1520)3/2^-$ and so on. In contrast, for the $\nu_\mu n \to \mu^- \pi^+\pi^- p$ case, the contribution from $\Delta(1232)3/2^+$ is barely seen, while higher resonance contributions are sizable above $W \gtrsim 1.45$ GeV, and their effects on the cross section reach a rather high Q^2, too. The single-pion productions are known to be a major background in the neutrino-oscillation experiments at $E_\nu \lesssim 1$ GeV. However, above $E_\nu \sim 1$ GeV, the cross section of the double-pion productions also becomes significant (the right panel of Fig. 7) because of the large branching ratios of higher N^* and Δ^* resonances for the decay into the three-body $\pi\pi N$ states. This implies the need of reliable information on the double-pion productions for next-generation oscillation experiments using multi-GeV neutrino beams.

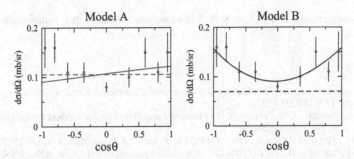

Fig. 5 Differential cross section at $W = 1672$ MeV for the $K^- p \rightarrow \eta \Lambda$ reaction. *Left (right)* panel is the result from Model A (Model B). *Solid curves* are the full results, while the dashed curves are the contribution from the S_{01} partial wave only. For Model B, the difference between the *solid* and *dashed curves* comes almost completely from the P_{03} partial wave dominated by the new narrow $J^P = 3/2^+$ Λ resonance with $M_R = 1671^{+2}_{-8} - i(5^{+11}_{-2})$ MeV

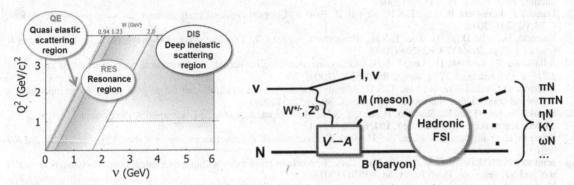

Fig. 6 *Left* Kinematical region relevant to neutrino-oscillation experiments. *Right* Schematic view of neutrino-nucleon reaction processes in resonance region

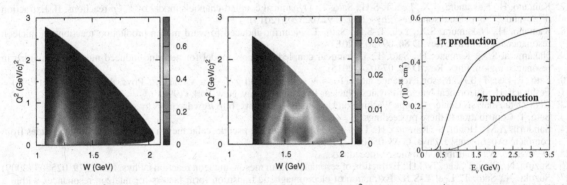

Fig. 7 *Left* $d\sigma/dW dQ^2$ for $\nu_\mu n \rightarrow \mu^- \pi N$ at $E_\nu = 2$ GeV. *Middle* $d\sigma/dW dQ^2$ for $\nu_\mu n \rightarrow \mu^- \pi^+ \pi^- p$ at $E_\nu = 2$ GeV. *Right* σ for $\nu_\mu n \rightarrow \mu^- \pi N$ (*black curve*) and $\nu_\mu n \rightarrow \mu^- \pi \pi N$ (*red curve*) as a function of E_ν. See Ref. [19] for the details

6 Summary

We have given an overview of our efforts for the light-quark baryon spectroscopy through a dynamical coupled-channels analysis of various meson-production reactions off the nucleon, induced by pion, photon, electron, and anti-kaon beams. In particular, our recent DCC analysis of $K^- p$ reactions has successfully extracted parameters associated with Λ^* and Σ^* resonances defined by poles of scattering amplitudes, and has shed light on a possible existence of new Λ^* and Σ^* resonances. However, our results also reveal that the current existing data for $K^- p$ reactions are still far from sufficient to constrain the extracted partial-wave amplitudes and Λ^* and Σ^* resonance parameters unambiguously. The help of hadron beam facilities such as J-PARC is highly desirable for further establishing Λ^* and Σ^* resonances. Finally, although we did not discuss in this contribution, our DCC approach has also been applied to meson spectroscopy [33,34]. We are putting more efforts into this direction, too.

Acknowledgments The author thanks T.-S. H. Lee, S. X. Nakamura, and T. Sato for their collaborations.

References

1. Dudek, J.J., et al.: (Hadron Spectrum Collaboration): resonances in coupled πK-ηK scattering from quantum chromodynamics. Phys. Rev. Lett. **113**, 182001 (2014)
2. Wu, J.-J., Lee, T.-S.H., Thomas, A.W., Young, R.D.: Finite-volume Hamiltonian method for coupled-channels interactions in lattice QCD. Phys. Rev. **90**, 055206 (2014)
3. Molina, R., Döring, M.: The pole structure of the $\Lambda(1405)$ in a recent QCD simulation. arXiv:1512.05831
4. Lee, T.-S. H., Wu, J.-J., Kamano, H.: From extraction of nucleon resonances to LQCD. arXiv:1602.01169
5. Matsuyama, A., Sato, T., Lee, T.-S.H.: Dynamical coupled-channel model of meson production reactions in the nucleon resonance region. Phys. Rep. **439**, 193 (2007)
6. Juliá-Díaz, B., Lee, T.-S.H., Matsuyama, A., Sato, T.: Dynamical coupled-channel model of πN scattering in the $W \leq 2$ GeV nucleon resonance region. Phys. Rev. C **76**, 065201 (2007)
7. Juliá-Díaz, B., Lee, T.-S.H., Matsuyama, A., Sato, T., Simith, L.C.: Dynamical coupled-channels effects on pion photoproduction. Phys. Rev. C **77**, 045205 (2008)
8. Durand, J., Juliá-Díaz, B., Lee, T.-S.H., Saghai, B., Sato, T.: Coupled-channels study of the $\pi^- p \to \eta n$ process. Phys. Rev. C **78**, 025204 (2008)
9. Kamano, H., Juliá-D iaz, B., Lee, T.-S.H., Matsuyama, A., Sato, T.: Dynamical coupled-channels study of $\pi N \to \pi\pi N$ reactions. Phys. Rev. C **79**, 025206 (2009)
10. Julia-Diaz, B., Kamano, H., Lee, T.-S.H., Matsuyama, A., Sato, T., Suzuki, N.: Dynamical coupled-channels analysis of $H^1(e, e'\pi)N$ reactions. Phys. Rev. C **80**, 025207 (2009)
11. Kamano, H., Juliá-D iaz, B., Lee, T.-S.H., Matsuyama, A., Sato, T.: Double and single pion photoproduction within a dynamical coupled-channels model. Phys. Rev. C **80**, 065203 (2009)
12. Suzuki, N., Juliá-Díaz, B., Kamano, H., Lee, T.-S.H., Matsuyama, A., Sato, T.: Disentangling the dynamical origin of P_{11} nucleon resonances. Phys. Rev. Lett. **104**, 042302 (2010)
13. Kamano, H., Nakamura, S.X., Lee, T.-S.H., Sato, T.: Extraction of P_{11} resonances from πN data. Phys. Rev. C **81**, 065207 (2010)
14. Kamano, H., Nakamura, S.X., Lee, T.-S.H., Sato, T.: Nucleon resonances within a dynamical coupled-channels model of πN and γN reactions. Phys. Rev. C **88**, 035209 (2013)
15. Kamano, H.: Impact of $\pi N \to \pi\pi N$ data on determining high-mass nucleon resonances. Phys. Rev. C **88**, 045203 (2013)
16. Kamano, H., Nakamura, S.X., Lee, T.-S.H., Sato, T.: Dynamical coupled-channels model of $K^- p$ reactions: determination of partial-wave amplitudes. Phys. Rev. C **90**, 065204 (2014)
17. Kamano, H., Nakamura, S.X., Lee, T.-S.H., Sato, T.: Dynamical coupled-channels model of $K^- p$ reactions. II. Extraction of Λ^* and Σ^* hyperon resonances. Phys. Rev. C **92**, 025205 (2015)
18. Kamano, H., Nakamura, S.X., Lee, T.-S.H., Sato, T.: Neutrino-induced forward meson-production reactions in nucleon resonance region. Phys. Rev. D **86**, 097503 (2012)
19. Nakamura, S.X., Kamano, H., Sato, T.: Dynamical coupled-channels model for neutrino-induced meson productions in resonance region. Phys. Rev. D **92**, 074024 (2015)
20. Sato, T., Lee, T.-S.H.: Meson exchange model for πN scattering and $\gamma N \to \pi N$ reaction. Phys. Rev. C **54**, 2660 (1996)
21. Feshbach, H.: Theoretical Nuclear Physics, Nuclear Reactions. Wiley, New York (1992)
22. CNS Data Analysis Center, George Washington University University. http://gwdac.phys.gwu.edu
23. Sato, T: Contribution to these proceedings
24. Sandorfi, A.M., Hoblit, S., Kamano, H., Lee, T.-S.H.: Determining pseudoscalar meson photoproduction amplitudes from complete experiments. J. Phys. G **38**, 053001 (2011)
25. D'Angerlo, A.: Contribution to these proceedings
26. Suzuki, N., Sato, T., Lee, T.-S.H.: Extraction of resonances from meson–nucleon reactions. Phys. Rev. C **79**, 025205 (2009)
27. Suzuki, N., Sato, T., Lee, T.-S.H.: Extraction of electromagnetic transition form factors for nucleon resonances within a dynamical coupled-channels model. Phys. Rev. C **82**, 045206 (2010)
28. Olive, K.A.: Particle data group collaboration. Chin. Phys. C **38**, 090001 (2014)
29. Starostin, A., et al.: (Crystal Ball Collaboration): Measurement of $K^- p \to \eta \Lambda$ near threshold. Phys. Rev. C **64**, 055205 (2001)
30. Alvarez-Ruso, L., Hayato, Y., Nieves, J.: Progress and open questions in the physics of neutrino cross sections at intermediate energies. New J. Phys. **16**, 075015 (2014)
31. Nakamura, S.X., Hayato, Y., Hirai, M., Kamano, H., Kumano, S., Sakuda, M., Saito, K., Sato, T.: Toward construction of the unified lepton–nucleus interaction model from a few hundred MeV to GeV region. AIP Conf. Proc. **1663**, 120010 (2015)
32. http://nuint.kek.jp/index_e.html
33. Kamano, H., Nakamura, S.X., Lee, T.-S.H., Lee, Sato, T.: Unitary coupled-channels model for three-mesons decays of heavy mesons. Phys. Rev. D **84**, 114019 (2011)
34. Nakamura, S.X., Kamano, H., Lee, T.-S.H., Lee, Sato, T.: Extraction of meson resonances from three-pions photo-production reactions. Phys. Rev. D **86**, 114012 (2012)

Few-Body Syst (2016) 57:1041–1050
DOI 10.1007/s00601-016-1146-5

Peter Kroll

Hard Exclusive Pion Leptoproduction

Received: 26 January 2016 / Accepted: 30 July 2016 / Published online: 17 August 2016
© Springer-Verlag Wien 2016

Abstract In this talk it is reported on an analysis of hard exclusive leptoproduction of pions within the handbag approach. It is argued that recent measurements of this process performed by HERMES and CLAS clearly indicate the occurrence of strong contributions from transversely polarized photons. Within the handbag approach such $\gamma_T^* \to \pi$ transitions are described by the transversity GPDs accompanied by twist-3 pion wave functions. It is shown that the handbag approach leads to results on cross sections and single-spin asymmetries in fair agreement with experiment. Predictions for other pseudoscalar meson channels are also briefly discussed.

1 Introduction

The handbag approach to hard exclusive leptoproduction of mesons offers a partonic description of these processes in the generalized Bjorken regime of large photon virtuality, Q^2, and large photon–proton center of mass energy, W, but small invariant momentum transfer, $-t$. The theoretical basis of the handbag approach is provided by the factorization theorems [1,2] which say that the process amplitudes for longitudinally polarized virtual photons, γ_L^*, are represented as convolutions of hard partonic subprocess amplitudes and soft hadronic matrix elements, so-called generalized parton distributions (GPDs), which encode the soft physics. The subprocess amplitudes are likewise convolutions of perturbatively calculable hard scattering kernels and meson wave functions. For pion production in particular the helicity amplitudes for $\gamma_L^* \to \pi$ transitions read

$$\mathcal{M}_{0+,0+} = \frac{e_0}{Q}\sqrt{1-\xi^2}\Big[\langle\tilde{H}\rangle - \frac{\xi^2}{1-\xi^2}\langle\tilde{E}\rangle\Big], \quad \mathcal{M}_{0-,0+} = \frac{e_0}{Q}\frac{\sqrt{-t'}}{2m}\xi\langle\tilde{E}\rangle, \tag{1}$$

where

$$\langle K \rangle = \int_{-1}^{1} dx \sum_{\lambda} \mathcal{H}_{0\lambda,0\lambda}(x,\xi,Q^2) K(x,\xi,t) \tag{2}$$

denotes the convolution of a subprocess amplitude, $\mathcal{H}$, with a GPD K. The nucleon mass is denoted by m. The skewness, ξ, is related to Bjorken-x by $\xi = x_B/(1-x_B)$ up to corrections of order $1/Q^2$. In (1) the usual abbreviation $t' = t - t_0$ is employed where the minimal value of $-t$ corresponding to forward scattering, is $t_0 = -4m^2\xi^2/(1-\xi^2)$. Helicities are labeled by their signs or by zero; they appear in the familiar order: pion, outgoing nucleon, photon, in-going nucleon.

Power corrections to the leading-twist result (1) are theoretically not under control. It is therefore not clear at which values of Q^2 and W the amplitudes (1) can be applied. The onset of the leading-twist dominance has

---

This article belongs to the special issue "Nucleon Resonances".

P. Kroll (✉)
Fachbereich Physik, Universität Wuppertal, 42097 Wuppertal, Germany
E-mail: pkroll@uni-wuppertal.de

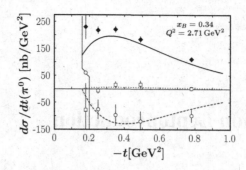

Fig. 1 The unseparated (long.–transv., transv.–transv.) cross section. Data [4] are shown as *diamonds* (*squares*, *circles*). Statistical and systematic errors are added in quadrature. The theoretical results are taken from [5]

to be found out by comparison with experiment. An example of power corrections is set by the amplitudes for transversely polarized photons, γ_T^*, which are suppressed by $1/Q$ as compared to the $\gamma_L^* \to \pi$ amplitudes. Still, as experiments tell us, the $\gamma_T^* \to \pi$ amplitudes play an important role in hard exclusive pion leptoproduction. The first experimental evidence for strong contribution from $\gamma_T^* \to \pi$ transitions came from the spin asymmetry, A_{UT}, measured with a transversely polarized target by the HERMES collaboration for π^+ production [3]. Its $\sin\phi_s$ modulation[1] unveils a particularly striking behavior: It is rather large and does not show any indication of a turnover towards zero for $t' \to 0$. In this limit $A_{UT}^{\sin\phi_s}$ is in control of an interference term of two helicity non-flip amplitudes

$$A_{UT}^{\sin\phi_s} \propto \mathrm{Im}\Big[\mathcal{M}_{0-,++}^* \mathcal{M}_{0+,0+} \Big]. \tag{3}$$

Both the amplitudes are not forced by angular momentum conservation to vanish in the limit $t' \to 0$. Hence, the small $-t'$ behavior of $A_{UT}^{\sin\phi_s}$ entails a sizeable $\gamma_T^+ \to \pi$ amplitude.

A second evidence for large contributions from transversely polarized photons comes from the CLAS measurement [4] of the π^0 electroproduction cross sections. As can be seen from Fig. 1 the transverse–transverse interference cross section is negative and amounts to about 50 % of the unseparated cross section in absolute value. Neglecting the double-flip amplitude $\mathcal{M}_{0-,-+}$ ($\propto t'$ for $t' \to 0$) and introducing the combinations

$$\mathcal{M}_{0+,\mu+}^{N,U} = \frac{1}{2}\Big[\mathcal{M}_{0+,\mu+} \pm \mathcal{M}_{0+,-\mu+} \Big], \tag{4}$$

one can write the transverse–transverse interference cross section as

$$\frac{d\sigma_{TT}}{dt} \simeq -\frac{1}{\kappa}\Big[|\mathcal{M}_{0+,++}^N|^2 - |\mathcal{M}_{0+,++}^U|^2 \Big] \tag{5}$$

where κ is a phase-space factor. Thus, from the CLAS data one learns that $|\mathcal{M}_{0+,++}^N| \gg |\mathcal{M}_{0+,++}^U|$ and is also large in comparison to the amplitudes (1).

In passing I remark that the combinations (4) are special cases of natural (N) and unnatural parity (U) combinations. They satisfy the symmetry relations

$$\mathcal{M}_{0\nu',-\mu\nu}^{N,U} = \mp(-1)^\mu \mathcal{M}_{0\nu',\mu\nu}^{N,U} \tag{6}$$

as a consequence of parity conservation.[2]

2 Transversity

In Fig. 2 a typical Feynman graph for pion production is depicted with helicity labels for the amplitude $\mathcal{M}_{0-,++}$. Angular momentum conservation forces both the subprocess amplitude and the nucleon-parton matrix element to vanish $\propto \sqrt{-t'}$ for $t' \to 0$ for any contribution to $\mathcal{M}_{0-,++}$ from the usual helicity non-flip GPDs $\widetilde{H}$ and

---

[1] The angle ϕ_s defines the orientation of the target spin vector.

[2] Parity conservation leads to the relation $\mathcal{M}_{0-\nu',-\mu-\nu} = -(-1)^{\mu-\nu+\nu'} \mathcal{M}_{0\nu',\mu\nu}$. An analogous relation holds for the subprocess amplitudes.

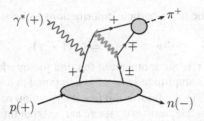

Fig. 2 A typical lowest-order Feynman graph for pion leptoproduction. The *signs* indicate the helicities of the involved particles

$\tilde{E}$. This result is in conflict with the HERMES data on $A_{UT}^{\sin \phi_s}$. However, there is a second set of GPDs, the helicity-flip or transversity ones, H_T, $\tilde{H}_T$, E_T, $\tilde{E}_T$ [6,7] for which the emitted and reabsorbed partons have opposite helicities. In [5,8] (see also [9]), it has been suggested that contributions from the transversity GPDs are responsible for the above mentioned experimental phenomena. Assuming handbag factorization for the $\gamma_T^* \to \pi$ amplitudes, they read ($\mu = \pm 1$)

$$M_{0+\mu+} = e_0 \frac{\sqrt{-t'}}{4m} \int_{-1}^{1} dx \left\{ (\mathcal{H}_{0+\mu-} - \mathcal{H}_{0-\mu+})(\bar{E}_T - \xi \tilde{E}_T) \right.$$
$$\left. + (\mathcal{H}_{0+\mu-} + \mathcal{H}_{0-\mu+})(\tilde{E}_T - \xi E_T) \right\}$$

$$M_{0-\mu+} = e_0 \sqrt{1-\xi^2} \int_{-1}^{1} dx \left\{ \mathcal{H}_{0-\mu+} \left[H_T + \frac{\xi}{1-\xi^2}(\tilde{E}_T - \xi E_T) \right] \right.$$
$$\left. + (\mathcal{H}_{0+\mu-} - \mathcal{H}_{0-\mu+}) \frac{t'}{4m^2} \tilde{H}_T \right\}. \tag{7}$$

With the help of parity conservation one can easily convince oneself that the nucleon helicity non-flip amplitude consists of a natural and a unnatural parity part. The natural parity part is related to the GPD $\bar{E}_T \equiv 2\tilde{H}_T + E_T$ with corrections[3] of order ξ^2 from $\tilde{E}_T$. The unnatural part is proportional to ξ. The amplitude $M_{0-\mu+}$ is more complicated. There is a contribution that does not have a definite parity behavior. It is dominated by H_T with corrections of order ξ^2 from E_T and $\tilde{E}_T$. It contributes to M_{0-++} while its contribution to M_{0--+} is suppressed by t/Q^2 from the double-flip subprocess amplitude $H_{0-,-+}$. In addition there is a natural parity contribution to $\mathcal{M}_{0-,\mu+}$ from $\tilde{H}_T$ which is suppressed by $t'/(4m^2)$. Thus, the $\gamma_T^* \to \pi$ amplitudes advocated for in [5,8]

$$\mathcal{M}_{0+,\mu+} = e_0 \frac{\sqrt{-t'}}{4m} \int_{-1}^{1} dx \mathcal{H}_{0-,++} \bar{E}_T$$

$$\mathcal{M}_{0-,++} = e_0 \sqrt{1-\xi^2} \int_{-1}^{1} dx \mathcal{H}_{0-,++} H_T$$

$$\mathcal{M}_{0-,-+} = 0 \tag{8}$$

are justified within the handbag approach at least at small ξ and $-t'$. There is only one subprocess amplitude common to the transverse amplitudes. These amplitudes meet the main features of the experimental data discussed in Sect. 1.

3 The Subprocess Amplitude $H_{0-,++}$

As can be seen from Fig. 2 quark and antiquark forming the pion have the same helicity. Hence, a twist-3 pion wave function is required. There are two twist-3 distribution amplitudes, a pseudoscalar one, Φ_P and a

---

[3] The GPD $\tilde{E}_T$ is an odd function of ξ as a consequence of time reversal invariance [7].

pseudotensor one, Φ_σ. Assuming the three-particle contributions to be strictly zero, one obtains the twist-3 distribution amplitudes [10]

$$\Phi_P \equiv 1 \quad \Phi_\sigma = 6\tau(1-\tau) \tag{9}$$

from the equation of motion. Here τ is the momentum fraction the quark in the pion carries with respect to the pion momentum. The subprocess amplitude $\mathcal{H}_{0-,++}$ is computed to lowest-order of perturbation theory (a typical Feynman graph is shown in Fig. 2). It turns out that the pseudotensor term is proportional to t/Q^2 and consequently neglected. The pseudoscalar term is non-zero at $t=0$ but infrared singular. In order to regularize this singularity the modified perturbative approach is used in [5,8] in which quark transverse momentum, $k_\perp$, are retained in the subprocess while the emission and reabsorption of partons from the nucleons is assumed to happen collinear to the nucleon momenta [11]. In this approach the subprocess amplitude reads

$$\mathcal{H}_{0-,++} = \frac{2}{\pi^2} \frac{C_F}{\sqrt{2N_C}} \mu_\pi \int d\tau d^2\mathbf{k}_\perp \Psi_{\pi P}(\tau, k_\perp) \alpha_s(\mu_R)$$

$$\times \left[\frac{e_u}{x - \xi + i\epsilon} \frac{1}{\bar{\tau}\frac{x-\xi}{2\xi}Q^2 - k_\perp^2 + i\epsilon} \right.$$

$$\left. - \frac{e_d}{x + \xi - i\epsilon} \frac{1}{-\tau\frac{x+\xi}{2\xi}Q^2 - k_\perp^2 + i\epsilon} \right]. \tag{10}$$

Here,[4] $\Psi_{\pi P}$ is a light-cone wave function for the pion which is parametrized as[5]

$$\Psi_{\pi P} = \frac{16\pi^{3/2}}{\sqrt{2N_C}} f_\pi a_P^3 |\mathbf{k}_\perp| \exp\left[-a_P^2 k_\perp^2\right]. \tag{11}$$

Its associated distribution amplitude is the pseudoscalar one given in (9). For the transverse-size parameter, a_P, the value 1.8 GeV is adopted and f_π (=0.132 GeV) denotes the pion decay constant. The parameter μ_π in (10) is related to the chiral condensate

$$\mu_\pi = \frac{m_\pi^2}{m_u + m_d} \tag{12}$$

where m_π is the pion mass while m_u and m_d denote current quark masses. In [5,8] a value[6] of 2 GeV^2 is taken for μ_π. The contributions from transversely polarized photons which are of twist-3 accuracy, are parametrically suppressed by μ_π/Q as compared to the $\gamma_L^* \to \pi$ amplitudes (1). However, for values of Q accessible in current experiments μ_π/Q is of order 1.

The amplitude (10) is Fourier transformed to the impact parameter space (after partial fraction decomposition) and the integrand is multiplied by a Sudakov factor, $\exp[-S]$, which represents gluon radiation in next-to-leading-log approximation using resummation techniques and having recourse to the renormalization group [15]. The Sudakov factor cuts off the b-integration at $b_0 = 1/\Lambda_{\rm QCD}$. In the modified perturbative approach the renormalization and factorization scales are $\mu_R = \max[\tau Q, (1-\tau)Q, 1/b]$ and $\mu_F = 1/b$, respectively. In [5,8] the modified approach is analogously applied to the $\gamma_L^* \to \pi$ amplitudes. The Sudakov factor guarantees the emergence of the twist-2 result for $Q^2 \to \infty$.

4 The Pion Pole

A special feature of π^+ production is the appearance of the pion pole. As has been shown in [16] the pion pole is part of the GPD $\tilde{E}$

$$\tilde{E}_{\rm pole}^u = -\tilde{E}_{\rm pole}^d = \Theta(\xi - |x|) \frac{F_P(t)}{4\xi} \Phi_\pi\left(\frac{x+\xi}{2\xi}\right) \tag{13}$$

---

[4] This is the subprocess amplitude for π^+ production. For the case of the π^0 the quark charges have to be taken out; they appear in the flavor combination of the GPDs.

[5] It may seem appropriate to use an $l_3 = \pm 1$ wave function for a particle moving along the 3-direction. Such a momentum space wave function is proportional to $k_\perp^\pm = k_\perp^1 \pm i k_\perp^2$ [14]. However, its collinear reduction leads to the pseudotensor distribution amplitude; the pseudoscalar one is lacking in this wave function.

[6] Taking the quark masses from [12], one finds $\mu_\pi = 2.6^{+0.52}_{-0.15}$ GeV while QCD sum rule analyses, e.g. [13], favor the value 1.6 ± 0.2 GeV.

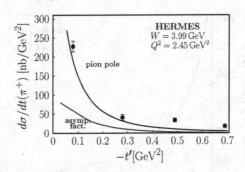

Fig. 3 The unseparated π^+ cross section versus $-t'$. The *lines* represent the pion pole contribution (16) with F_π^{pert} replaced by the experimental value and the leading-twist result. Data are taken from [3]

where F_P is the pole contribution to the pseudoscalar form factor of the nucleon which, with the help of PCAC and the Goldberger–Treiman relation, can be written as

$$F_{\dot{P}}(t) = \int\limits_{-1}^{1} dx\left[\tilde{E}^u - \tilde{E}^d\right] = -2m\frac{f_\pi}{F_\pi(Q^2)}\frac{\varrho_\pi}{t - m_\pi^2} \tag{14}$$

with

$$\varrho_\pi = \sqrt{2}g_{\pi NN}F_{\pi NN}(t)F_\pi(Q^2). \tag{15}$$

The coupling of the exchanged pion to the nucleons is described by the coupling constant $g_{\pi NN}(=13.1 \pm 0.2)$ and a form factor parametrized as a monopole with a parameter $\Lambda_N(=0.44 \pm 0.07)$; F_π denotes the electromagnetic form factor of the pion. The convolution of $\tilde{E}_{\text{pole}}$ with the subprocess amplitude $\mathcal{H}_{0\lambda,0\lambda}$ can be worked out analytically. The result leads to the following pole contribution to the longitudinal cross section

$$\frac{d\sigma_L^{\text{pole}}}{dt} = 4\pi\frac{\alpha_{em}}{\kappa}\frac{-t}{(t - m_\pi^2)^2}Q^2\varrho_\pi^2. \tag{16}$$

However, in this calculation the pion form factor is only the leading-order perturbative contribution to it which is known to underestimate the experimental form factor [17] substantially, and consequently the π^+ cross section, see Fig. 3. In [5,8] the perturbative result for F_π was therefore replaced by its experimental value. This prescription is equivalent to computing the pion pole contribution as a one-particle exchange.[7] With this replacement one sees that the pole term controls the π^+ cross section at small $-t'$, see Fig. 3. A detailed discussion of the pion pole contribution can be found in [18]. It also plays a striking role in electroproduction of ω mesons [19].

5 Phenomenology

In order to make predictions and comparison with experiment the GPDs are needed. In [5,8,11] the GPDs are constructed with the help of the double-distribution representation. According to [20] a double distribution is parametrized as a product of a zero-skewness GPD and an appropriate weight function that generates the skewness dependence. The zero-skewness GPD itself is composed of its forward limit, $K(x, \xi = t = 0) = k(x)$, multiplied by an exponential in t with a profile function, $f(x)$, parametrized in a Regge-like manner

$$f(x) = -\alpha' \ln x + B \tag{17}$$

at small $-t$. The GPD $\tilde{H}$ at $\xi = 0$ is taken from a recent analysis of the nucleon form factors [21] while $\tilde{E}$ is neglected. The forward limit of the transversity GPD H_T is given by the transversity parton distributions, known from an analysis of the azimuthal asymmetry in semi-inclusive deep inelastic lepton-nucleon scattering

---

[7] For $Q^2 \gg -t$ the virtuality of the exchanged pion can be neglected. The $\gamma^* \to \pi\pi^*$ vertex is therefore the pion form factor of the pion.

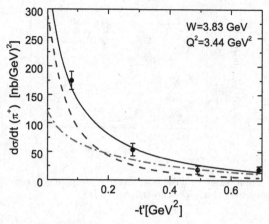

Fig. 4 The unseparated π^+ cross section. Data taken from [3]. The *solid (dashed, dash-dotted) curve* represents the results of [8] for the unseparated (longitudinal, transverse) cross section

and inclusive two-hadron production in electron–positron annihilations [22]. It turns out, however, that this parametrization underestimates H_T. Its moments are about 40 % smaller than lattice QCD results [23] and it leads to a very deep dip in the forward π^0 cross section which is at variance with experiment [4]. Therefore, the normalization of H_T is adjusted to the lattice QCD moments but the x-dependence of the transversity distributions is left unchanged. The forward limit of $\bar{E}_T$ is parametrized like the usual parton densities

$$\bar{e}_T(x) = N x^{-\alpha} (1-x)^\beta \tag{18}$$

with $\alpha = 0.3$ for both u and d quarks and $\beta^u = 4$, $\beta^d = 5$. The normalization as well as the parameters α' and B for each of the transversity GPDs are estimated by fits to the HERMES data on the π^+ cross section [3] and to the lattice QCD results [24] on the moments of $\bar{E}_T$.

An example of the results for the π^+ cross section is shown in Fig. 4, typical results for π^0 production in Fig. 1. At small $-t'$ the π^+ cross section is under control of the pion pole as discussed in Sect. 4. The contribution from $\tilde{H}$ to the longitudinal cross section is minor. The transverse cross section, although suppressed by μ_π^2/Q^2, is rather large and even dominates for $-t' \gtrsim 0.2$ GeV^2. It is governed by H_T, the contribution from $\bar{E}_T$ is very small. This fact can easily be understood considering the relative sign of u and d quark GPDs. For H_T they have opposite signs but the same sign for $\bar{E}_T$. Moreover, $\bar{E}_T^u$ and $\bar{E}_T^d$ are of similar size.[8] Since the GPDs contribute to π^+ production in the flavor combination

$$K^{\pi^+} = K^u - K^d \tag{19}$$

it is obvious that the contributions from $\bar{E}_T^u$ and $\bar{E}^d$ cancel each other to a large extent in contrast to those from H_T.

For π^0 production the situation is reversed since the GPDs appear now in the flavor combination

$$K^{\pi^0} = \frac{1}{\sqrt{2}} \left(e_u K^u - e_d K^d \right). \tag{20}$$

Therefore, $\bar{E}_T^u$ and $\bar{E}_T^d$ add while there is a partial cancellation between H_T^u and H_T^d. As can be seen from Eqs. (6) and (8) the $\bar{E}_T$-contribution is of the natural-parity type and, hence, makes up the transverse–transverse interference cross section (5), see Fig. 1. The transverse cross section, $d\sigma_T/dt$ receives contribution from both H_T and $\bar{E}_T$. However, the sum

$$\frac{d\sigma_T}{dt} + \frac{d\sigma_{TT}}{dt} \simeq \frac{1}{2\kappa} |\mathcal{M}_{0-,++}|^2 \tag{21}$$

is only fed by H_T. According to [5,8]

$$\frac{d\sigma_L}{dt} \ll \frac{d\sigma_T}{dt}. \tag{22}$$

---

[8] In [25] it has been speculated that $\bar{E}_T$ is linearly related to the Boer–Mulders function. Indeed both the functions show the same pattern [26]. This pattern is also supported by large-N_c considerations [27].

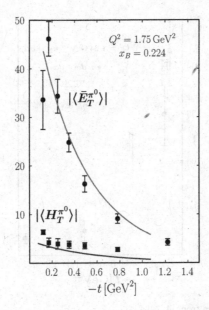

Fig. 5 The convolutions of $\bar{E}_T$ and H_T extracted from the CLAS data on π^0 leptoproduction [4] and the estimates from [5,8]

Hence, to a good approximation the transverse cross section equals the unseparated one. The prediction (22) is consistent with the very small longitudinal–transverse interference cross section, see Fig. 1. It is to be emphasized that this prediction is what is expected in the limit $Q^2 \to 0$ and not for $Q^2 \to \infty$. With the approximation $d\sigma_T \simeq d\sigma$ one can directly extract the convolutions of $H_T^{\pi^0}$ and $\bar{E}_T^{\pi^0}$ from the data on the π^0 cross sections. Results for the convolutions are displayed in Fig. 5 at $Q^2 = 1.75$ GeV^2 and $x_B = 0.224$ and compared to the estimates made in [5,8]. There is a further test of the set of amplitudes (8): The 'constant' modulation of the asymmetry A_{LL} measured with longitudinally polarized beam and target by the CLAS collaboration [29] for π^0 production, is related to the cross sections by

$$\frac{A_{LL}^{\text{const}}}{\sqrt{1-\varepsilon}} \frac{d\sigma}{dt} \simeq \frac{d\sigma_T}{dt} + \frac{d\sigma_{TT}}{dt} \simeq \frac{d\sigma}{dt} + \frac{d\sigma_{TT}}{dt} \tag{23}$$

(ε denotes the ratio of the longitudinal and transversal photon fluxes). A violation of this relation would indicate contributions from other transversity GPDs, especially from $\widetilde{H}_T$ [see (7)]. In Fig. 6 the right and left hand sides of (23) are compared to each other. Within admittedly large errors there is agreement except, perhaps, at the largest values of $-t$. Of course small contributions from other transversity GPDs cannot be excluded.

Data on π^0 production off neutrons as is planned to measure by the Jefferson Laboratory Hall A collaboration, will improve the flavor separation of the GPDs since they now appear in the combination

$$K_{\text{neutron}}^{\pi^0} = \frac{1}{\sqrt{2}} \left(e_u K^d - e_d K^u \right). \tag{24}$$

In analogy to the proton case (see Fig. 6) one may extract the convolutions of H_T and $\bar{E}_T$ from future data. According to the present estimate $\bar{E}_T$ for u and d quarks have about the same size. If this is correct the neutron/proton ratio of $d\sigma_{TT}$ is about one. On the other hand, for $d\sigma_T + d\sigma_{TT}$ the neutron/proton ratio is expected to be much smaller than one because of the properties of H_T.

The transversity GPDs play a similarly prominent role in leptoproduction of other pseudoscalar mesons. In Fig. 7 predictions for the η cross sections are shown. Except in the proximity of the forward direction the unseparated cross section for η production is considerably smaller than the π^0 one. In fact ratio $d\sigma(\eta)/d\sigma(\pi^0)$ amounts to about a third in fair agreement with the preliminary CLAS data on η production [28]. In the small $-t'$ region where the GPD H_T takes the lead, the η/π^0 ratio amounts to about 1 [5]. The behavior of the ratio found in [5] is in sharp contrast with earlier speculations [30] that the ratio is larger than 1 for all t'. One can easily understand the results for the η/π^0 ratio by considering again the relative signs and sizes of the u and d-quark GPDs. Under the plausible assumption $K^s = K^{\bar{s}}$ only the u and d-quark GPDs in the combination[9]

---

[9] Due to this assumption the flavor singlet and octet combinations are related by $K^{(1)} = \sqrt{2} K^{(8)}$.

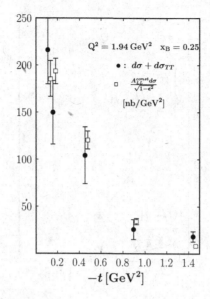

Fig. 6 Testing the relation (23). Data taken from [4,29]

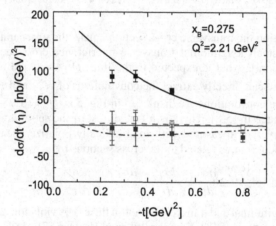

Fig. 7 The η production cross sections. Predictions are from [5], preliminary data are from [28]

$$K^\eta \simeq \frac{1}{\sqrt{6}}\left(e_u K^u + e_d K^d\right). \qquad (25)$$

contribute to η production. With regard to the different signs in (25) and (20) it is evident that H_T plays a more important role for η than for π^0 production while for $\bar{E}_T$ the situation is reversed with the consequence of a large η/π^0 ratio for $t' \to 0$ and a small ratio otherwise. In the evaluation of the η cross section $\eta - \eta'$ mixing is to be taken into account [31]. The η cross sections are shown in Fig. 7. The transverse–transverse cross section is now very small because of the strong cancellation between $\bar{E}_T^u$ and $\bar{E}_T^d$.

The handbag approach can straightforwardly be generalized to the production of Kaons [5]. Some results on the cross sections for various pseudoscalar meson channels are shown in Fig. 8.

6 Summary

In this article the present status of the analysis of hard exclusive leptoproduction of pions and other pseudoscalar mesons within the handbag approach is reviewed. The present GPD parametrizations are to be considered as estimates which however reproduces the main features of the data. A detailed fit to all pion data is pending. The surprising result is the dominance of $\gamma_T^* \to \pi$ transitions. The leading-twist contribution is small, in particular for π^0 production. The ultimate justification of this observation would be a measurement of the

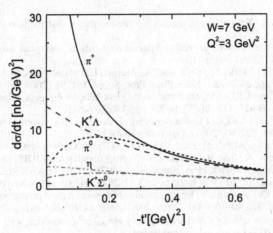

Fig. 8 Predictions for the unseparated cross sections of various pseudoscalar meson channels [5]

unseparated cross sections. The JLab Hall A collaboration has done this for π^0 production. The experiment is under analysis.

The statement which has been mentioned in many papers and talks, that from pion leptoproduction we learn about the GPDs $\tilde{H}$ and $\tilde{E}$ which was state of the art 10–15 years ago, is to be revised: from pion leptoproduction we learn about contributions from transversely polarized photons and in particular about the transversity GPDs H_T and $\bar{E}_T$.

References

1. Radyushkin, A.V.: Asymmetric gluon distributions and hard diffractive electroproduction. Phys. Lett. B **385**, 333 (1996). [arXiv:hep-ph/9605431]
2. Collins, J.C., Frankfurt, L., Strikman, M.: Factorization for hard exclusive electroproduction of mesons in QCD. Phys. Rev. D **56**, 2982 (1997). [arXiv:hep-ph/9611433]
3. Airapetian, A., et al. [HERMES Collaboration]: Single-spin azimuthal asymmetry in exclusive electroproduction of pi+ mesons on transversely polarized protons. Phys. Lett. B **682**, 345 (2010). [arXiv:0907.2596 [hep-ex]]
4. Bedlinskiy, I. et al. [CLAS Collaboration]: Exclusive π^0 electroproduction at $W > 2$ GeV with CLAS. Phys. Rev. C **90**(2), 025205 (2014) [Phys. Rev. C **90**(3), 039901 (2014)]. [arXiv:1405.0988 [nucl-ex]]
5. Goloskokov, S.V., Kroll, P.: Transversity in hard exclusive electroproduction of pseudoscalar mesons. Eur. Phys. J. A **47**, 112 (2011). [arXiv:1106.4897 [hep-ph]]
6. Hoodbhoy, P., Ji, X.D.: Helicity flip off forward parton distributions of the nucleon. Phys. Rev. D **58**, 054006 (1998). [arXiv:hep-ph/9801369]
7. Diehl, M.: Generalized parton distributions with helicity flip. Eur. Phys. J. C **19**, 485 (2001). [arXiv:hep-ph/0101335]
8. Goloskokov, S.V., Kroll, P.: An attempt to understand exclusive pi+ electroproduction. Eur. Phys. J. C **65**, 137 (2010). [arXiv:0906.0460 [hep-ph]]
9. Ahmad, S., Goldstein, G.R., Liuti, S.: Nucleon Tensor charge from exclusive pi**o electroproduction. Phys. Rev. D **79**, 054014 (2009). [arXiv:0805.3568 [hep-ph]]
10. Braun, V.M., Filyanov, I.E.: Conformal invariance and pion wave functions of nonleading twist. Z. Phys. C **48**, 239 (1990) [Sov. J. Nucl. Phys. **52**, 126 (1990)] [Yad. Fiz. **52**, 199 (1990)]
11. Goloskokov, S.V., Kroll, P.: The Role of the quark and gluon GPDs in hard vector-meson electroproduction. Eur. Phys. J. C **53**, 367 (2008). [arXiv:0708.3569 [hep-ph]]
12. Olive, K.A., et al. [Particle Data Group Collaboration]: Review of particle physics. Chin. Phys. C **38**, 090001 (2014)
13. Ball, P.: Theoretical update of pseudoscalar meson distribution amplitudes of higher twist: the nonsinglet case. JHEP **9901**, 010 (1999). [arXiv:hep-ph/9812375]
14. Ji, X.D., Ma, J.P., Yuan, F.: Classification and asymptotic scaling of hadrons' light cone wave function amplitudes. Eur. Phys. J. C **33**, 75 (2004). [arXiv:hep-ph/0304107]
15. Li, H.N., Sterman, G.F.: The perturbative pion form-factor with Sudakov suppression. Nucl. Phys. B **381**, 129 (1992)
16. Mankiewicz, L., Piller, G., Radyushkin, A.: Hard exclusive electroproduction of pions. Eur. Phys. J. C **10**, 307 (1999). [arXiv:hep-ph/9812467]
17. Blok, H.P., et al. [Jefferson Lab Collaboration]: Charged pion form factor between Q^2=0.60 and 2.45 GeV^2. I. Measurements of the cross section for the ^{1}H($e, e'\pi^+$)n reaction. Phys. Rev. C **78**, 045202 (2008). [arXiv:0809.3161 [nucl-ex]]
18. Favart, L., Guidal, M., Horn, T., Kroll, P.: Deeply virtual meson production on the nucleon. Eur. Phys. J. A **52**(6), 158 (2016). arXiv:1511.04535 [hep-ph]
19. Goloskokov, S.V., Kroll, P.: The pion pole in hard exclusive vector-meson leptoproduction. Eur. Phys. J. A **50**(9), 146 (2014). [arXiv:1407.1141 [hep-ph]]

20. Musatov, I.V., Radyushkin, A.V.: Evolution and models for skewed parton distributions. Phys. Rev. D **61**, 074027 (2000). [arXiv:hep-ph/9905376]

21. Diehl, M., Kroll, P.: Nucleon form factors, generalized parton distributions and quark angular momentum. Eur. Phys. J. C **73**(4), 2397 (2013). [arXiv:1302.4604 [hep-ph]]

22. Anselmino, M., Boglione, M., D'Alesio, U., Kotzinian, A., Murgia, F., Prokudin, A., Melis, S.: Update on transversity and Collins functions from SIDIS and e+ e- data. Nucl. Phys. Proc. Suppl. **191**, 98 (2009). [arXiv:0812.4366 [hep-ph]]

23. Gockeler, M., et al. [QCDSF and UKQCD Collaborations]: Quark helicity flip generalized parton distributions from two-flavor lattice QCD. Phys. Lett. B **627**, 113 (2005). [arXiv:hep-lat/0507001]

24. Gockeler, M., et al. [QCDSF and UKQCD Collaborations]: Transverse spin structure of the nucleon from lattice QCD simulations. Phys. Rev. Lett. **98**, 222001 (2007). [arXiv:hep-lat/0612032]

25. Burkardt, M.: Hadron tomography. AIP Conf. Proc. **915**, 313 (2007). [arXiv:hep-ph/0611256]

26. Barone, V., Melis, S., Prokudin, A.: The Boer–Mulders effect in unpolarized SIDIS: an analysis of the COMPASS and HERMES data on the cos 2 phi asymmetry. Phys. Rev. D **81**, 114026 (2010). [arXiv:0912.5194 [hep-ph]]

27. Schweitzer, P., Weiss, C.: Chiral-odd GPDs in large-N_c QCD. **2015**, 041 (2015). arXiv:1601.03016 [hep-ph]

28. Kubarovsky, V. [for the CLAS Collaboration]: Deeply virtual pseudoscalar meson production at jefferson lab and transversity GPDs. Int. J. Mod. Phys. Conf. Ser. **40**, 1660051 (2016). arXiv:1601.04367 [hep-ex]

29. Kim, A., et al.: Target and Double spin asymmetries of deeply virtual π^0 production with a longitudinally polarized proton target and CLAS. arXiv:1511.03338 [nucl-ex] (**unpublished**)

30. Eides, M.I., Frankfurt, L.L., Strikman, M.I.: Hard exclusive electroproduction of pseudoscalar mesons and QCD axial anomaly. Phys. Rev. D **59**, 114025 (1999). [arXiv:hep-ph/9809277]

31. Feldmann, T., Kroll, P., Stech, B.: Mixing and decay constants of pseudoscalar mesons. Phys. Rev. D **58**, 114006 (1998). [arXiv:hep-ph/9802409]

Few-Body Syst (2016) 57:949–953
DOI 10.1007/s00601-016-1132-y

Toru Sato

Electromagnetic Transition form Factor of Nucleon Resonances

Received: 14 February 2016 / Accepted: 1 June 2016 / Published online: 16 June 2016
© Springer-Verlag Wien 2016

Abstract A dynamical coupled channel model for electron and neutrino induced meson production reactions is developed. The model is an extension of our previous reaction model to describe reactions at finite Q^2. The electromagnetic transition form factors of the first $(3/2^+, 3/2)$ and $(3/2^-, 1/2)$ resonances extracted from partial wave amplitude are discussed.

1 Introduction

Studies of meson electroproduction reactions at momentum transfer Q^2 will reveal the space–time structure of excited states of nucleon [1,2]. Because the widths of nucleon resonances are large and the non-resonant mechanisms play important role, properties of nucleon resonance can be only obtained from an analysis covering all of available polarization and cross section data of meson production, such as πN, ηN and KY. Such theoretical efforts are in progress by many groups using isobar models or dynamical reaction models [1].

We concentrate on a dynamical coupled channel model of the meson production reactions. In [4] (SL), pion and photon induced pion production reactions around the $\Delta(1232)$ region have been studied. The model describe rather well pion production reactions [3,4]. The role of pion loop on the deformation of $N\Delta$ transition such as the quadruple form factors has been predicted in the model. In our dynamical model, model parameters of the strong interaction are determined from the analysis of the pion induced reactions. Then so called 'bare' electromagnetic form factors of NN^* transition are left to be determined from the analysis of the γN and $N(e, e')$ reactions. In [4], we have assumed a simple Q^2 dependence of form factor with a few parameters. Those parameters were determined from a few data available at that time. In the later work [5], as extensive data of pion electroproduction become available, we could fit the data at each Q^2 data without assuming a functional form of 'bare' form factors.

In SL model, only πN Fock space is taken into account, which is reasonable approximation for $\Delta(1232)$. However, to investigate higher mass resonances, the reaction model must be extended for $\pi N, \eta N, \pi\pi N$ channels. This has been done for πN reaction in Ref. [6] (JLMS) and for $N(e, e'\pi)N$ reaction in Ref. [7]. Though, pion induced reaction should be able to determine the strong interaction part of the reaction model in principle, a sensitivity of each resonance to photo reaction is different from that to pion induced reaction. A combined analysis of pion and photon induced reactions, though it involves much efforts for the analysis, would be important to refine the model. Recently combined analysis of pion and photon induced reactions up to $W < 2.2$ GeV has been done taken into account the kaon-hyperons Fock space in addition to $\pi N, \eta N, \pi\pi N$ [8] (ANL-Osaka).

This article belongs to the special issue "Nucleon Resonances".

T. Sato (✉)
Department of Physics, Osaka University, Toyonaka, Osaka 560-0043, Japan
Tel.: +81-66850-5345
E-mail: tsato@phys.sci.osaka-u.ac.jp

The neutrino experiments to investigate neutrino properties utilize neutrinos in a wide energy range from MeV to GeV region. A model for neutrino reactions on nucleon in the resonance region has been developed in [9] based on the ANL-Osaka dynamical model. This is the first dynamical model describing neutrino induced πN, ηN, KY and $\pi \pi N$ production reactions in a coupled channel reaction model. The model covers $W < 2\,\text{GeV}$ and $Q^2 < 3\,(\text{GeV/c})^2$ region. In this report, we use this model of meson production reaction to extract the transition form factors of nucleon resonances. The form factors are obtained from the residue of the meson production amplitude at resonance pole. A method of analytic continuation of amplitudes is described in [10,11].

In Sect. 2, we briefly explain our model of single pion electroproduction. The description of the dynamical model and the determination of the parameters for strong interaction and photo reactions are given in [12]. In Sect. 3, the explanation on the transition form factor and our results on $(3/2^+, 3/2)P_{33}$ and $(3/2^-, 1/2)D_{13}$ transition form factors are discussed.

2 Analysis of Pion Electroproduction

The partial wave amplitude for the virtual photon γ^* induced meson(M) and the baryon(B) production reaction $(\gamma^* + N \rightarrow M + B)$ is given as [9]

$$T_{MB,JN}(\lambda, k, q; W, Q^2) = t_{MB,JN}(\lambda, k, q; W, Q^2) + t^R_{MB,JN}(\lambda, k, q; W, Q^2). \tag{1}$$

The first term $t_{MB,JN}$ and the second term $t^R_{MB,JN}$ are the non-resonant and the resonant amplitude, respectively. The invariant mass, four momentum square of virtual photon and polarization of virtual photon are denoted as W, Q^2 and λ. k and q are the magnitude of spatial momentum of final and initial nucleon. The resonant amplitude can be written as

$$t^R_{MB,JN}(\lambda, k, q; W, Q^2) = \sum_{m,n} \bar{\Gamma}_{MB,N^*_m}(k, W)[G_{N^*}(W)]_{m,n} \bar{\Gamma}_{N^*_n, JN}(\lambda, q, W, Q^2), \tag{2}$$

where the summation is taken over the considered 'bare' N^* and Δ excited states. The N^* Green's function $G_{N^*}(W)$ is given as

$$[G_{N^*}(W)^{-1}]_{m,n} = (W - m_0)\delta_{m,n} - \Sigma(W)_{m,n}. \tag{3}$$

Here m_0 is the mass of 'bare' state. Σ is self-energy of the meson loop, which has off diagonal element. The dressed $\gamma^* + N \rightarrow N^*$ vertex is written as

$$\bar{\Gamma}_{N^*_n, JN}(\lambda, q, W, Q^2) = \Gamma_{N^*_n, JN}(\lambda, q, Q^2) + \sum_{M',B'} \int p^2 dp\, \Gamma_{N^*_n, M'B'}(p) G_{M'B'}(W) t_{M'B', JN}(\lambda, k, q; W, Q^2)$$

$$\tag{4}$$

The first term $\Gamma_{N^*_n, JN}(\lambda, q, Q^2)$ denotes a 'bare' N^* excitation current $< N^*_n |j_{em} \cdot \epsilon_\lambda| N >$, which is determined by analyzing meson electroproduction data. The second term is meson loop correction of the $\gamma^* N N^*$ vertex.

We have studied the data of single pion electroproduction off the proton from the CLAS collaboration. The kinematical region we have studied are $Q^2 < 3\,(\text{GeV/c})^2$ and $W < 2\,\text{GeV}$ as shown Fig. 1. It is noticed that for the $W > 1.4\,GeV$ and very low Q^2 region and the $W > 1.7\,\text{GeV}$ and $Q^2 < 2\,(\text{GeV/c})^2$, the data is missing. In those kinematical region, we have fitted the inclusive cross section from empirical mode by Christy and Bosted [13].

The data of four virtual photon cross sections $d\sigma_T/d\Omega_\pi + \epsilon d\sigma_L/d\Omega_\pi, d\sigma_{LT}/d\Omega_\pi, d\sigma_{TT}/d\Omega_\pi, d\sigma_{LT'}/d\Omega_\pi$ are analyzed. We have determined the helicity amplitudes of 'bare' states at each Q^2 corresponding to the data. The results of Ref. [9] based on ANL-Osaka model are shown in Fig. 2. The left and the right panel shows the the virtual photon cross section $d\sigma_T/d\Omega_\pi + \epsilon d\sigma_L/d\Omega_\pi$ at $Q^2 = 0.4\,(\text{GeV/c})^2$ for the neutral pion production $p(e, e\pi^0)p$ and positive pion production $p(e, e'\pi^+)n$, respectively. Our model gives reasonable description of the single pion electroproduction in the Q^2 and W range we have examined. The model also gives fair description of the inclusive cross section. At higher $W \sim 2\,\text{GeV}$ and $Q^2 \sim 2\,(\text{GeV/c})^2$, the inclusive cross section of the dynamical coupled channel model connets smoothly to the parton description.

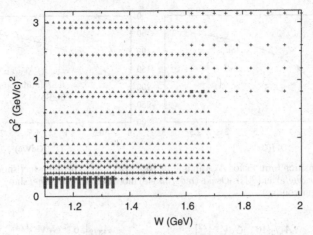

Fig. 1 Kinematical region covered by available single pion electroproduction data from the CLAS collaboration taken from Ref. [9]. The *red triangle* (*blue cross*) points indicate the kinematical points where data for $p(e, e'\pi^0)p$ $(p(e, e'\pi^+)n$ are available. At Green square points, data for both π^0 and π^+ production are available (color figure online)

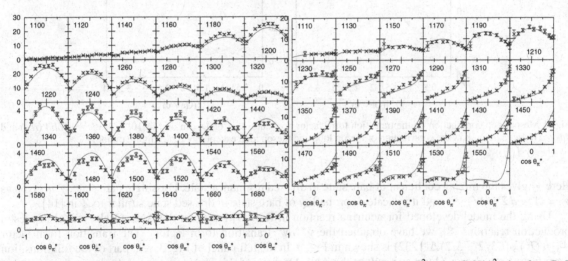

Fig. 2 Differential cross section of single pion production $d\sigma_T/d\Omega + \epsilon d\sigma_L/d\Omega (\mu b/sr)$ at $Q^2 = 0.4 \, (\text{GeV/c})^2$ taken from Ref. [9]. The *left* and the *right panel* show the neutral and positive pion production reaction from ANL-Osaka model. The *number* in each *panel* indicates center of mass energy W (MeV)

3 $\gamma^* NN^*$ Transition form Factor

The transition form factors are obtained from the residue of the partial wave amplitude Eq. (1) at the resonance pole. A method of analytic continuation of the partial wave amplitudes from the physical to the unphysical sheet within the dynamical coupled channel model is explained in [10,11]. The resonance poles are found in the resonant term $t_{MB,JN}^R$. The resonance energy E_R is determined by

$$det[G_{N^*}^{-1}(W = E_R)] = 0. \tag{5}$$

Near the resonance energy, the partial wave amplitude can be expressed as

$$t_{MB,JN}^R(\lambda, k, q; W, Q^2) = \frac{\bar{\Gamma}_{MB}^R \bar{\Gamma}_{JN}^R}{W - E_R} + \cdots \tag{6}$$

We obtain the transition form factor from the residue at the resonance pole as,

$$<N^*|J^{em} \cdot \epsilon|N> = \bar{\Gamma}_{JN}^R = \sum_j \chi_j \bar{\Gamma}_{j,JN}^R. \tag{7}$$

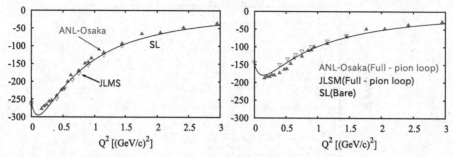

Fig. 3 Q^2 dependence of transition form factor $A_{3/2}$ in unit of $10^{-3}\,GeV^{-1/2}$. The *left panel* shows the results from SL (*solid curve*), JLSM (*blue inverted triangle*) and ANL-Osaka (*red triangle*) models. The *right panel* shows transition form factor except πN loop correction (color figure online)

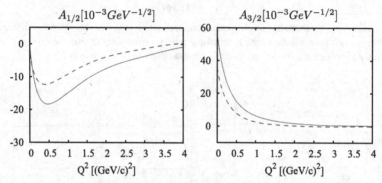

Fig. 4 Meson loop contribution on the transition form factor $A_{3/2}$ and $A_{1/2}$ of D_{13} in unit of $10^{-3}\,GeV^{-1/2}$. *Solid (red)* and *dashed (blue) curves* show the contribution of meson loop and meson loop except πN (color figure online)

Here χ_i gives the mixing coeficient of bare state in dressed resonance state. For single bare state, χ is given as $\chi = (1 - d\Sigma/dW)^{-1/2}$ and it indicates a probaility of bare state in dressed state similar to Z in [14].

Using the model developed for neutrino reaction in Ref. [9] based on the ANL-Osaka model of meson production reactions [8], we have obtained the γ^*NN^* transition form factors. The transition form factor $A_{3/2}(Q^2)$ of $(3/2^+, 3/2)\Delta(1232)$ is shown in Fig. 3. In the left panel of Fig. 3, real part of the full transition form factors are shown. Our new result of the ANL-Osaka model is shown in triangle(red) together with that of the JLMS model in inverted-triangle(blue) and the SL model in solid curve(black). Although the included meson-baryon Fock space is different for each of three reaction models, the obtained transition form factors agree well.

We have studied the contents of the transition form factor in the coupled channel model. We have switched off the contribution of $M'B' = \pi N$ loop in the second term of Eq. (4). Here, this is bare form factor in the SL model, since only πN space is taken into account. It is noticed that a small πN loop effect may still present with this procedure because of the coupled channel effect in the non-resonant interaction. The results are shown in the right panel of Fig. 3. The contributions of meson loop except pion for the three models agrees well again. As far as the $A_{3/2}$ form factor of pronounced resonance $\Delta(1232)$, the total transition form factor and pion loop contribution is stable against the variation of Fock space of meson-baryon channel included.

The transition form factors $A_{3/2}$ and $A_{1/2}$ of $D_{13}(1505)$ are shown in Fig. 4 and Fig. 5 as an example of higher energy resonances in ANL-Osaka model. In Fig. 4, the full meson-loop contribution(solid red curve) and the meson-loop except πN(dashed blue curve) are shown. The meson loop other than pion has large contribution in this transition.

The full $A_{1/2}$ and $A_{3/2}$ form factors are shown in Fig. 5. The real and the imaginary part of the transition form factors are shown in the filled triangle(red) and the empty triangle(red), respectively. For comparison, the real part of the form factor in JLMS model is shown in the inverted triangle(blue). The form factors in the ANL-Osaka model agrees qualitatively with those of JLMS. The imaginary part of the form factors is found to be small. At low Q^2 below $0.4\,(GeV/c)^2$, the fitted form factors are not smooth Q^2 dependence. It is noticed, however, in those Q^2 region around $W \sim 1.5\,GeV$, the inclusive cross sections could be more influential than the single pion electroproduction cross sections in our current fitting.

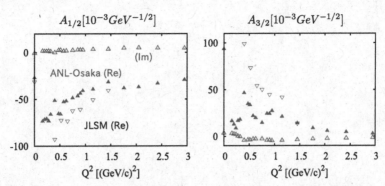

Fig. 5 Q^2 dependence of transition form factors $A_{3/2}$ and $A_{1/2}$ of D_{13} in unit of $10^{-3} \, GeV^{-1/2}$. The *filled triangle* (*red*) and *empty triangle* (*red*) show the real and the imaginary part of the form factor in ANL-Osaka model. The *inverted-triangle* (*blue*) shows that of JLSM model (color figure online)

4 Summary

The ANL-Osaka dynamical coupled channel model is extended to describe the electron and neutrino induced meson production reactions. The electromagnetic $\gamma^* N N^*$ transition form factors are extracted from the residue of the partial wave amplitude at resonance pole. Extracted full form factors and the pion-loop contributions for $\gamma^* N \Delta_{33}$ transition from SL, JLMS and ANL-Osaka models agree well with each other even though the meson-baryon Fock space employed is different for each of the models. Form factors of $(3/2^-, 1/2) D_{13}$ are analyzed. In general, contribution of meson cloud decreases as Q^2 increases and meson-baryon state other than pion-nucleon were important. Further analysis of pion electroproduction dedicated to study the N^* form factors in ANL-Osaka model is in progress.

Acknowledgments The author thanks to Drs. H. Kamano, S. X. Nakamura and T. -S. H. Lee for continuous collaboration to complete the analysis. This work was supported by the JSPS KAKENHI Grant No. 25105010.

References

1. Aznauryan, I.G., et al.: Studies of nucleon resonance structure in exclusive meson electroproduction. Int. J. Mod. Phys. E **22**, 1330015 (2013)
2. Cloet, I., Roberts, C.D.: Explanation and prediction of observables using continuum strong QCD. Prog. Part Nucl. Phys. **77**, 1 (2014)
3. Sato, T., Lee, T.-S.H.: Meson-exchange model for πN scattering and $\gamma N \to \pi N$ reaction. Phys. Rev. C **54**, 2660 (1996)
4. Sato, T., Lee, T.-S.H.: Dynamical study of the Δ excitation in $N(e, e'\pi)$ reactions. Phys. Rev. C **63**, 055201 (2001)
5. Juliá-Díaz, B., Lee, T.-S.H., Sato, T., Smith, L.C.: Extraction and interpretation of $\gamma N \to \Delta$ form factors within a dynamical model. Phys. Rev. C **75**, 015205 (2007)
6. Juliá-Díaz, B., Lee, T.-S.H., Matsuyama, A., Sato, T.: Dynamical coupled-channels model of πN scattering in the $W \leq 2$GeV nucleon resonance region. Phys. Rev. C **76**, 065201 (2007)
7. Juliá-Díaz, B., Kamano, H., Lee, T.-S.H., Matsuyama, A., Sato, T., Suzuki, N.: Dynamical coupled-channels analysis of $^1H(e, e'\pi)N$ reactions. Phys. Rev. C **80**, 025207 (2009)
8. Kamano, H., Nakamura, S.X., Lee, T.-S.H., Sato, T.: Nucleon resonances within a dynamical coupled-channels model of πN and γN. Phys. Rev. C **88**, 035209 (2013)
9. Nakamura, S.X., Kamano, H., Sato, T.: Dynamical coupled channels model for neutrino-induced meson productions in resonance region. Phys. Rev. D **92**, 074024 (2015)
10. Suzuki, N., Sato, T., Lee, T.-S.H.: Extraction of resonances from meson-nucleon reactions. Phys. Rev. C **79**, 025205 (2009)
11. Suzuki, N., Sato, T., Lee, T.-S.H.: Extraction of electromagnetic transition form factors for nucleon resonances within a dynamical coupled-channels model. Phys. Rev. C **82**, 045206 (2010)
12. Kamano H.: Light-quark baryon spectroscopy within ANL-Osaka dynamical coupled-channels approach. Contribution to these Proceedings. arXiv:1602.02511
13. Christy, M.E., Bosted, P.E.: Empirical fit to precision inclusive electron-proton cross sections in the resonance region. Phys. Rev. C **81**, 055213 (2010)
14. Théberge, T., Thomas, A.W., Miller, G.A.: Pionic corrections to the MIT bag model: the (3,3) resonance. Phys. Rev. D **22**, 2838 (1980)

Few-Body Syst (2016) 57:1087–1093
DOI 10.1007/s00601-016-1158-1

Lothar Tiator

Pion Electroproduction and Siegert's Theorem

Received: 12 September 2016 / Accepted: 13 September 2016 / Published online: 4 October 2016
© Springer-Verlag Wien 2016

Abstract Nucleon to Resonance transition form factors are discussed within the MAID model for pion electroproduction on the nucleon. For low Q^2 the consequences of Siegert's theorem are presented and medium to large violations of the Long Wavelength Limit at the pseudo-threshold are observed for the phenomenological parametrizations of the longitudinal transition form factors of different nucleon resonances.

1 Introduction

Pion electroproduction is the main source for investigations of the transition form factors of the nucleon to excited N^* and Δ baryons. After early measurements of the G_M^* form factor of the $N\Delta$ transition already in the 1960s, in the 1990s a large program was running at Mainz, Bonn, Bates and JLab in order to measure the very small E/M ratio of the $N\Delta$ transition and the Q^2 dependence of the E/M and S/M ratios in order to get information on the internal quadrupole deformations of the nucleon and the Δ. In parallel a large progress was achieved in various kinds of quark models that gave predictions to NN^* and $N\Delta^*$ transition form factors. Only at JLab both the energy and the photon virtuality were available to measure those form factors for a set of nucleon resonances up to $Q^2 \approx 5$ GeV^2. Two recent review articles on the electromagnetic excitation of nucleon resonances give a very good overview over experiment and theory and latest developments [1,2].

Whereas the transverse form factors are constrained by photoproduction results at $Q^2 = 0$, the longitudinal form factors remain very uncertain below $Q^2 \approx 0.5$ GeV^2. An application of the Siegert theorem in the Long Wavelength Limit (LWL), however, can constrain the longitudinal form factors at low Q^2. For pion photoproduction the LWL strictly holds in the unphysical region at the so-called pseudo-threshold, $Q^2_{pt} = -(W - M_N)^2$, which is energy dependent and closest to the physical region at pion threshold. But also in the Δ resonance region, consequences of this constraint can be observed in a curvature of the S/M ratio at the lowest measured Q^2 values. For higher resonances, the experimental data are not close enough to the pseudo-threshold, where the Siegert theorem exactly holds. Nevertheless, the LWL will have an important influence on the parametrization of the longitudinal form factors.

In addition, the Siegert theorem is also a good test of background parametrizations in isobar models. A full dynamical approach should automatically fulfill the Siegert theorem. Isobar models, where the background is often described by field theoretical Lagrangians with Born terms and vector meson exchanges usually use empirical electromagnetic form factors. Even if the models are constructed in a gauge invariant way, local Gauge invariance is often violated [3], which could be tested in the LWL.

This article belongs to the special issue "Nucleon Resonances".

L. Tiator (✉)
Universität Mainz, Mainz, Germany
E-mail: tiator@kph.uni-mainz.de

2 MAID

In the spirit of a dynamical approach to pion photo- and electroproduction, the t-matrix of the unitary isobar model MAID is set up by the ansatz [1,4]

$$t_{\gamma\pi}(W) = t_{\gamma\pi}^B(W) + t_{\gamma\pi}^R(W),$$ (1)

with a background and a resonance t-matrix, each of them constructed in a unitary way. Of course, this ansatz is not unique. However, it is a very important prerequisite to clearly separate resonance and background amplitudes within a Breit–Wigner concept also for higher and overlapping resonances.

For a specific partial wave $\alpha = \{j, l, \ldots\}$, the background t-matrix is set up by a potential multiplied by the pion-nucleon scattering amplitude in accordance with the K-matrix approximation,

$$t_{\gamma\pi}^{B,\alpha}(W, Q^2) = v_{\gamma\pi}^{B,\alpha}(W, Q^2)\,[1 + i t_{\pi N}^\alpha(W)],$$ (2)

where only the on-shell part of pion-nucleon rescattering is maintained and the off-shell part from pion-loop contributions is neglected. Whereas this approximation would fail near the threshold for γ, π^0, it is well justified in the resonance region because the main contribution from pion-loop effects is absorbed by the nucleon resonance dressing.

The background potential $v_{\gamma\pi}^{B,\alpha}(W, Q^2)$ is described by Born terms obtained with an energy-dependent mixing of pseudovector-pseudoscalar πNN coupling and t-channel vector meson exchanges. The mixing parameters and coupling constants are determined by an analysis of nonresonant multipoles in the appropriate energy regions [5]. In the latest version MAID2007 [4], the S, P, D, and F waves of the background contributions are unitarized as explained above, with the pion-nucleon elastic scattering amplitudes, $t_{\pi N}^\alpha = [\eta_\alpha \exp(2i\delta_\alpha) - 1]/2i$, described by phase shifts δ_α and the inelasticity parameters η_α taken from the GWU/SAID analysis [6].

For the resonance contributions Breit–Wigner forms for the resonance shape are assumed, following Ref. [5]

$$t_{\gamma\pi}^{R,\alpha}(W, Q^2) = \bar{\mathcal{A}}_\alpha^R(W, Q^2)\, \frac{f_{\gamma N}(W)\,\Gamma_{tot}(W)\,M_R\,f_{\pi N}(W)}{M_R^2 - W^2 - i M_R\,\Gamma_{tot}(W)}\, e^{i\phi_R(W)},$$ (3)

where $f_{\pi N}(W)$ is the usual Breit–Wigner factor describing the decay of a resonance with total width $\Gamma_{tot}(W)$, partial πN width $\Gamma_{\pi N}(W)$, and spin j,

$$f_{\pi N}(W) = C_{\pi N} \left[\frac{1}{(2j+1)\pi} \frac{\kappa(W)}{q(W)} \frac{M_N}{M_R} \frac{\Gamma_{\pi N}(W)}{\Gamma_{\text{tot}}^2(W)} \right]^{1/2}.$$ (4)

The energy dependence of the partial widths and of the γNN^* vertex can be found in Ref. [4]. The phase $\phi_R(W)$ in Eq. (3) is introduced to adjust the total phase such that the Fermi-Watson theorem is fulfilled below two-pion threshold.

While the original version of MAID included only the seven most important nucleon resonances with only transverse e.m. couplings in most cases, MAID2007 describes all 13 four-star resonances below $W = 2$ GeV. In a forthcoming update of MAID, these list of dominant resonances is no more sufficient, due to the high accuracy of the data and the availability of many polarization observables with single and double polarization. With these data also the weaker three-star and two-star resonances can be analyzed and will be included in the model. Currently an update of EtaMAID is in progress using recently measured high-quality cross sections and polarization observables from Mainz and Bonn, as well as the high-quality photon beam asymmetry measurements from GRAAL. Preliminary results show that many more N^* resonances have ηN branching rations of the order around 1 % than previously known. The quality of the data now allows analyses of resonances even below 1 %. A famous case is the $D_{13}(1520)3/2^-$ resonance, which plays a very important role in η photoproduction, despite having a branching ratio of only around 0.1 % [7].

Table 1 Parameters for the $N\Delta$ transition form factors G_M^*, G_E^*, G_C^* given by Eqs. (5, 6)

	g_α^0	β_α	γ_α		g_α^0	β_α	γ_α		g_α^0	β_α	γ_α	d_α
M1	3.00	0.0095	0.23	E2	0.0637	-0.0206	0.16	C2	0.1240	0.120	0.23	4.9

The normalization values at the photon point ($Q^2 = 0$), g_α^0 and d_α are dimensionless, the parameters β and γ in GeV^{-2}

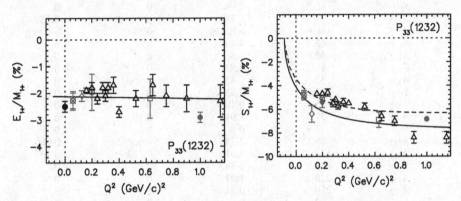

Fig. 1 The Q^2 dependence of the E/M and S/M ratios of the $\Delta(1232)$ excitation for low Q^2. The data are from Mainz, Bonn, Bates and JLab. For details see Ref. [1]. The behavior of the S/M ratio at low Q^2 and in particular for $Q^2 < 0$ in the unphysical region is a consequence of the Siegert theorem. The dashed line for S/M is a prediction according Eq. (14)

3 Transition Form Factors

In most cases, the resonance couplings $\bar{\mathcal{A}}_\alpha^R(W, Q^2)$ are assumed to be independent of the total energy. However, an energy dependence may occur if the resonance is parameterized in terms of the virtual photon three-momentum $k(W, Q^2)$, e.g., in MAID2007 for the $\Delta(1232)$ resonance. For all other resonances a simple Q^2 dependence is assumed for $\bar{\mathcal{A}}_\alpha(Q^2)$. These resonance couplings are taken as constants for a single-Q^2 analysis, e.g., for photoproduction ($Q^2 = 0$) but also at any fixed $Q^2 > 0$, whenever sufficient data with W and θ variation are available. Independently from this single-Q^2 analysis, also a Q^2-dependent analysis, with a simple ansatz using polynomials and exponentials, was performed. In MAID2007 the Q^2 dependence of the e.m. $N\Delta$ transition form factors is parameterized as follows:

$$G_{E,M}^*(Q^2) = g_{E,M}^0 (1 + \beta_{E,M} Q^2) e^{-\gamma_{E,M} Q^2} G_D(Q^2) , \tag{5}$$

$$G_C^*(Q^2) = g_C^0 \frac{1 + \beta_C Q^2}{1 + d_C Q^2/(4M_N^2)} \frac{2M_\Delta}{\kappa_\Delta} e^{-\gamma_C Q^2} G_D(Q^2), \tag{6}$$

where $G_D(Q^2) = 1/(1 + Q^2/0.71 \text{ GeV}^2)^2$ is the dipole form factor and the parameters are given in Table 1. In order to fulfill the Siegert theorem (see below), the parametrization of the Coulomb amplitude has been changed accordingly [4]. The E/M and S/M ratios for low Q^2 are shown in Fig. 1 together with single-Q^2 data of different sources, see Ref. [1,2]

Alternatively the couplings can be parameterized as functions of Q^2 by an ansatz like

$$\bar{\mathcal{A}}_\alpha(Q^2) = \bar{\mathcal{A}}_\alpha(0)(1 + a_1 Q^2 + a_2 Q^4 + a_4 Q^8) e^{-b_1 Q^2} . \tag{7}$$

For such an ansatz the parameters $\bar{\mathcal{A}}_\alpha(0)$ are determined by a fit to the world database of photoproduction, and the parameters a_i and b_1 are obtained from a combined fitting of all the electroproduction data at different Q^2. The latter procedure is called the Q^2-dependent fit. In MAID the photon couplings $\bar{\mathcal{A}}_\alpha(0)$ are input parameters, directly related to the helicity couplings $A_{1/2}$, $A_{3/2}$, and $S_{1/2}$ of nucleon resonance excitation. Relations between these helicity form factors, Sachs form factors G_M^*, G_E^*, G_C^* and Dirac form factors F_1, F_2, F_3 can be found in Refs. [1,4].

In Table 2 the parameters obtained from the Q^2-dependent fit to the resonances above the $\Delta(1232)$ are listed.

Table 2 MAID parametrization of the transition form factors, Eq. (7), for proton targets

N^*, Δ^*		$\bar{A}_\alpha(0)$	a_1	a_2	a_4	b_1
P_{11} (1440)	$A_{1/2}$	−61.4	0.871	−3.516	−0.158	1.36
	$S_{1/2}$	4.2	40.	0	1.50	1.75
S_{11} (1535)	$A_{1/2}$	66.4	1.608	0	0	0.70
	$S_{1/2}$	−2.0	23.9	0	0	0.81
D_{13} (1520)	$A_{1/2}$	−27.4	8.580	−0.252	0.357	1.20
	$A_{3/2}$	160.6	−0.820	0.541	−0.016	1.06
	$S_{1/2}$	−63.5	4.19	0	0	3.40
S_{31} (1620)	$A_{1/2}$	65.6	1.86	0	0	2.50
	$S_{1/2}$	16.2	2.83	0	0	2.00
S_{11} (1650)	$A_{1/2}$	33.3	1.45	0	0	0.62
	$S_{1/2}$	−3.5	2.88	0	0	0.76
D_{15} (1675)	$A_{1/2}$	15.3	0.10	0	0	2.00
	$A_{3/2}$	21.6	1.91	0.18	0	0.69
	$S_{1/2}$	1.1	0	0	0	2.00
F_{15} (1680)	$A_{1/2}$	−25.1	3.780	−0.292	0.080	1.25
	$A_{3/2}$	134.3	1.016	0.222	0.237	2.41
	$S_{1/2}$	−44.0	3.783	0	0	1.85
D_{33} (1700)	$A_{1/2}$	226.	1.91	0	0	1.77
	$A_{3/2}$	210.	0.88	1.71	0	2.02
	$S_{1/2}$	2.1	0	0	0	2.00
P_{13} (1720)	$A_{1/2}$	73.0	1.89	0	0	1.55
	$A_{3/2}$	−11.5	10.83	−0.66	0	0.43
	$S_{1/2}$	−53.0	2.46	0	0	1.55

$\bar{A}_\alpha(0)$ is given in units of 10^{-3} GeV$^{-1/2}$ and the coefficients a_1, a_2, a_4, b_1 in units of GeV^{-2}, GeV^{-4}, GeV^{-8}, GeV^{-2}, respectively

4 Siegert Theorem for Meson Electroproduction

The Siegert theorem was originally formulated for applications in nuclear physics in 1937 [8], and was later in the 1970s intensively applied in calculations with light nuclei for meson exchange contributions [9]. Later in the 1990s it was reconsidered by Naus [10] in terms of local gauge invariance and applied to the calculations of the electric (E2) and charge (C2) quadrupole excitations of the $\Delta(1232)$ resonance [4,11,12].

This dependence is rather model-independent, because it reflects the behavior of the multipoles at physical threshold (pion momentum $\mathbf{q} \to 0$) and pseudothreshold (Siegert limit, photon momentum $\mathbf{k} \to 0$). The longitudinal and Coulomb multipoles are related by gauge invariance, $\mathbf{k} \cdot \mathbf{J} = \omega_\gamma \rho$, which leads to

$$|\mathbf{k}| \, L_{\ell\pm}^I(W, Q^2) = \omega_\gamma \, S_{\ell\pm}^I(W, Q^2). \tag{8}$$

Since the photon c.m. energy ω_γ vanishes for $Q^2 = Q_0^2 = W^2 - M_N^2$, the longitudinal multipole must have a zero at that momentum transfer, $L_{\ell\pm}^I(W, Q_0^2) = 0$. Furthermore, gauge invariance implies that the longitudinal and Coulomb multipoles take the same value in the real photon limit, $L_{\ell\pm}^I(W, Q^2 = 0) = S_{\ell\pm}^I(W, Q^2 = 0)$. Finally, the multipoles obey the following model-independent relations at physical threshold ($\mathbf{q} \to 0$) and pseudothreshold ($\mathbf{k} \to 0$):

$$
\begin{aligned}
(E_{\ell+}^I, L_{\ell+}^I) &\to k^\ell q^\ell \quad (\ell \geq 0) \\
(M_{\ell+}^I, M_{\ell-}^I) &\to k^\ell q^\ell \quad (\ell \geq 1) \\
(L_{\ell-}^I) &\to kq \quad (\ell = 1) \\
(E_{\ell-}^I, L_{\ell-}^I) &\to k^{\ell-2} q^\ell \quad (\ell \geq 2).
\end{aligned}
\tag{9}
$$

According to Eq. (8) the Coulomb amplitudes acquire an additional factor k at pseudothreshold, i.e., $S_{\ell\pm}^I \sim k L_{\ell\pm}^I$. The photon momentum in the *cm* frame is given by

$$k(W, Q^2) = \frac{\sqrt{((W - M_N)^2 + Q^2)(W + M_N)^2 + Q^2)}}{2W} \tag{10}$$

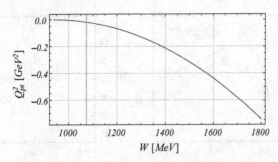

Fig. 2 The W dependence of the pseudo-threshold, where the Siegert theorem strictly holds. At πN threshold, the pseudo-threshold value is $Q^2_{\text{pt}} = -m^2_\pi = -0.018$ GeV^2, at $W = 1535$ MeV, $Q^2_{\text{pt}} = -0.356$ GeV^2

and the limit $k = 0$ is reached at $Q^2 = Q^2_{\text{pt}} = -(W - M_N)^2$ (pseudo-threshold), see Fig. 2, and because no direction is defined for $\mathbf{k} = 0$, the electric and longitudinal multipoles are no longer independent at this point,

$$E^I_{\ell+}/L^I_{\ell+} \to 1 \quad \text{and} \quad E^I_{\ell-}/L^I_{\ell-} \to -\ell/(\ell-1) \quad \text{if } k \to 0. \tag{11}$$

$$k_\gamma(W, Q^2_{pt}) = 0 \quad \to \quad Q^2_{pt} = -(W - M_N)^2 \tag{12}$$

In the case of the $N\Delta$ multipoles, Eq. (11) yields the following relation in the limit $\mathbf{k} \to 0$: $L^{3/2}_{1+} \to E^{3/2}_{1+} \to \mathcal{O}(k)$ and consequently $S^{3/2}_{1+} = kE^{3/2}_{1+}/\omega_\gamma \to \mathcal{O}(k^2)$. Although the pseudo-threshold is reached at the unphysical point $Q^2_{\text{pt}} = -(M_\Delta - M_N)^2 \approx -0.084$ GeV^2, it still influences the multipoles near $Q^2 = 0$ because of the relatively small excitation energy of the $\Delta(1232)$, see Fig. 1. In particular it leads to the following relation for $Q^2 \to Q^2_{\text{pt}}$

$$R_{SM} = \frac{S^{(3/2)}_{1+}}{M^{(3/2)}_{1+}} = \frac{k}{\omega_\gamma}\frac{E^{(3/2)}_{1+}}{M^{(3/2)}_{1+}} \to \frac{k}{M_\Delta - M_N} R_{EM}. \tag{13}$$

With increasing value of Q^2, the Siegert relation fails to describe the experimental data. Moreover, it contains a singularity at $\omega_\gamma = 0$, which occurs in $\Delta(1232)$ electroproduction already at $Q^2 = 0.64$ GeV^2. However, we obtain a good overall description by using the idea of Ref. [12] that the ratio R_{SM} is related to the (elastic) form factors of the neutron,

$$R_{SM}(Q^2) = \frac{M_N k_\Delta(Q^2) G^n_E(Q^2)}{2 Q^2 G^n_M(Q^2)}. \tag{14}$$

This relation gives the necessary proportionality to the photon momentum at small Q^2, describes the experimental value of the ratio over a wide range of Q^2, and yields an asymptotic behavior consistent with the prediction of perturbative QCD that R_{SM} should approach a constant for $Q^2 \to \infty$. This leads to the following simple parametrization

$$R_{SM}(Q^2) = -\frac{k_\Delta(Q^2)}{8M_N}\frac{a}{1+d\tau}, \tag{15}$$

with $\tau = Q^2/(4M^2_N)$, and the parameters a and d to be determined by a fit to the data. On the basis of this ansatz, the Coulomb coupling has been modified as follows:

$$\bar{\mathcal{A}}^\Delta_S(W, Q^2) = A^0_S \frac{1 + \beta_S Q^2}{1 + d\tau} \frac{k^2}{k_W k^\Delta_W} e^{-\gamma_S Q^2} G_D(Q^2), \tag{16}$$

with parameters given in Table 1. This leads to the multipole ratio

$$R_{SM}(Q^2) = \frac{A^0_S}{A^0_M}\frac{1}{1+d\tau}\left(\frac{1 + \beta_S Q^2}{1 + \beta_M Q^2}\right)\frac{k_\Delta}{k^\Delta_W}. \tag{17}$$

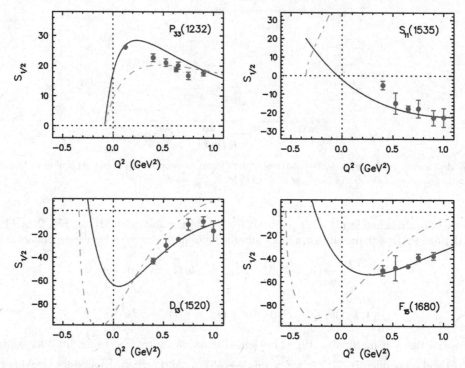

Fig. 3 Electric and longitudinal transition form factors in the Siegert limit. The red dash-dotted lines show the predictions for the longitudinal form factors, when the LWL is applied to the electric form factors. For the $\Delta(1232)$ this is fulfilled, for the $N(1520)3/2^-$ and $N(1680)5/2^+$ it is moderately violated, but for the $N(1535)1/2^-$ the Siegert theorem is extremely violated

By construction this ratio vanishes in the Siegert limit, $Q^2 \to Q_{pt}^2$, and approaches a (negative) constant for $Q^2 \to \infty$ in agreement with perturbative QCD. However, a word of caution has to be added at this point. The polynomials and gaussians used to fit the data in the range of low and intermediate virtualities, $Q^2 < 10\,\text{GeV}^2$, should not be expected to yield realistic extrapolations to the higher values of Q^2. The correct Siegert limit is even more important for pion S-wave production in the threshold region, in which case the pseudo-threshold comes as close as $Q_{pt}^2 = -m_\pi^2 \approx -0.02\,\text{GeV}^2$. The term describing the pion cloud contribution has therefore been parameterized as follows:

$$ L_{0+}^{\text{corr}}(W, Q^2) = \frac{\omega_\gamma}{\omega_{pt}} e^{-\beta(Q^2 - Q_{pt}^2)} E_{0+}^{\text{corr}}(W, Q^2), \tag{18} $$

where $\omega_{pt}^2 = -Q_{pt}^2 = (W - M_N)^2$. From a fit to π^0 electroproduction data near threshold, we obtain $\beta = 10$ GeV^{-2}.

In Fig. 3 we show the longitudinal transition form factors $S_{1/2}(Q^2)$ for resonance excitations of $\Delta(1232)3/2^+$, $N(1535)1/2^-$, $N(1520)3/2^-$ and $N(1680)5/2^+$. The red points show the single-Q^2 data obtained in the MAID2007 analysis [4] and the solid blue lines are the empirical MAID parametrizations, which are extended into the unphysical region until pseudo-threshold. The red dash-dotted lines give the predictions of Siegert's theorem applied to the MAID parametrizations of the electric form factors. At the pseudo-threshold these form factors must vanish and even the slope of the form factors are still determined by the theorem. This is almost fulfilled for the $\Delta(1232)$ with only a small discrepancy in the slope at pseudo-threshold. For the form factors of $N(1520)$ and $N(1680)$ the Siegert theorem is moderately violated, the blue curves miss the zero crossing, but the slopes are very similar to the predictions. This violation and also the violation in the slope of the $N\Delta$ form factor can easily be corrected by slight adjustments in the parametrizations. For the final $N(1535)$ form factor, however, the Siegert theorem is heavily violated by a completely different slope with opposite sign. In this case a sign change of the longitudinal form factor would solve this problem, however, this is excluded by both the MAID and the JLab/CLAS analyses. Very recently, an article by Gilberto Ramalho appeared on the archive, that gives in an improved empirical parametrization a possible explanation with an additional turnover to meet the Siegert limit without destroying the good agreement with the single-Q^2 data for $Q^2 > 0.3\,\text{GeV}^2$ [13].

5 Summary

In this paper the extraction of electromagnetic resonance excitation form factors from pion electroproduction is shortly summarized and transition form factors at low Q^2 are presented and confronted with the Siegert theorem in the Long Wavelength Limit. In the unphysical region this theorem constrains the longitudinal form factors for nucleon resonances (except $P_{11} 1/2^+$ states which do not have an electric form factor) by the electric transition form factors, which are much better known at low Q^2 because of the constraints to photoproduction data at $Q^2 = 0$. For the $\Delta(1232)$ the onset of the LWL is clearly visible in the existing data, for other resonances it gives an important constraint for the phenomenological parametrizations. The biggest violation of Siegert's theorem is observed for the nucleon to $N(1535)1/2^-$ transition.

Acknowledgments I want to thank Gilberto Ramalho for encouraging discussions. This work was supported in parts by the Deutsche Forschungsgemeinschaft (SFB 1044).

References

1. Tiator, L., Drechsel, D., Kamalov, S.S., Vanderhaeghen, M.: Electromagnetic excitation of nucleon resonances. Eur. Phys. J. ST **198**, 141 (2011). [arXiv:1109.6745 [nucl-th]]
2. Aznauryan, I.G., Burkert, V.D.: Electroexcitation of nucleon resonances. Prog. Part. Nucl. Phys. **67**, 1 (2012). [arXiv:1109.1720 [hep-ph]]
3. Haberzettl, H., Nakayama, K., Krewald, S.: Gauge-invariant approach to meson photoproduction including the final-state interaction. Phys. Rev. C **74**, 045202 (2006)
4. Drechsel, D., Kamalov, S. S., Tiator, L.: Unitary isobar model–MAID2007. Eur. Phys. J. A **34** (2007) 69. http://www.kph.uni-mainz.de/MAID/
5. Drechsel, D., Hanstein, O., Kamalov, S.S., Tiator, L.: A unitary isobar model for pion photo-and electroproduction on the proton up to 1 GeV. Nucl. Phys. A **645**, 145 (1999)
6. Arndt, R. A., Strakovsky, I. I., Workman, R. L.: Updated resonance photodecay amplitudes to 2 GeV. Phys. Rev. **C53** (1996) 430-440; (SP99 solution of the GW/SAID analysis); http://gwdac.phys.gwu.edu/
7. Tiator, L., Drechsel, D., Knochlein, G., Bennhold, C.: Analysis of resonance multipoles from polarization observables in η photoproduction. Phys. Rev. C **60**, 035210 (1999)
8. Siegert, A.J.F.: Note on the interaction between nuclei and electromagnetic radiation. Phys. Rev. **52**, 787 (1937)
9. Arenhovel, H.: Exchange currents in electric transitions and the role of Siegert's theorem: a case study in Deuteron photo-disintegration. Phys. Z A **302**, 25 (1981)
10. Naus, H.W.L.: Local gauge invariance implies Siegert's hypothesis. Phys. Rev. C **55**, 1580 (1997)
11. Buchmann, A.J., Meyer, U., Faessler, A., Hernandez, E.: N $\to$ Δ (1232) E 2 transition and Siegert's theorem. Phys. Rev. C **58**, 2478 (1998)
12. Buchmann, A.J.: Electromagnetic N $\to$ Δ transition and neutron form factors. Phys. Rev. Lett. **93**, 212301 (2004)
13. G. Ramalho, arXiv:1602.03832 [hep-ph] and private communication

Part V
Nucleon Resonances and Strong QCD Dynamics from Dyson Schwinger Equation Approaches

Few-Body Syst (2016) 57:955–963
DOI 10.1007/s00601-016-1133-x

Bruno El-Bennich · Gastão Krein · Eduardo Rojas ·
Fernando E. Serna

Excited Hadrons and the Analytical Structure of Bound-State Interaction Kernels

Received: 18 February 2016 / Accepted: 2 June 2016 / Published online: 28 June 2016
© Springer-Verlag Wien 2016

Abstract We highlight Hermiticity issues in bound-state equations whose kernels are subject to a highly asymmetric mass and momentum distribution and whose eigenvalue spectrum becomes complex for radially excited states. We trace back the presence of imaginary parts in the eigenvalues and wave functions to truncation artifacts and suggest how they can be eliminated in the case of charmed mesons. The solutions of the gap equation in the complex plane, which play a crucial role in the analytic structure of the Bethe–Salpeter kernel, are discussed for several interaction models and qualitatively and quantitatively compared to analytic continuations by means of complex-conjugate pole models fitted to real solutions.

1 Introduction

A time-honored tool to study and understand the structure of matter and its constituents, be it solids, molecules, atoms or nuclei, is spectroscopy. Indeed, if we revisit the history of atoms, it becomes clear that spectroscopy has been an invaluable and very efficient tool to reveal their *quantified* nature via spectral lines and to understand their constituent content. Eventually, the discoveries made thanks to spectroscopy led to the foundation of quantum theory and its refinement to the development of Quantum Electrodynamics with its glorious explanation, amongst other phenomena, of the Lamb shift [1–4].

A like-minded approach has been pursued in particle and hadron physics since the early experiments at the Stanford Linear Accelerator which ultimately led to the quark picture of hadrons. The quark model was eventually promoted to a non-Abelian gauge theory we nowadays believe to describe the strong interactions between the quarks, namely Quantum Chromodynamics (QCD) [5]. In recent years this program has been

---

This article belongs to the special issue "Nucleon Resonances".

B. El-Bennich (✉)
Laboratório de Física Teórica e Computacional, Universidade Cruzeiro do Sul, Rua Galvão Bueno 868, São Paulo,
SP 01506-000, Brazil
E-mail: bruno.bennich@cruzeirodosul.edu.br

G. Krein · F. E. Serna
Instituto de Física Teórica, Universidade Estadual Paulista, Rua Dr. Bento Teobaldo Ferraz, 271 – Bloco II, São Paulo,
SP 01140-070, Brazil
E-mail: gkrein@ift.unesp.br

F. E. Serna
E-mail: fernandoserna@ift.unesp.br

E. Rojas
Instituto de Física, Universidad de Antioquia, Calle 70, no. 52-21, Medellín, Colombia
E-mail: rojas@gfif.udea.edu.co

pursued and strongly extended with meson electroproduction experiments [6] at Jefferson Lab where its Continuous Electron Beam Accelerator Facility (CEBAF) recently delivered the first batch of 12 GeV electrons to Hall D. In particular, the extensive study of mesons and baryon ground states, the mass differences and level ordering with their radially excited states and parity partners as well as with exotics shed important light on the constituent structure of the hadrons. In some cases it will help to distinguish between very different or controversial pictures of a hadron's composition, for instance the competing descriptions of scalar mesons and the $X(3872)$ as either molecular bound states of lighter mesons and tetraquarks; we note that quantum field theory does not prevent these hadrons to be a superposition of many-quark states *and* flavored meson loops.

At a fundamental level, we are not merely interested in spectroscopy as a means to inform us about the mass spectrum and level orderings of hadrons. Confinement and asymptotic freedom are the prime paradigms of QCD and we expect the experimental data to teach us something about the interaction that holds the quarks together in bound states and in particular at larger distances or lower momenta. This is because if confinement is related to the analytic properties of QCD's Schwinger functions, we should gain more insights into its mechanism by mapping out the infrared behavior of the theory's coupling constant. This task cannot be completed by perturbative analyses, yet an adequate nonperturbative continuum approach is provided by Dyson–Schwinger equations (DSE) [7–10]. More precisely, one compares the effect of DSE predictions for the quark's mass functions embedded in bound-state calculations with the hadronic mass spectrum as well as form factors, wherefrom one obtains valuable information on the long-range behavior of the strong interaction [11–17].

We remind that excited hadron properties are considerably more sensitive to this long-range behavior than those of ground states [18–23]. Therefore, excited mesons and nucleons [6] provide an extremely valuable source of improving our understanding of QCD in the strong coupling regime, additional to the insight gained from light mesons [24–32], nucleons [33–40] and heavy-light mesons characterized by a disparate range of energy scales [41–48].

In here, we focus on nontrivial issues which arise out of artifacts of a given truncation scheme of the gap and bound-state equations. The latter is given by the Bethe–Salpeter equation (BSE). In particular the simplest truncation that satisfies the axialvector Ward–Green–Takahashi identity (WGTI), the rainbow-ladder truncation, produces a non-Hermitian interaction kernel once one departs from (relatively) symmetric mass constellations, such as pions, kaons, η, ρ and quarkonia. We describe in Sect. 2 under which conditions this non-Hermiticity manifests itself and argue that kernels beyond the rainbow-ladder truncation are indispensable once the physical bound states are not protected by the WGTI, as it is the case for Goldstone bosons, or the gauge-fermion vertex in the simplified infinite-mass limit becomes invalid. Calculations of BSE beyond the leading symmetry-preserving truncation are underway and as a digression we discuss in Sect. 3 the complex-conjugate pole approach that has been employed to represent the quark propagators in the complex plane [49] for approaches in and beyond the rainbow-ladder truncation.

2 Interaction Kernel and Hermiticity Issues

As we work in continuum QCD, the interactions of a quark-antiquark pair are described by a BSE which is treated as an eigenvalue problem. As an example, consider the relativistic bound-state equation of a pseudoscalar $J^P = 0^-$ meson with relative $\bar{q}q$ momentum p and total momentum P omitting flavor and Dirac indices for simplicity:

$$\Gamma_{0^-}(p, P) = \int \frac{d^4k}{(2\pi)^4}\, \mathcal{K}(p, k, P) \left[S(k + \eta_+ P)\, \Gamma_{0^-}(k, P)\, S(k - \eta_- P) \right]. \tag{1}$$

The dressed quark propagators $S(k \pm \eta_\pm P)$ are solutions of the DSE for a given flavor with $\eta_+ + \eta_- = 1$, where $\eta_\pm$ are arbitrary partition parameters since numerical results are independent of the momentum distribution in a Poincaré invariant calculation. In rainbow-ladder truncation, the interaction kernel is given by,

$$\mathcal{K}(p, k, P) = -\frac{Z_2^2\, \mathcal{G}(q^2)}{q^2}\, \frac{\lambda^a}{2}\, \gamma_\mu T_{\mu\nu}(q)\, \frac{\lambda^a}{2}\, \gamma_\nu, \tag{2}$$

which introduces the transverse projection operator, $T_{\mu\nu}(q) := g_{\mu\nu} - q_\mu q_\nu/q^2$, with $q = p - k$, and where Z_2 is the wave-function renormalization constant and λ^a are the SU(3) color matrices in the fundamental representation.

A series of ansätze for the effective interaction $\mathcal{G}(q^2)$ has been studied [24–26,50], which serves to emulate the combined effect of the gluon and quark–gluon vertex dressing functions and reflects the historic evolution of the understanding of the infrared limit of the strong interaction. There have been various efforts to extend these models by including other tensor components of the vertex beyond the leading truncation [51–59]. An important feature of Eq. (2) is that it satisfies the axialvector WGTI [60] which warrants a massless pion in the chiral limit. The set of Eqs. (1) and (2) defines an eigenvalue problem with physical solutions at the mass-shell points, $P^2 = -M_i^2$, where M_i^2 is the bound-state mass of the ground and excited states; see discussions in Refs. [15,61].

Using a Euclidian nonorthogonal basis with respect to the Dirac trace, the Poincaré invariant solutions of Eq. (1) are generally given by:

$$\Gamma_{0^-}(p, P) = \gamma_5 \Big[i\, \mathbb{I}_D\, E_{0^-}(p, P) + \slashed{P} F_{0^-}(p, P) + \slashed{p}(p \cdot P)\, G_{0^-}(p, P) + \sigma_{\mu\nu} p_\mu P_\nu\, H_{0^-}(p, P) \Big], \tag{3}$$

The scalar functions $\mathcal{F}_{0^-}^\alpha(p, P) = \big\{ E_{0^-}(p, P), F_{0^-}(p, P), G_{0^-}(p, P), H_{0^-}(p, P) \big\}$ are the Lorentz invariant components of the Bethe–Salpeter amplitude and it can be shown [15] that the BSE rewritten in component form,

$$\lambda(P^2)\, \mathcal{F}_{0^-}^\alpha(p, P) = \int \frac{d^4k}{(2\pi)^4}\, \mathcal{K}^{\alpha\beta}(p, k, P)\, \mathcal{F}_{0^-}^\beta(k, P), \tag{4}$$

is an eigenvalue equation as indicated on the left-hand side of Eq. (4) by the eigenvalue function $\lambda(P^2)$. The latter has a solution for every value of P^2. The tensor $\mathcal{K}^{\alpha\beta}(p, k, P)$ is a projection of Eq. (1) with the interaction kernel in Eq. (2) [15,23]. The scalar functions $\mathcal{F}_{0^-}^\alpha(p, P)$ in Eq. (4) are expanded in Chebyshev polynomials of the second kind, $U_m(z_k)$ and $U_m(z_p)$, which are functions of the angles, $z_k = P \cdot k/(\sqrt{P^2}\sqrt{k^2})$ and $z_p = P \cdot p/(\sqrt{P^2}\sqrt{p^2})$, for example:

$$\mathcal{F}_{0^-}^\alpha(p, P) = \sum_{m=0}^\infty \mathcal{F}_{0^-}^{\alpha m}(p, P)\, U_m(z_p). \tag{5}$$

This expansion can of course also be applied to mesons with quantum numbers other than 0^{-+}.

One of the postulates of quantum mechanics states that the eigenvalues that describe a physical observable must be real, hence the necessity of Hermitian operators. Equivalently, in case of the BSE, the eigenvalue spectrum should be positive definite [62] owing to the Hermiticity requirement of physical operators. The eigenvalues in Eq. (4) describe a mass spectrum with physical solutions for $\lambda_0(M_0^2) = \lambda_1(M_1^2) = \lambda_2(M_2^2) = \cdots = \lambda_i(M_i^2) = 1$, where M_0^2 denotes the ground state and M_i^2, $i = 1, 2, \ldots$ are radial excitations with $M_i < M_{i+1}$, and are ordered as $\lambda_0(M_i^2) > \lambda_1(M_i^2) > \cdots > \lambda_i(M_i^2)$. As an example, the individual eigenvalue trajectories, $\lambda_0(M^2)$ and $\lambda_1(M^2)$, for the nucleon and its first excited state are visualized in Fig. 1 of Ref. [15].

Nonetheless, within the framework of the rainbow-ladder truncation using an infrared-massive and finite interaction in agreement with numerical results from lattice QCD and DSE for the gluon dressing function, we observe that the eigenvalues of heavy-light systems become complex with increasing mass difference of the antiquark–quark pairs [23]. More precisely, we find a complex-conjugate pair of eigenvalues for the first radially excited states of D and D_s mesons when we include higher Chebyshev polynomials, $U_{m>1}(z_p)$. The ground-state eigenvalues of the D mesons remain real. This is independent of the amount of Chebyshev moments employed and one finds that for $U_m(z_p)$, $m \geq 4$, the numerical solution always converges to the same eigenvalue. The appearance of imaginary parts in higher eigenvalues is not limited to mesons—we also find them in case of the Faddeev kernel that describes the three-quark correlation function with the quantum numbers of the nucleon. Its first radial excitation, hence the second eigenvalue the Faddeev kernel produces, was shown to be consistent with the Roper resonance [63]. However, the third eigenvalue which corresponds to the second excited state also consists of a complex-conjugate pair.

The above findings indicate that the rainbow-ladder kernels, which describe the repeated antiquark–quark interactions, are not Hermitian when one of the quarks becomes much heavier. Whereas the eigenvalues and Chebyshev moments of the kaon and its excited states are all real, we observe a threshold midway between the strange and charm masses where the eigenvalues acquire an imaginary component. Moreover, in the case of D and D_s mesons we observe that *odd* Chebyshev moments contribute and they acquire an imaginary part in the ground and excited states. The odd contributions are not surprising since K and D mesons are not eigenstates of the charge-conjugation operator and thus $\bar{\Gamma}(k, P) = \lambda_c \Gamma(k, P)$ does not imply $\lambda_c = \pm 1$ for the charge parity, whereas for equal-mass pseudoscalar mesons with $J^{PC} = 0^{-+}$, the constraint that the Dirac base satisfies $\lambda_c = +1$ imposes that $\mathcal{F}_{0^-}^\alpha(p, P)$ be *even* in the angular variable z_p. Yet, the angular dependence in the odd moments is the source of their complexity when the mass difference increases—the zeroth Chebyshev moment of the D mesons is angle independent and remains real. This does not occur for the kaon where flavor-symmetry and charge-parity breaking are still negligible.

Thus, the appearance of complex solutions of the BSE for heavy-light mesons is intimately related with the asymmetric momentum distribution within the bound state at a given truncation. Consider initially a pseudoscalar meson in the isovector channel with equal masses in the rainbow-ladder truncation. A consistent next-order truncation of the DSE and BSE implies vertex-corrections and crossed-box contributions in the BSE kernel [60]. A necessary consequence of the Goldstone theorem is that these contributions cancel, which is the reason for the successes of the rainbow-ladder truncation in many applications in hadron physics. Now, consider again a pseudoscalar isovector meson but with two very different quark masses, as in the case of D and B mesons. For heavy quarks, the vertex corrections in the DSE are insignificant and the vertex is almost bare, which is another reason the leading truncation is also successful for quarkonia. To see this, consider the Ball-Chiu ansatz for the quark–gluon vertex,

$$\Gamma_\mu^{BC}(k, p) = \tfrac{1}{2}\big[A(k) + A(p)\big]\gamma_\mu + \frac{A(k) - A(p)}{2(k^2 - p^2)}(\slashed{k} + \slashed{p})(k_\mu + p_\mu) - i\frac{B(k) - B(p)}{k^2 - p^2}(k_\mu + p_\mu)\mathbb{I}_D, \quad (6)$$

where $A(p^2)$ and $B(p^2)$ are the vector and scalar components, respectively, of the quark's gap equation solution: $S^{-1}(p) = iA(p^2)\slashed{p} + B(p^2)\mathbb{I}_D$. As the mass and wave functions, $M(p^2) = B(p^2)/A(p^2)$ and $Z(p^2) = 1/A(p^2)$, vary insignificantly for the b quark and modestly for the c quark, one has $A(k^2) \approx A(p^2) \simeq 1$ and $B(k^2) \approx B(p^2)$, where k is the incoming and p the outgoing quark momentum related by the gluon momentum $q = p - k$. In this heavy-mass limit, the tensor structures of the Ball-Chiu proportional to the dressing functions $\Delta B(k, p)$ and $\Delta A(k, p)$ vanish [64] and only the leading term survives:

$$\Gamma_\mu(k, p) \propto \frac{A(k) + A(p)}{2}\gamma_\mu \approx \gamma_\mu. \quad (7)$$

Hence, unlike the equal-mass case, the kernel of a heavy-light $\bar{Q}q$ meson contains very different corrections to either quark-gluon vertex. As a consequence, the repulsive crossed-box contribution and the two asymmetrically dressed vertices do not cancel perfectly. The rainbow-ladder truncation is therefore not a good approximation and one must take the unavoidable dressing of the light quark seriously, which adds to the angular complexity of the $\bar{Q}q$ scattering kernel with terms proportional to z_k discarded in the lowest truncation. It is plausible that this additional angular dependence of the kernel will cancel imaginary contributions to Chebyshev moments by which the rainbow-ladder truncation is plagued and thereby restore its Hermiticity.

A different case is the quark-diquark Faddeev equation employed in Refs. [33,34,63], which relies on more model input and its kernel cannot be described by a simple ladder approximation. More precisely, the fact that the Faddeev amplitude contains a diquark in the $\bar{3}_c$ channel that is free at spacelike momenta but pole-free on the timelike axis implies a qq scattering kernel beyond ladder truncation [60]. It is however possible that the appearance of an imaginary component in the third eigenvalue can also be traced back to missing angular dependence in the Faddeev kernel that originates in three-body interactions or additional diquark contributions not included in Ref. [63].

3 Analytical Quark Structure: Numerical DSE Solutions and Complex-Conjugate Poles

As we have argued above, while the rainbow-ladder truncation is successful in studies of meson spectroscopy and decays for a broad range of ground-state mesons, the approximation inexorably runs into trouble when applied to excited mesons states, heavy-light mesons and certain mass splittings, such as the a_1-ρ mass difference [53]. As mentioned, progress in calculation of the quark's DSE beyond the rainbow-ladder truncation has been made and is underway [12,51–58]. More effort is required, though, as practical computations of the

Bethe-Salpeter amplitude imply the knowledge of the quark propagator in the complex plane,[1] where poles produced by a given interaction pose severe numerical difficulties. This also occurs for vertex ansätze beyond the rainbow-ladder ansatz. An economic expedient to represent these propagators by analytical expressions [49] is based on a complex-conjugate pole model and has been successfully used in calculations of pion distribution amplitudes and elastic form factors:

$$S(p) = \sum_i^n \left[\frac{z_i}{i\not{p} + m_i} + \frac{z_i^*}{i\not{p} + m_i^*} \right], \quad m_i, z_i \in \mathbb{C}. \tag{8}$$

In essence, the expression in Eq. (8) does not produce poles on the real timelike axis and has no Källén-Lehmann representation; therefore the quark cannot appear in the Hilbert space of observable states. This is consistent with confinement, though in real QCD many more poles or cuts may characterize the analytic structure of the quark's dressing functions. The numerical DSE solutions for the quark can be fitted with $n = 3$ complex-conjugate poles, where the pole locations depend on the model for the gluon- and vertex-dressing functions and whether the parameters m_i and z_i are fitted to the DSE solutions on the real axis or to a numerical solution in a parabola on the complex plane.

We here present the results of such fits for three different cases, namely the quark propagator obtained in rainbow-ladder truncation with two different interaction parameter sets listed in Table 1 of Ref. [23] (corresponding to model 1 and 2 which best reproduce ground and excited states, respectively) and with the Ball-Chiu vertex. As mentioned previously, the quark propagators, $S(k \pm \eta_\pm P)$, involve complex arguments for their respective momenta in the Euclidean formulation of the BSE (1). It was argued that a good model to reproduce the quark propagator on the real spacelike axis and to analytically continue it on the complex plane is given by the expression in Eq. (8). Nonetheless, using Cauchy's integral method, one can also directly solve the quark DSE on the complex plane with a given model for the gluon dressing function [23,65,66].

It is thus of interest to study whether this complex solution is in qualitative and quantitative agreement with the pole-model fit to the real-axis solution and whether the analytic structure of the quark-dressing functions off the real axis matters at all for hadronic observables. The latter point deals with the physical content of these poles—if any of the fits or extrapolations to the complex plane yield quark propagators that produce in conjunction with the BSE exactly the same values for meson observables, may one simply view them as models that work well, very much like the models [34,42,64] based on entire functions?

The complex-conjugate pole parameters, following Eq. (8) ($i = 1, 2, 3$), for the three different interaction ansätze and vertices are found with a best least-squares fit (all entries are in GeV^2):

	Re m_1^2	Im m_1^2	Re m_2^2	Im m_2^2	Re m_3^2	Im m_3^2
Model 1 [23]	−0.24	0.53	−1.35	−1.37	−0.35	−1.71
Model 2 [23]	−0.39	0.83	−2.70	−2.92	−0.96	−3.90
Ball-Chiu	−0.13	0.05	−0.57	0.70	−1.07	−1.63

The dressed-quark propagator is the solution of a gap equation and has the general form,

$$S(p) = -i\,\sigma_V(p^2)\,\not{p} + \sigma_S(p^2)\mathbb{I}_D = \left[i\,A(p^2)\,\not{p} + B(p^2)\,\mathbb{I}_D \right]^{-1}. \tag{9}$$

To visualize the poles in the above table we plot the real component of the scalar function $\sigma_s(p^2)$ for a light quark in Fig. 1. The upper row of graphs corresponds to the interaction parametrization [50] in rainbow-ladder approximation that best reproduces the ground-state spectrum with $\omega D = (0.8 \text{ GeV})^3$, $\omega = 0.4$ GeV, and the one that yields the most satisfying excited-state spectrum with $\omega D = (1.1 \text{ GeV})^3$, $\omega = 0.6$ GeV. Here, ω plays the role of the interaction width while D accounts for its strength. Both parameters are chosen for a fixed value of ωD and describe effectively the combined effect of the gluon and vertex dressing functions in the leading truncation and therefore determine the support and location of maximal strength of the interaction. Comparing both graphs, one observes a set of two conjugate-complex poles that are located in the vicinity of the real axis. The two other pairs lie more remote from the real axis, but in the excited-state parametrization this distance increases considerably and the poles shift towards much larger timelike squared momenta. When the gap equation is solved with the beyond rainbow-ladder Ball-Chiu ansatz for the vertex, one of the complex-conjugate

---

[1] See discussion in Sect. 3 of Ref. [23] and references therein.

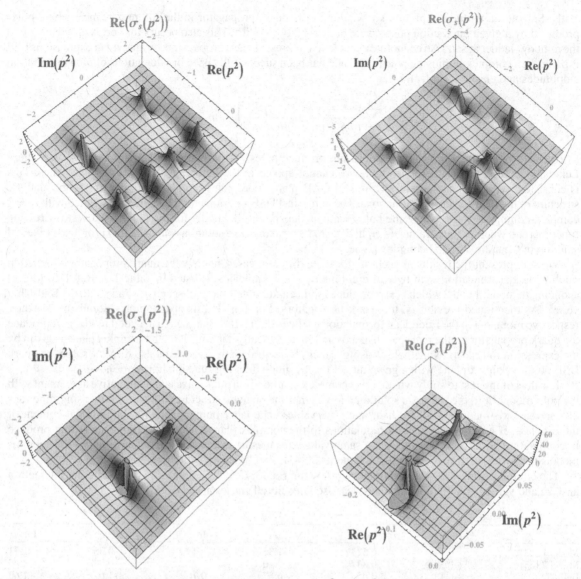

Fig. 1 Scalar function σ_s of the light quark propagator $S(p) = -i\,\not{p}\,\sigma_V(p^2) + \sigma_S(p^2)$ for the three cases: model 1 in Ref. [23] with $\omega D = (0.8\,\text{GeV})^3$, $\omega = 0.4\,\text{GeV}$, in rainbow-ladder truncation (top-left graph); model 2 in Ref. [23] with $\omega D = (1.1\,\text{GeV})^3$, $\omega = 0.6\,\text{GeV}$, in rainbow-ladder truncation (top-right graph); Ball-Chiu vertex with $\omega D = (0.55\,\text{GeV})^3$, $\omega = 0.5\,\text{GeV}$ (bottom-left and -right graph). In all cases the interaction of Ref. [50] is used with a renormalized running quark mass $m(19\,\text{GeV}) = 3.4\,\text{MeV}$

pole pairs moves indistinguishably close to the real axis, whereas the pole pair most distant from the real axis shifts toward larger timelike squared momenta as depicted in the bottom-left graph of Fig. 1. In the bottom-right graph, we plot the enlarged vicinity of the complex-conjugate poles closest to the real axis for better visibility. The left panel of Fig. 2 depicts the difference of the real part of the excited state's mass function, $\text{Re}\,B_{\text{Model}\,2}(p^2)$, and the ground state's mass function, $\text{Re}\,B_{\text{Model}\,1}(p^2)$: $\Delta B(p^2) = \text{Re}\,B_{\text{Model}\,2}(p^2) - \text{Re}\,B_{\text{Model}\,1}(p^2)$. Upon inspection of the real function $\Delta B(p^2)$ in the complex plane it becomes clear that the most dramatic differences are localized in the timelike region close to the complex-conjugate poles which characterize the different parameterizations of the given interaction [50]. Below the zero level, represented by a green-colored plane, $\Delta B(p^2)$ becomes slightly negative but especially in the spacelike region the difference is vanishing. For further comparison, we reproduce in the right panel of Fig. 2 again the function $\sigma_s(p^2)$ obtained with the Maris-Tandy interaction [24] from which it can be inferred that in a fit with three conjugate-complex poles one pole pair lies again closer to the real axis. Roughly, this solution gives rise to a pole structure that resembles the ground-state parametrization in the top-left corner of Fig. 1.

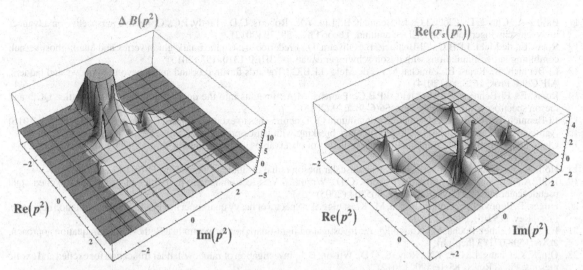

Fig. 2 Left panel: Plot of the function $\Delta B(p^2) = \operatorname{Re} B_{\text{Model 2}}(p^2) - \operatorname{Re} B_{\text{Model 1}}(p^2)$, where the green horizontal plane indicates the zero level; see text for details. Right panel: complex-conjugate pole representation of σ_s obtained with the Maris–Tandy model [24], the parameter values $\omega = 0.4$ GeV, $\omega D = 0.372$ GeV^3 and a running quark mass renormalized with $m(19$ GeV$) = 3.4$ MeV

4 Final Remarks

At the present point it is difficult to draw firm conclusions about the analytic structure of the quark propagators we just presented. It is evident that either complex-conjugate pole representation is too simplistic: we checked that while they yield the same numerical results for light-meson observables as the full DSE solution in a parabola on the complex plane [23,65,66], they do so only for a restrained set of hadronic data. It is reasonable to assume that a realistic solution, such as the one produced by fully unquenched QCD, is considerably more complicated. We note that the pole model also works well for charm quarks and we verify that the D and D_s masses and decay constants are almost indistinguishable from the ones obtained with the numerical solution in a complex parabola. The effect of all 12 tensor structures derived from longitudinal *and* transverse Slavnov–Taylor identities (STI) [67] in the quark-gap equation is currently being studied (Ahmed et al., in preparation), which calls for a full comparison of the analytical structure of the quark propagators in the rainbow-ladder approximation with that including all vertex tensor structures.

Acknowledgments B. E. thanks the organizers of the ECT* workshop "Nucleon Resonances From Photoproduction to High Photon Virtualities" for their kind invitation and local support. The work mentioned in this contribution was made possible by: FAPESP Grant Nos. 2013/01907-0 and 2013/16088-4 (São Paulo State); CNPq Grant Nos. 305894/2009-9, 458371/2014-9 and 305852/2014-0 (Brazil). E. R. acknowledges support by *Patrimonio Autónomo Fondo Nacional de Financiamiento para la Ciencia, la Tecnología y la Innovación, Francisco José de Caldas* and by *Sostenibilidad-UDEA 2014–2015* (Colombia).

References

1. Tomonaga, S.: On a relativistically invariant formulation of the quantum theory of wave fields. Prog. Theor. Phys. **1**, 27 (1946)
2. Schwinger, J.S.: On Quantum electrodynamics and the magnetic moment of the electron. Phys. Rev. **73**, 416 (1948); *ibid*.**82**, 664 (1951)
3. Feynman, R.P.: The Theory of positrons. Phys. Rev. **76**, 749 (1949); *ibid*.**76**, 769 (1949)
4. Dyson, F.J.: The Radiation theories of Tomonaga, Schwinger, and Feynman. Phys. Rev. **75**, 486 (1949)
5. Fritzsch, H., Gell-Mann, M., Leutwyler, H.: Advantages of the color cctet gluon picture. Phys. Lett. B **47**, 365 (1973)
6. Aznauryan, I.G., et al.: Studies of Nucleon Resonance Structure in Exclusive Meson Electroproduction. Int. J. Mod. Phys. E **22**, 1330015 (2013)
7. Dyson, F.J.: The S matrix in quantum electrodynamics. Phys. Rev. **75**, 1736 (1949)
8. Schwinger, J.S.: On the Green's functions of quantized fields. 1., Proc. Nat. Acad. Sci. **37**, 452 (1951); *ibid*.**37**, 455 (1951)
9. Roberts, C.D., Williams, A.G.: Dyson-Schwinger equations and their application to hadronic physics. Prog. Part. Nucl. Phys. **33**, 477 (1994)
10. Alkofer, R., von Smekal, L.: The Infrared behavior of QCD Green's functions: Confinement dynamical symmetry breaking, and hadrons as relativistic bound states. Phys. Rept. **353**, 281 (2001)

11. Bashir, A., Chang, L., Cloët, I.C., El-Bennich, B., Liu, Y.X., Roberts, C.D., Tandy, P.C.: Collective perspective on advances in dyson-schwinger equation QCD. Commun. Theor. Phys. **58**, 79 (2012)

12. Rojas, E., de Melo, J.P.B.C., El-Bennich, B., Oliveira, O., Frederico, T.: On the quark-gluon vertex and quark-ghost kernel: combining lattice simulations with dyson-schwinger equations. JHEP **1310**, 193 (2013)

13. El-Bennich, B., Rojas, E., Paracha, M.A., de Melo, J.P.B.C.: Towards flavored bound states beyond rainbows and ladders. AIP Conf. Proc. **1625**, 80 (2014)

14. Rojas, E., El-Bennich, B., De Melo, J.P.B.C., Paracha, M.A.: Insights into the quark-gluon vertex from lattice QCD and meson spectroscopy. Few Body Syst. **56**(6–9), 639 (2015)

15. El-Bennich, B., Rojas, E.: Contemporary continuum QCD approaches to excited hadrons. EPJ Web Conf. **113**, 05003 (2016)

16. Aguilar, A.C., Papavassiliou, J.: Chiral symmetry breaking with lattice propagators. Phys. Rev. D **83**, 014013 (2011)

17. Cloët, I.C., Roberts, C.D.: Explanation and prediction of observables using continuum strong QCD. Prog. Part. Nucl. Phys. **77**, 1 (2014)

18. Höll, A., Krassnigg, A., Roberts, C.D.: Pseudoscalar meson radial excitations. Phys. Rev. C **70**, 042203 (2004)

19. Höll, A., Krassnigg, A., Maris, P., Roberts, C.D., Wright, S.V.: Electromagnetic properties of ground and excited state pseudoscalar mesons. Phys. Rev. C **71**, 065204 (2005)

20. Hilger, T., Popovici, C., Gómez-Rocha, M., Krassnigg, A.: Spectra of heavy quarkonia in a Bethe-Salpeter-equation approach. Phys. Rev. D **91**(3), 034013 (2015)

21. Hilger, T., Gómez-Rocha, M., Krassnigg, A.: Investigating light-quarkonium spectra in a Bethe-Salpeter-equation approach, arXiv:1508.07183 [hep-ph]

22. Qin, S.X., Chang, L., Liu, Y.X., Roberts, C.D., Wilson, D.J.: Investigation of rainbow-ladder truncation for excited and exotic mesons. Phys. Rev. C **85**, 035202 (2012)

23. Rojas, E., El-Bennich, B., de Melo, J.P.B.C.: Exciting flavored bound states. Phys. Rev. D **90**, 074025 (2014)

24. Maris, P., Roberts, C.D., Tandy, P.C.: Pion mass and decay constant. Phys. Lett. B **420**, 267 (1998)

25. Maris, P., Roberts, C.D.: Pi- and K meson Bethe-Salpeter amplitudes. Phys. Rev. C **56**, 3369 (1997)

26. Maris, P., Tandy, P.C.: Bethe-Salpeter study of vector meson masses and decay constants. Phys. Rev. C **60**, 055214 (1999)

27. El-Bennich, B., de Melo, J.P.B.C., Loiseau, B., Dedonder, J.-P., Frederico, T.: Modeling electromagnetic form-factors of light and heavy pseudoscalar mesons. Braz. J. Phys. **38**, 465 (2008)

28. Chang, L., Cloët, I.C., El-Bennich, B., Klähn, T., Roberts, C.D.: Exploring the light-quark interaction. Chin. Phys. C **33**, 1189 (2009)

29. da Silva, E.O., de Melo, J.P.B.C., El-Bennich, B., Filho, V.S.: Pion and kaon elastic form factors in a refined light-front model. Phys. Rev. C **86**, 038202 (2012)

30. El-Bennich, B., de Melo, J.P.B.C., Frederico, T.: A combined study of the pion's static properties and form factors. Few Body Syst. **54**, 1851 (2013)

31. de Melo, J.P.B.C., El-Bennich, B., Frederico, T.: The photon-pion transition form factor: incompatible data or incompatible models? Few Body Syst. **55**, 373 (2014)

32. Eichmann, G., Alkofer, R., Cloët, I.C., Krassnigg, A., Roberts, C.D.: Perspective on rainbow-ladder truncation. Phys. Rev. C **77**, 042202 (2008)

33. Oettel, M., Alkofer, R., von Smekal, L.: Nucleon properties in the covariant quark diquark model. Eur. Phys. J. A **8**, 553 (2000)

34. Cloët, I.C., Eichmann, G., El-Bennich, B., Klähn, T., Roberts, C.D.: Survey of nucleon electromagnetic form factors. Few Body Syst. **46**, 1 (2009)

35. Eichmann, G., Alkofer, R., Krassnigg, A., Nicmorus, D.: Nucleon mass from a covariant three-quark Faddeev equation. Phys. Rev. Lett. **104**, 201601 (2010)

36. Eichmann, G., Fischer, C.S.: Unified description of hadron-photon and hadron-meson scattering in the Dyson-Schwinger approach. Phys. Rev. D **85**, 034015 (2012)

37. Eichmann, G., Nicmorus, D.: Nucleon to delta electromagnetic transition in the dyson-schwinger approach. Phys. Rev. D **85**, 093004 (2012)

38. Segovia, J., Cloët, I.C., Roberts, C.D., Schmidt, S.M.: Nucleon and Δ elastic and transition form factors. Few Body Syst. **55**, 1185 (2014)

39. Segovia, J.: Elastic and transition form factors in DSEs. Few Body Syst. **57**(6), 461 (2016)

40. Sanchis-Alepuz, H., Fischer, C.S., Kubrak, S.: Pion cloud effects on baryon masses. Phys. Lett. B **733**, 151 (2014)

41. El-Bennich, B., Ivanov, M.A., Roberts, C.D.: Flavourful hadronic physics. Nucl. Phys. Proc. Suppl. **199**, 184 (2010)

42. El-Bennich, B., Ivanov, M.A., Roberts, C.D.: Strong $D^* \to D\pi$ and $B^* \to B\pi$ couplings. Phys. Rev. C **83**, 025205 (2011)

43. El-Bennich, B., Krein, G., Chang, L., Roberts, C.D., Wilson, D.J.: Flavor SU(4) breaking between effective couplings. Phys. Rev. D **85**, 031502 (2012)

44. El-Bennich, B., Roberts, C.D., Ivanov, M.A.: Heavy-quark symmetries in the light of nonperturbative QCD approaches, POS(QCD-TNT-II) 018, arXiv:1202.0454 [nucl-th] (2012)

45. El-Bennich, B., Furman, A., Kamiński, R., Leśniak, L., Loiseau, B., Moussallam, B.: CP violation and kaon-pion interactions in $B \to K\pi^+\pi^-$ decays. Phys. Rev. D **79**, 094005 (2009) [Erratum: Phys. Rev. D **83**, 039903 (2011)]

46. Gómez-Rocha, M., Hilger, T., Krassnigg, A.: Effects of a dressed quark-gluon vertex in pseudoscalar heavy-light mesons. Phys. Rev. D **92**(5), 054030 (2015)

47. Serna, F.E., Brito, M.A., Krein, G.: Symmetry-preserving contact interaction model for heavy-light mesons. AIP Conf. Proc. **1701**, 100018 (2016)

48. Krein, G.: Dressed perturbation theory: perturbative approach to Dyson-Schwinger and Bethe-Salpeter equations. PoS QCD TNT-III, 021 (2013)

49. Bhagwat, M., Pichowsky, M.A., Tandy, P.C.: Confinement phenomenology in the Bethe-Salpeter equation. Phys. Rev. D **67**, 054019 (2003)

50. Qin, S.X., Chang, L., Liu, Y.X., Roberts, C.D., Wilson, D.J.: Interaction model for the gap equation. Phys. Rev. C **84**, 042202 (2011)

51. Matevosyan, H.H., Thomas, A.W., Tandy, P.C.: Quark-gluon vertex dressing and meson masses beyond ladder-rainbow truncation. Phys. Rev. C **75**, 045201 (2007)
52. Chang, L., Roberts, C.D.: Sketching the Bethe-Salpeter kernel. Phys. Rev. Lett. **103**, 081601 (2009)
53. Chang, L., Roberts, C.D.: Tracing masses of ground-state light-quark mesons. Phys. Rev. C **85**, 052201 (2012)
54. Fischer, C.S., Williams, R.: Probing the gluon self-interaction in light mesons. Phys. Rev. Lett. **103**, 122001 (2009)
55. Qin, S.X., Chang, L., Liu, Y.X., Roberts, C.D., Schmidt, S.M.: Practical corollaries of transverse Ward-Green-Takahashi identities. Phys. Lett. B **722**, 384 (2013)
56. Aguilar, A.C., Binosi, D., Ibañez, D., Papavassiliou, J.: New method for determining the quark-gluon vertex. Phys. Rev. D **90**(6), 065027 (2014)
57. Binosi, D., Chang, L., Papavassiliou, J., Roberts, C.D.: Bridging a gap between continuum-QCD and ab initio predictions of hadron observables. Phys. Lett. B **742**, 183 (2015)
58. Binosi, D., Chang, L., Papavassiliou, J., Qin, S.X., Roberts, C.D.: Symmetry preserving truncations of the gap and Bethe-Salpeter equations. Phys. Rev. D **93**(9), 096010 (2016) arXiv:1601.05441 [nucl-th]
59. Sanchis-Alepuz, H., Williams, R.: Probing the quark-gluon interaction with hadrons. Phys. Lett. B **749**, 592 (2015).
60. Bender, A., Roberts, C.D., Von Smekal, L.: Goldstone theorem and diquark confinement beyond rainbow ladder approximation. Phys. Lett. B **380**, 7 (1996)
61. Krassnigg, A., Roberts, C.D.: DSEs, the pion, and related matters. Fizika B **13**, 143 (2004)
62. Ahlig, S., Alkofer, R.: (In)consistencies in the relativistic description of excited states in the Bethe-Salpeter equation Ann. Phys. **275**, 113 (1999)
63. Segovia, J., El-Bennich, B., Rojas, E., Cloët, I.C., Roberts, C.D., Xu, S.S., Zong, H.S.: Completing the picture of the Roper resonance. Phys. Rev. Lett. **115**(17), 171801 (2015)
64. Ivanov, M.A., Körner, J.G., Kovalenko, S.G., Roberts, C.D.: B- to light-meson transition form-factors. Phys. Rev. D **76**, 034018 (2007)
65. Fischer, C.S., Watson, P., Cassing, W.: Probing unquenching effects in the gluon polarisation in light mesons. Phys. Rev. D **72**, 094025 (2005)
66. Krassnigg, A.: Excited mesons in a Bethe-Salpeter approach. PoS Confin. **8**, 075 (2008)
67. He, H.X.: Transverse Symmetry Transformations and the Quark-Gluon Vertex Function in QCD. Phys. Rev. D **80**, 016004 (2009)

Few-Body Syst (2016) 57:965–973
DOI 10.1007/s00601-016-1134-9

Gernot Eichmann

Progress in the Calculation of Nucleon Transition form Factors

Received: 10 February 2016 / Accepted: 1 June 2016 / Published online: 28 June 2016
© Springer-Verlag Wien 2016

Abstract We give a brief account of the Dyson–Schwinger and Faddeev-equation approach and its application to nucleon resonances and their transition form factors. We compare the three-body with the quark–diquark approach and present a quark–diquark calculation for the low-lying nucleon resonances including scalar, axialvector, pseudoscalar and vector diquarks. We also discuss the timelike structure of transition form factors and highlight the advantages of form factors over helicity amplitudes.

1 Introduction

Understanding the spectrum and structure of nucleon resonances has been among the main challenges in hadron physics ever since the discovery of the Roper resonance. Much experimental information on the photo- and electroexcitations of baryonic resonances has been collected at Jefferson Lab, MAMI, ELSA, and other facilities [1–4]; especially Jefferson Lab with the CLAS detector in Hall B has contributed a large amount to the electroproduction world data in recent years. Transition form factors over a wide kinematic domain are now available for a number of nucleon resonances and provide a unique window into the nonperturbative structure of Quantum Chromodynamics (QCD).

There is still an abundance of open questions concerning the nature of resonances. Are the three quarks in a baryon spatially equally distributed or do they cluster into diquarks? What is the importance of molecular components that are generated by meson–baryon interactions, and how does the 'pion cloud' reveal itself in form factors? How are confinement and spontaneous chiral symmetry breaking manifest, and what is the microscopic origin of vector-meson dominance?

To address these questions, it is desirable to establish a consistent microscopic description of π, ρ, nucleon and nucleon resonance properties within QCD. We will focus on the Dyson–Schwinger equation (DSE) framework, whose basic promise is to calculate such observables from the nonperturbative structure of QCD's Green functions [5–7]. While there has been progress with regard to meson spectrum and structure properties as well as nucleon and Δ form factors (among other areas), its application to excited baryons is still at an early stage. In the following we give a brief account of the approach on its way towards calculating nucleon resonances and transition form factors.

This article belongs to the special issue "Nucleon Resonances".

This work is supported by the German Science Foundation DFG under project number DFG TR-16.

G. Eichmann (✉)
Institut für Theoretische Physik, Justus-Liebig-Universität Giessen, Heinrich-Buff-Ring 16, 35392 Giessen, Germany
E-mail: gernot.eichmann@theo.physik.uni-giessen.de
Tel.: +49 641 9933342
Fax: +49 641 9933309

2 The Covariant Faddeev Approach

To illustrate the basic ideas, consider an electromagnetic nucleon or nucleon-to-resonance transition form factor. Its typical shape will be kindred to that in Fig. 1: its spacelike behavior ($Q^2 > 0$) is experimentally extracted from eN scattering or, in the case of a resonance, pion photo- and electroproduction. Experimental information on timelike nucleon form factors above threshold comes from the reaction $e^+e^- \to N\overline{N}$ or its inverse. For nucleon resonances, the region below threshold is indirectly accessible via the Dalitz decays $N^* \to Ne^+e^-$ that contribute to the dilepton 'cocktail' in NN and heavy-ion collisions. The characteristic features in this regime are the vector-meson bumps that are produced when the photon fluctuates into ρ and ω. The bump landscape in Fig. 1 is drawn from the pion form factor, experimentally measured via $e^+e^- \to \pi^+\pi^-$, where this property is exposed due to the much smaller threshold ($2m_\pi < m_\rho$).

Is it reasonable to expect a similar behavior also for nucleon transition form factors, and how can one understand these features from the quark level? The current matrix element that encodes the form factors will be made of diagrams such as that in Fig. 1: the incoming baryon splits into its valence quarks which emit and reabsorb gluons, obtain a boost from the photon, and finally recombine into the outgoing baryon. The blobs represent the baryons' covariant Faddeev amplitudes, the quantum field-theoretical analogues of a baryon wave function. Fortunately, these complicated interactions can be absorbed into a few compact building blocks: the Faddeev amplitudes, the dressed quark propagator, and the dressed quark–photon vertex. The complete expression for the electromagnetic current matrix element is shown in Fig. 2; it also couples the photon to the two- and three-quark kernels and thereby satisfies electromagnetic gauge invariance. In turn, the Faddeev amplitudes must be solved from their corresponding Faddeev equations, which at the same time determine the masses of the baryons [8–10].

The quark propagator and kernels that enter the equations are not arbitrary but related to each other. The quark propagator is determined from its Dyson–Schwinger equation (DSE) in Fig. 3. The resulting quark mass function becomes momentum-dependent; it describes the transition from the input current-quark mass at large momenta to a nonperturbative, dressed 'constituent quark' mass of a few hundred MeV in the infrared. The analogous $q\bar{q}$ kernel that appears in a meson's Bethe–Salpeter equation (BSE) is connected to the quark propagator via chiral symmetry and the corresponding axial Ward–Takahashi identity (WTI), which pictorially amounts to 'cutting' dressed quark lines in the DSE; see [11] and references therein. The leading kernel

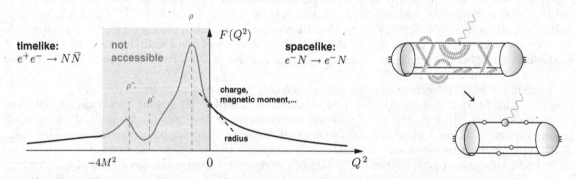

Fig. 1 *Left* Sketch of a generic form factor in the spacelike and timelike region. *Right* Representative (rainbow-ladder-like) contribution to a NN^* transition matrix element

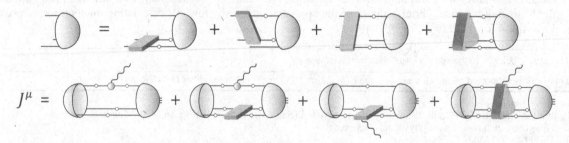

Fig. 2 Three-quark Faddeev equation (*top*) and electromagnetic current matrix element (*bottom*)

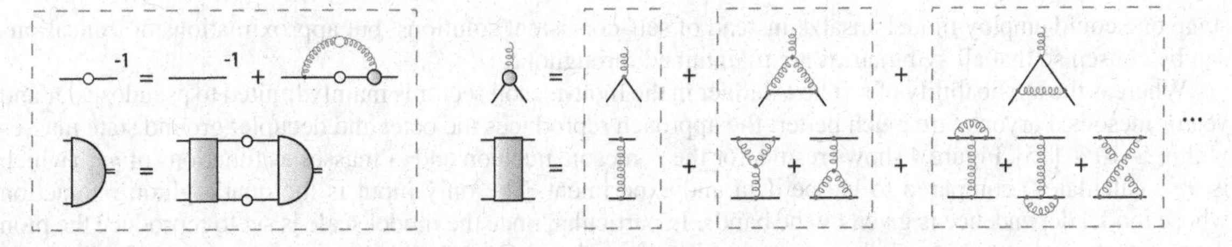

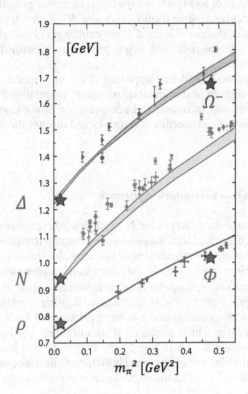

Fig. 3 Quark DSE, meson BSE, and symmetry-preserving relation between quark–gluon vertex and $q\bar{q}$ kernel

Fig. 4 ρ–meson [14], nucleon and Δ masses [8–10] calculated from their Bethe–Salpeter and Faddeev equations. *Stars* are PDG values and symbols with *error bars* are lattice data (see [8–10] for references)

contribution is a gluon exchange with a bare quark–gluon vertex ('rainbow-ladder'); the BSE will then produce a ladder of gluons upon iteration.

In combination with spontaneous chiral symmetry breaking, the axial WTI ensures that the pion is the massless Goldstone boson in the chiral limit [12]. Hence, if we solve the BSE for the pion from Fig. 3 we will get these features for free: the pion mass follows the Gell–Mann–Oakes–Renner relation for small quark masses ($m_\pi^2 \sim m_q$) and vanishes in the chiral limit where the current-quark mass m_q is zero. Another key ingredient is the quark–photon vertex: its inhomogeneous BSE (cf. Fig. 5) features the same $q\bar{q}$ kernel, so the resulting vertex respects the vector WTI which is a necessary prerequisite for electromagnetic gauge invariance and thus current conservation. In addition, the BSE automatically generates vector-meson poles in the transverse part of the vertex [13]. Since any form factor diagram ultimately couples the photon to the quarks through a dressed quark–photon vertex, this is the microscopic origin of the timelike resonance structure sketched in Fig. 1. (In rainbow-ladder ρ and ω are bound states without a width, hence one obtains a series of poles instead of bumps.)

Apparently there is a deep underlying connection between the ingredients of these equations. Figure 3 also makes clear that one cannot insert a crossed-ladder term without a consistent quark–gluon vertex; one cannot add a confinement potential or drop spin-orbit terms by hand; etc. Although in principle one is equipped with an exact set of equations, in practice one has to make concessions due to the complexity of the problem: at any stage one could employ model ansätze instead of self-consistent solutions, but approximations or truncations can be chosen so that all symmetries are maintained throughout.

Whereas the applicability of rainbow-ladder in the light-meson sector is mainly limited to pseudoscalar and vector mesons, baryons fare much better: the approach reproduces the octet and decuplet ground state masses within 5–10 % [15]. Figure 4 shows results for the ρ-meson, nucleon and Δ masses as functions of m_π^2 (which is also calculated) compared to lattice data and experiment. The only input is the quark–gluon interaction whose model dependence is given by the bands. In particular, once the model scale is set to reproduce the pion decay constant, there are no further parameters or approximations and all subsequent results are predictions.

Apart from mass spectra, a range of form factors have been calculated as well within this setup. Among them are nucleon, Δ and hyperon electromagnetic form factors, the $N \to \Delta\gamma$ transition, and nucleon axial form factors; see [16,17] and references therein. All these cases exhibit good overall agreement with experimental data (where available) and also lattice results at larger pion masses, with discrepancies at low Q^2 where pion-cloud effects become important.

We emphasize that the three-body Faddeev approach does not depend on explicit diquark degrees of freedom. Nevertheless, the assumption of dominant quark–quark correlations inside baryons is quite natural, and it can contribute much to our understanding and interpretation of nucleon transition form factors. In the following we will therefore review the properties of diquarks and discuss the diquark composition of nucleon resonances.

3 Nucleon Resonances in the Quark–Diquark Approach

Diquarks are not observable because they carry color. Nevertheless, there are several observations that support a quark–diquark interpretation of baryons. First, the presumably leading irreducible three-body force (the three-gluon vertex that connects three quarks) vanishes simply due the color traces [11], which seems to suggest that two-quark correlations are more important. Second, the interaction between two quarks is attractive in the color-$\bar{3}$ channel and supports the formation of diquarks as colored 'bound states' of quarks. Neglecting three-body terms, the Faddeev equation in Fig. 2 can be rewritten in a more familiar form that depends on the qq scattering matrix. Assuming that the latter is separable and can be approximated by a sum over diquark correlations, similarly to the $q\bar{q}$ scattering matrix which contains all meson poles, the Faddeev equation simplifies to a two-body problem—the quark–diquark BSE in Fig. 5. The baryon is then bound by quark exchange; gluons no longer appear explicitly but they are rather absorbed into the building blocks: the quark propagator, diquark amplitudes and diquark propagators. The electromagnetic current is constructed accordingly [18].

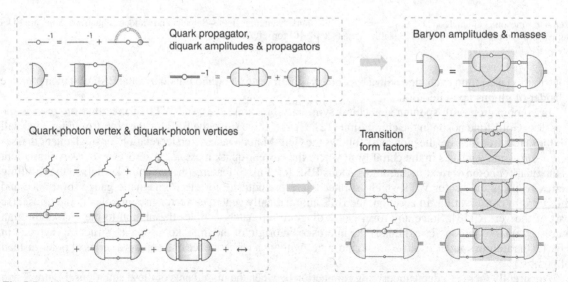

Fig. 5 Recipe for calculating transition form factors in the quark–diquark approach. The quark propagator, diquark amplitudes and diquark propagators form the input of the quark–diquark BSE. The current matrix element needs in addition the quark–photon vertex, the diquark–photon vertices and the seagull amplitudes. The ingredients are calculated from their DSEs and BSEs in rainbow-ladder truncation

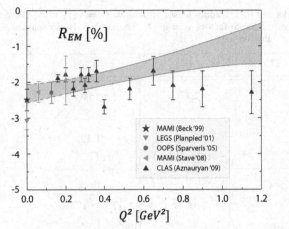 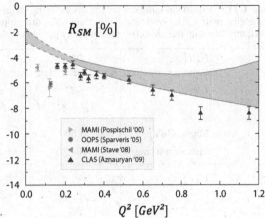

Fig. 6 Q^2-dependence of the electric and Coulomb quadrupole form-factor ratios R_{EM} and R_{SM} compared to experimental data [31]

Scalar and axialvector diquarks are the lightest ones, hence they are most important for describing the positive-parity nucleon and Δ baryons. Diquark masses have been calculated in rainbow-ladder [19] which produces actual diquark poles in the qq scattering matrix. Although their appearance is presumably a truncation artifact [20], it still suggests the presence of strong diquark correlations and allows one to compute diquark properties in close analogy to those of mesons from their BSEs.[1]

Quark-diquark models yield results for a range of baryon properties that are indeed similar to those obtained from the three-body equation. In addition to quark–diquark models based on relativistic quantum mechanics (see [21,22] and references therein), there are essentially three quantum-field theoretical variants of the approach that have been pursued in recent studies. One is the NJL/contact-interaction model from where qualitative statements and overall properties of the baryon spectrum have been extracted [24,25]. Another is the QCD-based model developed in [26,27] which implements more realistic models for the propagators, amplitudes and vertices and thereby also allows one to make quantitative predictions for form factors, such as in a recent calculation of the nucleon to Roper transition [28].

Here we will revisit and extend a third variant of the model, namely the approach of Ref. [29,30] which is outlined in Fig. 5. In that case the model input is replaced by DSE and BSE solutions, which provide a direct link to the microscopic building blocks since any modification in the fundamental quark–gluon interaction induces consistent changes up to the observables of interest. The calculated nucleon and Δ masses in that setup are indeed comparable to those in Fig. 4. As an example we quote the $N\gamma \to \Delta$ calculation of Ref. [31] in Fig. 6. The calculated form factor ratios R_{EM} and R_{SM} agree rather well with the experimental data, which can be partially traced to the presence of relativistic p waves in the nucleon and Δ amplitudes. On the other hand, the magnetic dipole form factor agrees with experiment only above $Q^2 \gtrsim 1$ GeV^2 and deviates by ~30% at $Q^2 = 0$, presumably due to missing pion-cloud effects. So far the approach has been applied to nucleon and Δ ground state properties, also including their elastic form factors and the $N \to \Delta\pi$ decay [29,30,32,33]. What are then the necessary steps to calculate negative-parity resonances such as the $N(1535)$ and $N(1520)$? Let us first collect some properties of the diquarks.

3.1 Diquark Properties

Diquarks are subject to the Pauli principle, so they must be totally antisymmetric under quark exchange. Concerning isospin, the diquark flavor wave functions are either antisymmetric ($I = 0$) or symmetric ($I = 1$). Since color is antisymmetric, the corresponding Dirac parts must be either antisymmetric ($I = 0$) or symmetric ($I = 1$) as well. Denoting the total onshell momentum by P (with $P^2 = -m^2$, where m is the respective

---

[1] Similarly, the four-body amplitude for a tetraquark is dominated by the lowest-lying two-body poles (the pseudoscalar mesons), which leads to a 'meson-molecule' interpretation for the light scalar mesons [23].

Table 1 $I(J^P)$ quantum numbers for the lowest-lying diquarks and their leading tensor structures. $C = \gamma_4\gamma_2$ is the charge-conjugation matrix. The entries in light (blue) color are suppressed due to factors $\omega = q \cdot P$. The rainbow-ladder masses are quoted together with the respective meson parity partners (color figure online)

$I(J^P)$	0^+	1^+	0^-	1^-
$I = 0$	$\gamma_5 C$	$\omega\,\gamma^\mu C$	C	$\gamma^\mu\gamma_5 C$
	$i\slashed{P}\gamma_5 C$	$\omega\,[\gamma^\mu, i\slashed{P}]\,C$	$\omega\,i\slashed{P}\,C$	$\omega\,[\gamma^\mu, i\slashed{P}]\,\gamma_5 C$
Mass [GeV]	0.80	–	1.01	1.12
Meson partner	$\pi\,(0^{-+})$	exotic (1^{-+})	scalar (0^{++})	$a_1\,(1^{++})$
$I = 1$	$\omega\,\gamma_5 C$	$\gamma^\mu C$	$\omega\,C$	$\omega\,\gamma^\mu\gamma_5 C$
	$\omega\,i\slashed{P}\gamma_5 C$	$[\gamma^\mu, i\slashed{P}]\,C$	$i\slashed{P}\,C$	$[\gamma^\mu, i\slashed{P}]\,\gamma_5 C$
Mass [GeV]	–	1.00	–	1.12
Meson partner	exotic (0^{--})	$\rho\,(1^{--})$	(0^{+-})	$b_1\,(1^{+-})$

diquark mass) and the relative momentum by q, this entails

$$\Gamma(q, P) = \begin{cases} -\Gamma^T(-q, P) \dots I = 0 \\ \Gamma^T(-q, P) \dots I = 1 \end{cases} \tag{1}$$

for the Dirac parts of the $J^P = 0^\pm$ diquark amplitudes, where T is a matrix transpose. The same relations hold for the $J^P = 1^\pm$ amplitudes with the replacement $\Gamma(q, P) \to \Gamma^\mu(q, P)$. This leads to the classification in Table 1, where we display the two leading (s-wave) tensor structures for each case. The $0^\pm$ amplitudes depend on four tensors and the $1^\pm$ amplitudes on eight, but the remaining tensor components are suppressed because they carry relative momentum and thus higher orbital angular momentum. The two isospin states must differ by $\omega = q \cdot P$ to ensure the correct symmetry (the dressing functions are all even in ω) and the appearance of ω induces further suppression.

One can draw a close analogy between diquarks and mesons: each diquark amplitude can be mapped onto a respective meson by removing the charge-conjugation matrix C. The upper and lower entries in Table 1 then correspond to opposite C-parities for mesons. After taking traces, the rainbow-ladder BSEs for diquarks only differ by a factor 2 from their meson parity partners—diquarks are 'less bound' than mesons. The scalar diquarks are the lightest ones ($\sim$800 MeV), followed by axialvector ($\sim$1 GeV), pseudoscalar and vector diquarks. The meson analogues of $I(J^P) = 1(0^+)$, $0(1^+)$ and $1(0^-)$ are exotic and therefore the respective diquark masses are larger. In contrast to their meson counterparts, diquarks are rather sensitive to the quark–gluon interaction: in Table 1 we quote the central values for the model employed herein but the masses can vary by $100 \dots 150$ MeV in both directions [19]. There is another consequence of the diquark–meson analogy: whereas pseudoscalar and vector mesons are well described in rainbow-ladder, scalar and axialvector mesons come out too light [19,34]—which can be remedied with more sophisticated truncations [35,36]. Hence, the pseudoscalar and vector diquarks should inherit this behavior and produce negative-parity baryons that are also too light and acquire repulsive shifts beyond rainbow-ladder [24,25].

3.2 Baryons from Quarks and Diquarks

To assess the diquark content for the various nucleon resonances, we note that nucleons ($I = 1/2$) can feature both diquark isospins whereas Δ baryons ($I = 3/2$) can only consist of $I = 1$ diquarks. Moreover, baryons with $J = 3/2$ should be dominated by $J = 1$ diquarks because those with $J = 0$ require orbital angular momentum; and positive/negative-parity baryons should be dominated by positive/negative-parity diquarks. Hence we arrive at Fig. 7, which shows the $J^P = 1/2^\pm$, $3/2^\pm$ nucleon resonances with at least two stars in the PDG. Each box corresponds to a given $I(J^P)$ channel, with one ground state and further radial excitations. The presumably dominant diquark channels (scalar, axialvector, pseudoscalar or vector) are shown in bold font.

I \ J^P	$\frac{1}{2}^+$	$\frac{3}{2}^+$	$\frac{1}{2}^-$	$\frac{3}{2}^-$	$\cdots$
$\frac{1}{2}$	P_{11} **N(940)** **N(1440)** N(1710) N(1880) **sc, av,** ps, v	P_{13} **N(1720)** N(1900) **sc, av,** ps, v	S_{11} **N(1535)** **N(1650)** N(1895) sc, av, **ps, v**	D_{13} **N(1520)** **N(1700)** N(1875) sc, av, ps, v	
$\frac{3}{2}$	P_{31} **Δ(1910)** **av,** v	P_{33} **Δ(1232)** **Δ(1600)** **Δ(1920)** **av,** v	S_{31} **Δ(1620)** Δ(1900) av, **v**	D_{33} **Δ(1700)** Δ(1940) av, **v**	

Fig. 7 Nucleon resonances below 2 GeV [37] and expectations for their dominant diquark content

Table 2 Rainbow-ladder masses and diquark contributions (see text) for various nucleon resonances

	N	$N(1535)$	$\Delta(1232)$	$\Delta(1620)$	$\Delta(1700)$	$\Delta(1910)$
Mass (GeV)	0.95	1.21	1.28	1.42	1.46	1.60
sc	1	0.45				
av	0.37	0.56	1	−0.40	−0.10	0.39
ps	0.02	1				
v	0.03	0.27	−0.02	1	1	1

To test whether these expectations hold we applied the recipe in Fig. 5 (the details of the calculation will be given elsewhere). After solving the quark DSE, the various diquark amplitudes and propagators are calculated from their BSEs and quark loop integrals [29,30]. They subsequently enter in the quark–diquark BSE from where the baryon masses are determined. The results for the ground states are collected in Table 2. The {sc, av, ps, v} entries provide a measure for the importance of the various diquark channels: they denote the magnitude of the quark–diquark amplitudes' dressing functions corresponding to the leading tensor structures at vanishing relative momentum, all normalized to the strongest component. So far we have neglected the $0(1^-)$ vector diquark which would only contribute to nucleons but not to Δ baryons; 'v' therefore refers to the $1(1^-)$ diquark only.

Table 2 shows that the pseudoscalar and vector diquark contributions are indeed strongly suppressed for the nucleon and $\Delta(1232)$. The $N(1535)$, on the other hand, is dominated by pseudoscalar diquarks but the other channels (sc, av, v) are all sizeable and similarly important. Note that the masses of all states that are dominated by pseudoscalar or vector diquarks are severely underestimated compared to experiment. This is a consequence of what was discussed above: in analogy to scalar and axialvector mesons, the mass scales for pseudoscalar and vector diquarks should obtain repulsive shifts beyond rainbow-ladder and so would the masses of the baryons that they constitute. In the contact-interaction model of Refs. [24,25] a fictitious coupling strength was introduced into the BSEs for those channels to mimic beyond-rainbow-ladder effects and reproduce the $\rho - a_1$ splitting. If we adopted the same strategy here, then Table 2 suggests that such a single parameter could indeed move the $N(1535)$, $\Delta(1620)$, $\Delta(1700)$ and $\Delta(1910)$ masses into the ballpark of their experimental values. So far we did not obtain convergent solutions for the $N(1520)$ and $N(1720)$, which could be an artifact of the omission of $0(1^-)$ diquarks, or it might also signal a breakdown of the quark–diquark description for these states.

$N\gamma \rightarrow N(1535)$ *transition.* We conclude by returning to Fig. 1. The timelike vector-meson structure is contained in the same quark–photon vertex that appears in all current matrix elements, so we should expect similar

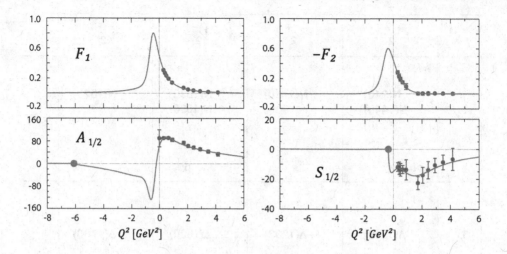

Fig. 8 CLAS data for the $p\gamma^* \to N(1535)$ transition form factors and helicity amplitudes [4], together with a toy parametrization including a ρ-meson bump. The helicity amplitudes carry units of 10^{-3} GeV$^{-1/2}$

features for nucleon transition form factors. These are traditionally discussed in terms of helicity amplitudes. In Fig. 8 we plot the JLab/CLAS data for the $N\gamma^* \to N(1535)$ transition form factors and corresponding helicity amplitudes. We define the form factors $F_1(Q^2)$ and $F_2(Q^2)$ from

$$J^\mu = i\,\bar{u}(p_f) \left[\frac{F_1(Q^2)}{m_N^2} t_{QQ}^{\mu\nu} \gamma^\nu + \frac{F_2(Q^2)}{2m_N} \frac{i}{2}[\gamma^\mu, \mathcal{Q}] \right] u(p_i), \qquad t_{AB}^{\mu\nu} = A \cdot B\, \delta^{\mu\nu} - B^\mu A^\nu \qquad (2)$$

to expose the transversality and analyticity properties of the offshell spin-1/2 transition vertex [38] from where the transition matrix element is derived (the standard Dirac-like form factor $F_1(Q^2)\,Q^2/m_N^2$ has a kinematic zero at the origin). For illustration we parametrize the form factors F_1 and F_2 by a simple ansatz including a single ρ-meson bump. Observe that the rich structure of the helicity amplitudes is essentially due to kinematic effects, including kinematic zeros at threshold and pseudothreshold $Q^2 = -(m_{N*} \pm m_N)^2$. Vice versa, it was noted in Ref. [39] that the Pauli-like form factor F_2 practically vanishes over a wide Q^2 domain and only rises at very low Q^2. This is in contrast to the usual multipole behavior of form factors and rather resembles what one would expect from chiral meson-cloud effects. Hence, to access the underlying properties of QCD from the data it is preferable to discuss the behavior of form factors, which are free of kinematic constraints, rather than that of helicity amplitudes.

Acknowledgments I would like to thank C. S. Fischer for a critical reading of the manuscript.

References

1. Klempt, E., Richard, J.-M.: Baryon spectroscopy. Rev. Mod. Phys. **82**, 1095 (2010)
2. Tiator, L., et al.: Electromagnetic excitation of nucleon resonances. Eur. Phys. J. Spec. Topics **198**, 141 (2011)
3. Aznauryan, I.G., et al.: Studies of nucleon resonance structure in exclusive meson electroproduction. Int. J. Mod. Phys. **E22**, 1330015 (2013)
4. Aznauryan, I., Burkert, V.: Electroexcitation of nucleon resonances. Prog. Part. Nucl. Phys. **67**, 1 (2012)
5. Roberts, C.D., Williams, A.G.: Dyson-Schwinger equations and their application to hadronic physics. Prog. Part. Nucl. Phys. **33**, 477 (1994)
6. Alkofer, R., von Smekal, L.: The Infrared behavior of QCD Green's functions. Phys. Rep. **353**, 281 (2001)
7. Fischer, C.S.: Infrared properties of QCD from Dyson–Schwinger equations. J. Phys. **G32**, R253 (2006)
8. Eichmann, G., Alkofer, R., Krassnigg, A., Nicmorus, D.: Nucleon mass from a covariant three-quark Faddeev equation. Phys. Rev. Lett. **104**, 201601 (2010)
9. Sanchis-Alepuz, H., et al.: Delta and Omega masses in a three-quark covariant Faddeev approach. Phys. Rev. **D84**, 096003 (2011)
10. Eichmann, G.: Nucleon electromagnetic form factors from the covariant Faddeev equation. Phys. Rev. **D84**, 014014 (2011)
11. Sanchis-Alepuz, H., Williams, R.: Hadronic observables from Dyson–Schwinger and Bethe–Salpeter equations. J. Phys. Conf. Ser. **631**(1), 012064 (2015)

12. Maris, P., Roberts, C.D., Tandy, P.C.: Pion mass and decay constant. Phys. Lett. **B420**, 267 (1998)
13. Maris, P., Tandy, P.C.: The Quark photon vertex and the pion charge radius. Phys. Rev. **C61**, 045202 (2000)
14. Maris, P., Tandy, P.C.: Bethe–Salpeter study of vector meson masses and decay constants. Phys. Rev. **C60**, 055214 (1999)
15. Sanchis-Alepuz, H., Fischer, C.S.: Octet and Decuplet masses: a covariant three-body Faddeev calculation. Phys. Rev. **D90**(9), 096001 (2014)
16. Alkofer, R., et al.: Electromagnetic baryon form factors in the Poincaré-covariant Faddeev approach. Hyperfine Interact. **234**(1), 149 (2015)
17. Sanchis-Alepuz, H., Fischer, C.S.: Hyperon elastic electromagnetic form factors in the space-like momentum region. arXiv:1512.00833 [hep-ph]
18. Oettel, M., Pichowsky, M., von Smekal, L.: Current conservation in the covariant quark diquark model of the nucleon. Eur. Phys. J. **A8**, 251 (2000)
19. Maris, P.: Effective masses of diquarks. Few Body Syst. **32**, 41 (2002)
20. Bender, A., Roberts, C.D., Von Smekal, L.: Goldstone theorem and diquark confinement beyond rainbow ladder approximation. Phys. Lett. **B380**, 7 (1996)
21. De Sanctis, M., Ferretti, J., Santopinto, E., Vassallo, A.: Electromagnetic form factors in the relativistic interacting quark-diquark model of baryons. Phys. Rev. **C84**, 055201 (2011)
22. Santopinto, E., Ferretti, J.: Strange and nonstrange baryon spectra in the relativistic interacting quark-diquark model with a Gürsey and Radicati-inspired exchange interaction. Phys. Rev. **C92**(2), 025202 (2015)
23. Eichmann, G., Fischer, C.S., Heupel, W.: The light scalar mesons as tetraquarks. Phys. Lett. **B753**, 282 (2016)
24. Roberts, H.L., et al.: Masses of ground and excited-state hadrons. Few Body Syst. **51**, 1 (2011)
25. Chen, C., et al.: Spectrum of hadrons with strangeness. Few Body Syst. **53**, 293 (2012)
26. Oettel, M., Hellstern, G., Alkofer, R., Reinhardt, H.: Octet and decuplet baryons in a covariant and confining diquark-quark model. Phys. Rev. **C58**, 2459 (1998)
27. Cloet, I.C., Eichmann, G., El-Bennich, B., Klahn, T., Roberts, C.D.: Survey of nucleon electromagnetic form factors. Few Body Syst. **46**, 1 (2009)
28. Segovia, J., et al.: Completing the picture of the Roper resonance. Phys. Rev. Lett. **115**(17), 171801 (2015)
29. Eichmann, G.: Hadron Properties from QCD Bound-State Equations. Ph.D. thesis, University of Graz (2009), arXiv:0909.0703 [hep-ph]
30. Eichmann, G., Cloet, I.C., Alkofer, R., Krassnigg, A., Roberts, C.D.: Toward unifying the description of meson and baryon properties. Phys. Rev. **C79**, 012202 (2009)
31. Eichmann, G., Nicmorus, D.: Nucleon to Delta electromagnetic transition in the Dyson–Schwinger approach. Phys. Rev. **D85**, 093004 (2012)
32. Nicmorus, D., Eichmann, G., Alkofer, R.: Delta and Omega electromagnetic form factors in a Dyson–Schwinger/Bethe–Salpeter approach. Phys. Rev. **D82**, 114017 (2010)
33. Mader, V., et al.: Hadronic decays of mesons and baryons in the Dyson–Schwinger approach. Phys. Rev. **D84**, 034012 (2011)
34. Krassnigg, A.: Survey of $J = 0, 1$ mesons in a Bethe–Salpeter approach. Phys. Rev. **D80**, 114010 (2009)
35. Chang, L., Roberts, C.D.: Sketching the Bethe–Salpeter kernel. Phys. Rev. Lett. **103**, 081601 (2009)
36. Williams, R., Fischer, C.S., Heupel, W.: Light mesons in QCD and unquenching effects from the 3PI effective action. arXiv:1512.00455 [hep-ph]
37. Olive, K.A., et al.: Review of particle physics. Chin. Phys. **C38**, 090001 (2014)
38. Eichmann, G., Ramalho, G.: (In preparation)
39. Ramalho, G., Tsushima, K.: A simple relation between the $\gamma N \rightarrow N(1535)$ helicity amplitudes. Phys. Rev. **D84**, 051301 (2011)

Few-Body Syst (2016) 57:993–1000
DOI 10.1007/s00601-016-1140-y

$\gamma_{\mathrm{v}} \mathrm{NN}^*$ Electrocouplings in Dyson–Schwinger Equations

Jorge Segovia

Received: 5 February 2016 / Accepted: 26 July 2016 / Published online: 13 August 2016
© Springer-Verlag Wien 2016

Abstract A symmetry preserving framework for the study of continuum Quantum Chromodynamics (QCD) is obtained from a truncated solution of the QCD equations of motion or QCD's Dyson–Schwinger equations (DSEs). A nonperturbative solution of the DSEs enables the study of, e.g., hadrons as composites of dressed-quarks and dressed-gluons, the phenomena of confinement and dynamical chiral symmetry breaking (DCSB), and therefrom an articulation of any connection between them. It is within this context that we present a unified study of Nucleon, Delta and Roper elastic and transition electromagnetic form factors, and compare predictions made using a framework built upon a Faddeev equation kernel and interaction vertices that possess QCD-like momentum dependence with results obtained using a symmetry-preserving treatment of a vector ⊗ vector contact-interaction. The comparison emphasises that experiment is sensitive to the momentum dependence of the running coupling and masses in QCD and highlights that the key to describing hadron properties is a veracious expression of DCSB in the bound-state problem.

1 Introduction

Nonperturbative QCD poses significant challenges. Primary amongst them is a need to chart the behaviour of QCD's running coupling and masses into the domain of infrared momenta. Contemporary theory is incapable of solving this problem alone but a collaboration with experiment holds a promise for progress. This effort can benefit substantially by exposing the structure of nucleon excited states and measuring the associated transition form factors at large momentum transfer [1]. Large momenta are needed in order to pierce the meson-cloud that, often to a significant extent, screens the dressed-quark core of all baryons [2,3]; and it is via the Q^2 evolution of form factors that one gains access to the running of QCD's coupling and masses from the infrared into the ultraviolet [4,5].

It is within the context just described that we have performed a simultaneous treatment of elastic and transition form factors involving the Nucleon, Delta and Roper baryons in Refs. [6–10]. In order to address the issue of charting the behaviour of the running coupling and masses in the strong interaction sector of the Standard Model, we use a widely-accepted leading-order (rainbow-ladder) truncation of QCD's Dyson–Schwinger equations [11–13] and compare results between a QCD-based framework and a confining, symmetry-preserving treatment of a vector ⊗ vector contact interaction.

A unified QCD-based description of elastic and transition form factors involving the nucleon and its resonances has acquired additional significance owing to substantial progress in the extraction of transition electrocouplings, $g_{\upsilon NN^*}$, from meson electroproduction data, obtained primarily with the CLAS detector at

---

This article belongs to the special issue "Nucleon Resonances".

J. Segovia (✉)
Physik-Department, Technische Universität München, James-Franck-Str. 1, 85748 Garching, Germany
E-mail: jorge.segovia@tum.de

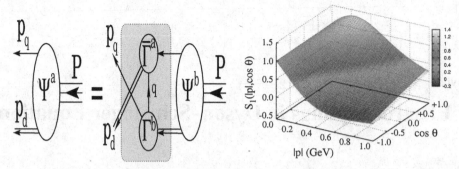

Fig. 1 *Left panel* Poincaré covariant Faddeev equation. Ψ is the Faddeev amplitude for a baryon of total momentum $P = p_q + p_d$, where $p_{q,d}$ are, respectively, the momenta of the quark and diquark within the bound-state. The *shaded area* demarcates the Faddeev equation kernel: *single line* dressed-quark propagator; Γ, diquark correlation amplitude; and *double line* diquark propagator. *Right panel* Dominant piece in the nucleon's eight-component Poincaré-covariant Faddeev amplitude: $S_1(|p|, \cos\theta)$. In the nucleon rest frame, this term describes that piece of the quark–scalar-diquark relative momentum correlation which possesses zero intrinsic quark–diquark orbital angular momentum, i.e. $L = 0$, before the propagator lines are reattached to form the Faddeev wave function. Referring to Fig. 1, $p = P/3 - p_q$ and $\cos\theta = p \cdot P/\sqrt{p^2 P^2}$. The amplitude is normalised such that its U_0 Chebyshev moment is unity at $|p| = 0$

the Thomas Jefferson National Accelerator Facility (JLab). The electrocouplings of all low-lying N^* states with mass <1.6 GeV have been determined via independent analyses of $\pi^+ n$, $\pi^0 p$ and $\pi^+ \pi^- p$ exclusive channels [14,15]; and preliminary results for the g_{vNN^*} electrocouplings of most high-lying N^* states with masses below 1.8 GeV have also been obtained from CLAS meson electroproduction data [1,16].

2 Baryon Structure

Dynamical chiral symmetry breaking (DCSB) is a theoretically-established feature of QCD and the most important mass generating mechanism for visible matter in the Universe, being responsible for approximately 98 % of the proton's mass. A fundamental expression of DCSB is the behaviour of the quark mass-function, $M(p)$. This appears in the dressed-quark propagator which may be obtained as a solution to the most famous and simple QCD's Dyson–Schwinger equation: the gap equation [13]. The nontrivial character of the mass function arises primarily because a dense cloud of gluons comes to clothe a low-momentum quark. It explains how an almost-massless parton-like quark at high energies transforms, at low energies, into a constituent-like quark with an effective mass of around 350 MeV.

DCSB ensures the existence of nearly-massless pseudo-Goldstone modes (pions). Another equally important consequence of DCSB is less well known. Namely, any interaction capable of creating pseudo-Goldstone modes as bound-states of a light dressed-quark and -antiquark, and reproducing the measured value of their leptonic decay constants, will necessarily also generate strong colour-antitriplet correlations between any two dressed quarks contained within a baryon. Although a rigorous proof within QCD cannot be claimed, this assertion is based upon an accumulated body of evidence, gathered in two decades of studying two- and three-body bound-state problems in hadron physics (the interested reader is referred to the discussion in Ref. [9] and to Refs. [17–31] cited therein). No realistic counter examples are known; and the existence of such diquark correlations is also supported by simulations of lattice QCD [32].

The existence of diquark correlations considerably simplifies analyses of the three valence-quark scattering problem and hence baryon bound states because it reduces that task to solving a Poincaré covariant Faddeev equation depicted in the left panel of Fig. 1. Two main contributions appear in the binding energy: (1) the formation of tight diquark correlations and (2) the quark exchange depicted in the shaded area of the left panel of Fig. 1.[1] This exchange ensures that diquark correlations within the baryon are fully dynamical: no quark holds a special place because each one participates in all diquarks to the fullest extent allowed by its quantum numbers. Attending to the quantum numbers of the nucleon and Roper, scalar-isoscalar and pseudovector-isotriplet diquark correlations are dominant. For the Δ-baryon, only the pseudovector-isotriplet ones are present.

---

[1] Whilst an explicit three-body term might affect fine details of baryon structure, the dominant effect of non-Abelian multi-gluon vertices is expressed in the formation of diquark correlations [33].

The quark + diquark structure of the nucleon is elucidated in the right panel of Fig. 1, which depicts the leading component of its Faddeev amplitude: with the notation of Ref. [8], $S_1(|p|, \cos\theta)$, computed using the Faddeev kernel described therein. The most interesting features are: (1) there is strong variation with respect to both arguments; (2) support is concentrated in the forward direction, $\cos\theta > 0$, so that alignment of p and P is favoured; and (3) in the antiparallel direction, $\cos\theta < 0$, the most probable situation is that in which $|p| = 0$, i.e. $p_q \sim P/3$, $p_d \sim 2P/3$.

3 The $\gamma^*N \to$ *Nucleon* Transition

The strong diquark correlations must be evident in many physical observables. We focus our attention on the flavour separated versions of the Dirac a Pauli form factors of the nucleon. The upper panels of Fig. 2 display the proton's flavour separated Dirac and Pauli form factors. The salient features of the data are: the d-quark contribution to F_1^p is far smaller than the u-quark contribution; $F_2^d/\kappa_d > F_2^u/\kappa_u$ on $x < 2$ but this ordering is reversed on $x > 2$; and in both cases the d-quark contribution falls dramatically on $x > 3$ whereas the u-quark contribution remains roughly constant. Our calculations are in semi-quantitative agreement with the empirical data.

It is natural to seek an explanation for the pattern of behaviour in the upper panels of Fig. 2. We have mentioned that the proton contains scalar and pseudovector diquark correlations. The dominant piece of its Faddeev wave function is $u[ud]$; namely, a u-quark in tandem with a $[ud]$ scalar correlation, which produces 62% of the proton's normalisation. If this were the sole component, then photon–d-quark interactions within the proton would receive a $1/x$ suppression on $x > 1$, because the d-quark is sequestered in a soft correlation, whereas a spectator u-quark is always available to participate in a hard interaction. At large $x = Q^2/M_N^2$, therefore, scalar diquark dominance leads one to expect $F^d \sim F^u/x$. Available data are consistent with this prediction but measurements at $x > 4$ are necessary for confirmation.

Consider now the ratio of proton electric and magnetic form factors, $R_{EM}(Q^2) = \mu_p G_E(Q^2)/G_M(Q^2)$, $\mu_p = G_M(0)$. A clear conclusion from the lower-left panel of Fig. 2 is that pseudovector diquark correlations have little influence on the momentum dependence of $R_{EM}(Q^2)$. Their contribution is indicated by the dotted (blue) curve, which was obtained by setting the scalar diquark component of the proton's Faddeev amplitude to zero and renormalising the result to unity at $Q^2 = 0$. As apparent from the dot-dashed (red) curve, the evolution of $R_{EM}(Q^2)$ with Q^2 is primarily determined by the proton's scalar diquark component. As we have explained above, in this component, the valence d-quark is sequestered inside the soft scalar diquark correlation so that the only objects within the nucleon which can participate in a hard scattering event are the valence u-quarks. The scattering from the proton's valence u-quarks is responsible for the momentum dependence of $R_{EM}(Q^2)$. However, the dashed (green) curve in the lower-left panel of Fig. 2 reveals something more, i.e. components of the nucleon associated with quark–diquark orbital angular momentum $L \geq 1$ in the nucleon rest frame are critical in explaining the data. Notably, the presence of such components is an inescapable consequence of the self-consistent solution of a realistic Poincaré-covariant Faddeev equation for the nucleon.

It is natural now to consider the proton ratio: $R_{21}(x) = xF_2(x)/F_1(x)$, $x = Q^2/M_N^2$, drawn in the lower-right panel of Fig. 2. As with R_{EM}, the momentum dependence of $R_{21}(x)$ is principally determined by the scalar diquark component of the proton. Moreover, the rest-frame $L \geq 1$ terms are again seen to be critical in explaining the data: the behaviour of the dashed (green) curve highlights the impact of omitting these components.

4 The $\gamma^*N \to$ *Delta* Transition

The electromagnetic $\gamma^*N \to \Delta$ transition is described by three Poincaré-invariant form factors [21]: magnetic-dipole, G_M^*, electric quadrupole, G_E^*, and Coulomb (longitudinal) quadrupole, G_C^*; that can be extracted in the Dyson–Schwinger approach by a sensible set of projection operators [22]. The following ratios

$$R_{EM} = -\frac{G_E^*}{G_M^*}, \qquad R_{SM} = -\frac{|\mathbf{Q}|}{2m_\Delta}\frac{G_C^*}{G_M^*}, \tag{1}$$

are often considered because they can be read as measures of the deformation of the hadrons involved in the reaction and how such deformation influences the structure of the transition current.

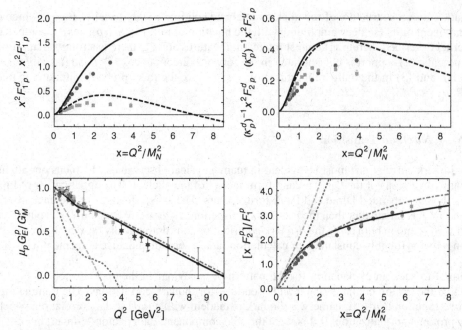

Fig. 2 *Upper-left panel* Flavour separation of the proton's Dirac form factor as a function of $x = Q^2/M_N^2$. The results have been obtained using a framework built upon a Faddeev equation kernel and interaction vertices that possess QCD-like momentum dependence. The *solid-curve* is the u-quark contribution, and the *dashed-curve* is the d-quark contribution. Experimental data taken from Ref. [20] and references therein: *circles* u-quark; and *squares* d-quark. *Upper-right panel* Same for Pauli form factor. *Lower-left panel* Computed ratio of proton electric and magnetic form factors. *Curves solid (black)* full result, determined from the complete proton Faddeev wave function and current; *dot-dashed (red)*—momentum-dependence of scalar-diquark contribution; *dashed (green)*— momentum-dependence produced by that piece of the scalar diquark contribution to the proton's Faddeev wave function which is purely S-wave in the rest-frame; *dotted (blue)*—momentum-dependence of pseudovector diquark contribution. All partial contributions have been renormalised to produce unity at $Q^2 = 0$. Data: *circles (blue)* [34]; *squares (green)* [17]; *asterisks (brown)* [18]; and *diamonds (purple)* [19]. *Lower-right panel* Proton ratio $R_{21}(x) = x F_2(x)/F_1(x)$, $x = Q^2/M_N^2$. The legend for the curves is the same than that of the *lower-left panel*. Experimental data taken from Ref. [20]. (colour figure online)

In considering the behaviour of the $\gamma^* N \rightarrow \Delta$ transition form factors, it is useful to begin by recapitulating upon a few facts. Note then that in analyses of baryon electromagnetic properties, using a quark model framework which implements a current that transforms according to the adjoint representation of spin-flavour $SU(6)$, one finds simple relations between magnetic-transition matrix elements [23,24]:

$$\langle p|\mu|\Delta^+\rangle = +\langle n|\mu|\Delta^0\rangle , \quad \langle p|\mu|\Delta^+\rangle = -\sqrt{2}\langle n|\mu|n\rangle ; \tag{2}$$

i.e., the magnetic components of the $\gamma^* p \rightarrow \Delta^+$ and $\gamma^* n \rightarrow \Delta^0$ are equal in magnitude and, moreover, simply proportional to the neutron's magnetic form factor. Furthermore, both the nucleon and Δ are S-wave states (neither is deformed) and hence $G_E^* \equiv 0 \equiv G_C^*$.

The second entry in Eq. (2) is consistent with perturbative QCD (pQCD) [25] in the following sense: both suggest that $G_M^{*p}(Q^2)$ should decay with Q^2 at the same rate as the neutron's magnetic form factor, which is dipole-like in QCD. It is often suggested that this is not the case empirically [26,31]. However, as argued elsewhere [6,7], such claims arise from a confusion between the form factors defined in the Ash et al. [27] and Jones et al. [21] conventions:

$$G_{M,Ash}^*(Q^2) = \frac{G_M^*(Q^2)}{[1 + Q^2/(m_\Delta + m_N)^2]^{1/2}}, \tag{3}$$

where $G_M^*(Q^2)$ is the Jones–Scadron form factor.

In addition, helicity conservation arguments within the context of pQCD enable one to make [25] the follow predictions for the ratios in Eq. (4):

$$R_{EM} \overset{Q^2 \to \infty}{=} 1 , \quad R_{SM} \overset{Q^2 \to \infty}{=} \text{constant.} \tag{4}$$

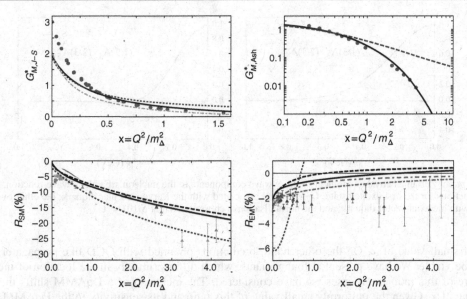

Fig. 3 *Upper-left panel* $G^*_{M,J-S}$ result obtained with QCD-based interaction (*solid, black*) and with contact-interaction (CI) (*dotted, blue*); The *green dot-dashed curve* is the dressed-quark core contribution inferred using SL-model [28]. *Upper-right panel* $G^*_{M,Ash}$ result obtained with QCD-based interaction (*solid, black*) and with CI (*dotted, blue*). *Lower-left panel* R_{SM} prediction of QCD-based kernel including dressed-quark anomalous magnetic moment (DqAMM) (*black, solid*), nonincluding DqAMM (*black, dashed*), and CI result (*dotted, blue*). *Lower-right panel* R_{EM} prediction obtained with QCD-kindred framework (*solid, black*); same input but without DqAMM (*dashed, black*); these results renormalised (by a factor of 1.34) to agree with experiment at $x = 0$ (*dot-dashed, red* zero at $x \approx 14$; and *dot-dash-dashed, red*, zero at $x \approx 6$); and CI result (*dotted, blue*). The data in the panels are from references that can be found in [8] (colour figure online)

These predictions are in marked disagreement with the outcomes produced by $SU(6)$-based quark models: $R_{EM} \equiv 0 \equiv R_{SM}$. More importantly, they are inconsistent with available data [26,31].

The upper-left panel of Fig. 3 displays the magnetic transition form factor in the Jones–Scadron convention. Our prediction obtained with a QCD-based kernel agrees with the data on $x \gtrsim 0.4$, and a similar conclusion can be inferred from the contact interaction result. On the other hand, both curves disagree markedly with the data at infrared momenta. This is explained by the similarity between these predictions and the bare result determined using the Sato–Lee (SL) dynamical meson-exchange model [28]. The SL result supports a view that the discrepancy owes to omission of meson-cloud effects in the DSEs' computations. An exploratory study of the effect of pion-cloud contributions to the mass of the nucleon and the Δ-baryon has been performed within a DSEs' framework in Ref. [29].

Presentations of the experimental data associated with the magnetic transition form factor typically use the Ash convention. This comparison is depicted in the upper-right panel of Fig. 3. One can see that the difference between form factors obtained with the QCD-kindred and CI frameworks increases with the transfer momentum. Moreover, the normalized QCD-kindred curve is in fair agreement with the data, indicating that the Ash form factor falls unexpectedly rapidly mainly for two reasons. First: meson-cloud effects provide up-to 35 % of the form factor for $x \lesssim 2$; these contributions are very soft; and hence they disappear quickly. Second: the additional kinematic factor $\sim 1/\sqrt{Q^2}$ that appears between Ash and Jones-Scadron conventions and provides material damping for $x \gtrsim 2$.

Our predictions for the ratios in Eq. (1) are depicted in the lower panels of Fig. 3. The lower-left panel displays the Coulomb quadrupole ratio. Both the prediction obtained with QCD-like propagators and vertices and the contact-interaction result are broadly consistent with available data. This shows that even a contact-interaction can produce correlations between dressed-quarks within Faddeev wave-functions and related features in the current that are comparable in size with those observed empirically. Moreover, suppressing the dressed-quark anomalous magnetic moment (DqAMM) in the transition current has little impact. These remarks highlight that R_{SM} is not particularly sensitive to details of the Faddeev kernel and transition current.

This is certainly not the case with R_{EM}. The differences between the curves displayed in the lower-right panel in Fig. 3 show that this ratio is a particularly sensitive measure of diquark and orbital angular momentum correlations. The contact-interaction result is inconsistent with data, possessing a zero that appears

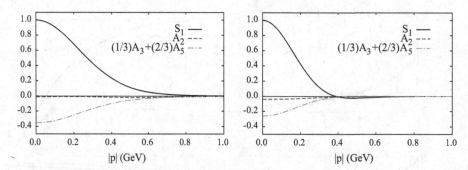

Fig. 4 *Left panel* Zeroth Chebyshev moment of all S-wave components in the nucleon's Faddeev wave function. *Right panel* Kindred functions for the first excited state. *Legend* S_1 is associated with the Baryon's scalar diquark; the other two curves are associated with the axial-vector diquark; and the normalisation is chosen such that $S_1(0) = 1$

at a rather small value of x. On the other hand, predictions obtained with QCD-like propagators and vertices can be viable. We have presented four variants, which differ primarily in the location of the zero that is a feature of this ratio in all cases we have considered. The inclusion of a DqAMM shifts the zero to a larger value of x. Given the uniformly small value of this ratio and its sensitivity to the DqAMM, we judge that meson-cloud affects must play a large role on the entire domain that is currently accessible to experiment.

5 The $\gamma^* N \rightarrow$ *Roper* Transition

Jefferson Lab experiments [15,30,31,35] have yielded precise nucleon-Roper ($N \rightarrow R$) transition form factors and thereby exposed the first zero seen in any hadron form factor or transition amplitude. It has also attracted much theoretical attention; but Ref. [10] provides the first continuum treatment of this problem using the power of relativistic quantum field theory. That study begins with a computation of the mass and wave function of the proton and its first radial excitation. The masses are (in GeV): $M_{\text{nucleon (N)}} = 1.18$ and $M_{\text{nucleon}-\text{excited (R)}} = 1.73$. These values correspond to the locations of the two lowest-magnitude $J^P = 1/2^+$ poles in the three-quark scattering problem. The associated residues are the Faddeev wave functions, which depend upon $(p^2, p \cdot P)$, where p is the quark–diquark relative momentum. Figure 4 depicts the zeroth Chebyshev moment of all S-wave components in that wave function. The appearance of a single zero in S-wave components of the Faddeev wave function associated with the first excited state in the three dressed-quark scattering problem indicates that this state is a radial excitation.

The empirical values of the pole locations for the first two states in the nucleon channel are [36]: 0.939 GeV and $1.36 - i\, 0.091$ GeV, respectively. At first glance, these values appear unrelated to those obtained within the DSEs framework. However, deeper consideration reveals [37,38] that the kernel in the Faddeev equation omits all those resonant contributions which may be associated with the meson-baryon final-state interactions that are resummed in dynamical coupled channels models in order to transform a bare-baryon into the observed state [3,36]. This Faddeev equation should therefore be understood as producing the dressed-quark core of the bound-state, not the completely-dressed and hence observable object. Crucial, therefore, is a comparison between the quark-core mass and the value determined for the mass of the meson-undressed bare-Roper in Ref. [36] which is 1.76 GeV.

The transition form factors are displayed in Fig. 5. The results obtained using QCD-derived propagators and vertices agree with the data on $x \gtrsim 2$. The contact-interaction result simply disagrees both quantitatively and qualitatively with the data. Therefore, experiment is evidently a sensitive tool with which to chart the nature of the quark–quark interaction and hence discriminate between competing theoretical hypotheses.

The mismatch between the DSE predictions and data on $x \lesssim 2$ is due to Meson-cloud contributions that are expected to be important on this domain. An inferred form of that contribution is provided by the dotted (green) curves in Fig. 5. These curves have fallen to just 20 % of their maximum value by $x = 2$ and vanish rapidly thereafter so that the DSE predictions alone remain as the explanation of the data. Importantly, the existence of a zero in F_2^* is not influenced by meson-cloud effects, although its precise location is.

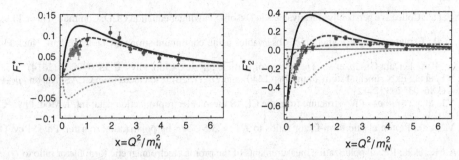

Fig. 5 *Left* Dirac transition form factor, $F_1^*(x)$, $x = Q^2/m_N^2$. *Solid (black) curve*, QCD-kindred prediction; *dot-dashed (red) curve*, contact-interaction result; *dotted (green) curve*, inferred meson-cloud contribution; and *dashed (blue) curve*, anticipated complete result. *Right* Pauli transition form factor, $F_2^*(x)$, with same legend. Data in *both panels*: *circles (blue)* [30]; *triangle (gold)* [35]; *squares (purple)* [15]; and *star (green)* [14] (colour figure online)

6 Conclusions

We have presented a unified study of nucleon, Delta and Roper elastic and transition form factors, and compare predictions made using a framework built upon a Faddeev equation kernel and interaction vertices that possess QCD-like momentum dependence with results obtained using a symmetry-preserving treatment of a vector ⊗ vector contact-interaction. The comparison emphasises that experiment is sensitive to the momentum dependence of the running coupling and masses in QCD and highlights that the key to describing hadron properties is a veracious expression of dynamical chiral symmetry breaking in the bound-state problem. Amongst our results, the following are of particular interest: The scaling behaviour of the electromagnetic ratios G_E^p/G_M^p and F_2^p/F_1^p is due to higher quark orbital angular momentum components in the nucleon wave function but also to strong diquark correlations. In fact, the presence of strong diquark correlations within the nucleon is sufficient to understand empirical extractions of the flavour-separated versions of Dirac and Pauli form factors. In connection with the $\gamma^*N \rightarrow \Delta$ transition, the momentum-dependence of the magnetic transition form factor, G_M^*, matches that of G_M^n once the momentum transfer is high enough to pierce the meson-cloud; and the electric quadrupole ratio is a keen measure of diquark and orbital angular momentum correlations, the zero in which is obscured by meson-cloud effects on the domain currently accessible to experiment. Finally, the Roper resonance is at heart of the nFucleon's first radial excitation, consisting of a dressed-quark core augmented by a meson cloud that reduces its mass by approximately 20 %. Our analysis shows that a meson-cloud obscures the dressed-quark core from long-wavelength probes, but that it is revealed to probes with $Q^2 \gtrsim 3m_N^2$.

Acknowledgments The material described in this contribution is drawn from work completed in collaboration with numerous excellent people, to all of whom I am greatly indebted. I would also like to thank V. Mokeev, R. Gothe, T.-S. H. Lee and G. Eichmann for insightful comments; and to express my gratitude to the organisers of the ECT* Workshop in Trento *Nucleon Resonances: From Photoproduction to High Photon Virtualities*, whose support helped my participation. I acknowledges financial support from the Alexander von Humboldt Foundation.

References

1. Aznauryan, I.G., et al.: Studies of nucleon resonance structure in exclusive meson electroproduction. Int. J. Mod. Phys. E **22**, 1330015 (2013)
2. Tiator, L., et al.: Electroproduction of nucleon resonances. Eur. Phys. J. A **19**, 55 (2004)
3. Kamano, H., et al.: Nucleon resonances within a dynamical coupled-channels model of πN and γN reactions. Phys. Rev. C **88**, 035209 (2013)
4. Cloët, I.C., et al.: Revealing dressed-quarks via the proton's charge distribution. Phys. Rev. Lett. **111**, 101803 (2013)
5. Chang, L., et al.: Pion electromagnetic form factor at spacelike momenta. Phys. Rev. Lett. **111**, 141802 (2013)
6. Segovia, J., et al.: Insights into the $\gamma^*N \rightarrow \Delta$ transition. Phys. Rev. C **88**, 032201 (2013)
7. Segovia, J., et al.: Elastic and transition form factors of the $\Delta(1232)$. Few Body Syst. **55**, 1–33 (2013)
8. Segovia, J., et al.: Nucleon and Δ elastic and transition form factors. Few Body Syst. **55**, 1185–1222 (2014)
9. Segovia, J., et al.: Understanding the nucleon as a Borromean bound-state. Phys. Lett. B **750**, 100–106 (2015)
10. Segovia, J., et al.: Completing the picture of the Roper resonance. Phys. Rev. Lett. **115**, 171801 (2015)
11. Chang, L., et al.: Selected highlights from the study of mesons. Chin. J. Phys. **49**, 955–1004 (2011)

12. Bashir, A., et al.: Collective perspective on advances in Dyson–Schwinger equation QCD. Commun. Theor. Phys. **58**, 79–134 (2012)
13. Cloët, I., et al.: Explanation and prediction of observables using continuum strong QCD. Prog. Part. Nucl. Phys. **77**, 1–69 (2014)
14. Olive, K.A., Particle Data Group, et al.: The review of particle physics. Chin. Phys. C **38**, 090001 (2014)
15. Mokeev, V.I., et al.: Experimental study of the $P_{11}(1440)$ and $D_{13}(1520)$ resonances from CLAS data on $ep \to e'\pi^+\pi^-p'$. Phys. Rev. C **86**, 035203 (2012)
16. Mokeev, V.I., et al.: Studies of N^* structure from the CLAS meson electroproduction data. Int. J. Mod. Phys. Conf. Ser. **26**, 1460080 (2013)
17. Punjabi, V., et al.: Proton elastic form-factor ratios to $Q^2 = 3.5\,\text{GeV}^2$ by polarization transfer. Phys. Rev. C **71**, 055202 (2005)
18. Puckett, A.J.R., et al.: Recoil polarization measurements of the proton electromagnetic form factor ratio to $Q^2 = 8.5\,\text{GeV}^2$. Phys. Rev. Lett. **104**, 242301 (2010)
19. Puckett, A.J.R., et al.: Final analysis of proton form factor ratio data at $Q^2 = 4.0, 4.8$ and $5.6\,\text{GeV}^2$. Phys. Rev. C **85**, 045203 (2012)
20. Cates, G.D., et al.: Flavor decomposition of the elastic nucleon electromagnetic form factors. Phys. Rev. Lett. **106**, 252003 (2011)
21. Jones, H.F., et al.: Multipole $\gamma N \Delta$ form-factors and resonant photoproduction and electroproduction. Ann. Phys. **81**, 1–14 (1973)
22. Eichmann, G., et al.: Nucleon to Delta electromagnetic transition in the Dyson–Schwinger approach. Phys. Rev. D **85**, 093004 (2012)
23. Beg, M.A.B., et al.: SU(6) and electromagnetic interactions. Phys. Rev. Lett. **13**, 514–517 (1964)
24. Buchmann, A.J.: Electromagnetic $N \to \Delta$ transition and neutron form-factors. Phys. Rev. Lett. **93**, 212301 (2004)
25. Carlson, Carl E.: Electromagnetic $N - \Delta$ transition at high Q^2. Phys. Rev. D **34**, 2704 (1986)
26. Aznauryan, I.: Results from the N^* program at JLab. J. Phys. Conf. Ser. **299**, 012008 (2011)
27. Ash, W., et al.: Measurement of the $\gamma N N^*$ form factor. Phys. Lett. B **24**, 165–168 (1967)
28. Julia-Diaz, B., et al.: Extraction and interpretation of $\gamma N \to \Delta$ form factors within a dynamical model. Phys. Rev. C **75**, 015205 (2007)
29. Sanchis-Alepuz, H., et al.: Pion cloud effects on baryon masses. Phys. Lett. B **733**, 151–157 (2014)
30. Aznauryan, I.G., et al.: Electroexcitation of nucleon resonances from CLAS data on single pion electroproduction. Phys. Rev. C **80**, 055203 (2009)
31. Aznauryan, I.G., et al.: Electroexcitation of nucleon resonances. Prog. Part. Nucl. Phys. **67**, 1–54 (2012)
32. Alexandrou, C., et al.: Evidence for diquarks in lattice QCD. Phys. Rev. Lett. **97**, 222002 (2006)
33. Eichmann, G., et al.: Nucleon mass from a covariant three-quark Faddeev equation. Phys. Rev. Lett. **104**, 201601 (2010)
34. Gayou, O., et al.: Measurements of the elastic electromagnetic form-factor ratio $mu_p G_{Ep}/G_{Mp}$ via polarization transfer. Phys. Rev. C **64**, 038202 (2001)
35. Dugger, M., et al.: π^+ photoproduction on the proton for photon energies from 0.725 to 2.875 GeV. Phys. Rev. C **79**, 065206 (2009)
36. Suzuki, N., et al.: Disentangling the dynamical origin of P_{11} nucleon resonances. Phys. Rev. Lett. **104**, 042302 (2010)
37. Eichmann, G., et al.: Perspective on rainbow-ladder truncation. Phys. Rev. C **77**, 042202 (2008)
38. Eichmann, G., et al.: Toward unifying the description of meson and baryon properties. Phys. Rev. C **79**, 012202 (2009)

Few-Body Syst (2016) 57:1059–1065
DOI 10.1007/s00601-016-1149-2

Si-xue Qin

Comments on Formulating Meson Bound-State Equations Beyond Rainbow-Ladder Approximation

Received: 16 February 2016 / Accepted: 6 August 2016 / Published online: 27 August 2016
© Springer-Verlag Wien 2016

Abstract We study mesons through solving the coupled system of the gap equation for the quark propagator and the Bethe–Salpeter equation for the meson wavefunction. The gap equation and Bethe–Salpeter equation are in fact members of infinitely coupled Dyson–Schwinger equations of Green functions of QCD. To make it solvable, the system must be truncated. The simplest rainbow-ladder truncation is widely used but shows drawbacks in many aspects. To improve the simplest truncation, we analyze symmetries of the fundamental theory and solve the corresponding Ward–Green–Takahashi identities. Then, the elements of the coupled system, i.e., the quark-gluon vertex and the quark-antiquark scattering kernel, can be constructed accordingly.

1 Introduction

In particle physics, hadrons, e.g., mesons and baryons, which contribute to the most visible mass of the universe, consist of fundamental blocks, i.e., quarks and gluons. Quarks and gluons are held together by the strong force which is described by quantum chromodynamics (QCD)—the strong interaction sector of the Standard Model. Mesons and baryons have two and three valence quarks, respectively. In recent years, new hadron states which contain more than three valence quarks have been discovered, for instance, the tetraquark and pentaquark states. On the other hand, numerous experiments have been devoted to the study of hadron structures, e.g., the measurements of the electromagnetic form factors of nucleon started around 50 years ago and a lot of experiments are investigating hadron structures in major facilities. For example, the 12 GeV upgrade at JLab is planned to provide new and high precision data on the nucleon generalized parton distribution functions and the pion electromagnetic form factor at high Q^2. Therefore, as the fundamental theory of the dynamics of quarks and gluons, QCD must answer two questions: what states are possible and how are they constituted?

The fascinating features of QCD, i.e., dynamical chiral symmetry breaking (DCSB) and confinement, are the key to understand hadron properties from the spectroscopy to the structure determination. But none of these features can be understood by perturbation QCD. The greatest challenge is to solve QCD nonperturbatively. Lattice QCD which directly evaluates the path integral of quantum fields on discrete spacetime lattice by Monte-Carlo techniques plays an important role for the study of nonperturbative QCD. Lattice QCD is a first-principal method because it recovers QCD in the continuum limit. Lattice QCD has already had impressive successes, e.g., the precise spectrum of ground state hadrons in the physical point—the calculated mass of the proton has an error less than 2 % of the empirical value [5]. However, lattice QCD still has its own difficulties. Therefore, it is necessary to work with an approach which can directly handle the degrees of freedom of quarks and gluons and also have a solid connection to QCD.

This article belongs to the special issue "Nucleon Resonances".

S. Qin (✉)
Physics Division, Argonne National Laboratory, Argonne, IL 60439, USA
E-mail: sqin@anl.gov

The Dyson–Schwinger equations (DSEs) [15] are an infinite tower of coupled integral equations which involve Green functions of all orders. The DSEs are also called equation of motion of quantum field theory (QFT). Note that the DSEs are nonperturbative in nature since Green functions all are fully dressed. To solve QCD nonperturbatively is equivalent to solving its DSEs. Because of the infinitely-coupling feature, it is unavoidable to truncate the DSEs at some point for any practical calculations. Truncation must be an eternal topic in the DSE framework. In this paper, we will limit ourselves to the meson case where we only face the lowest-order DSEs: the gap equation and the Bethe–Salpeter equation. Their coupled system involves three unknown elements: the gluon propagator, the quark-gluon vertex, and the quark-antiquark scattering kernel. Among the elements, the gluon propagator can be inputed with a solution of the DSE of the gauge sector [2] or a model inspired by lattice QCD [11]. By analyzing symmetries of the theory and solving corresponding Ward–Green–Takahashi identities (WGTIs), the main task of this paper is to construct the vertex and the kernel, of course, beyond the simplest approximation.

This paper is organized as follows: in Sect. 2 we introduce the DSE formalism which we are interested in and their form in the simplest approximation, i.e., the rainbow-ladder approximation; in Sect. 3 we discuss the structure of the quark-gluon vertex in Abelian approximation; in Sect. 4 we present a scheme to construct the quark-antiquark scattering kernel; at last we summarize and discuss potential applications.

2 Dyson–Schwinger Equations

The derivation of the Dyson–Schwinger equations (DSEs) has a standard procedure which can be found in a number of text books or review articles (see Ref. [15], and references therein). Here we will not go through details but briefly demonstrate how the DSEs emerge from the least action principle as the equation of motion of quantum field theory. Letting ϕ denote all fields of a given theory, the action $S[\phi]$ then is a functional of ϕ. Introducing external sources for the fields, the action is written as

$$S[\phi, j] = S[\phi] + \int d^4x \, j(x)\phi(x). \tag{1}$$

Then the QFT version of the least action principle is expressed as

$$0 = \left\langle \frac{\delta S[\phi, j]}{\delta \phi} \right\rangle := \int \mathcal{D}\phi \, \frac{\delta S[\phi, j]}{\delta \phi} \, \exp\left(iS[\phi, j]\right), \tag{2}$$

where the average over the path integral differs from the classic extreme condition $\delta S[\phi, j]/\delta \phi = 0$. Eq. (2) is the mother equation of the DSEs, since equations of all order Green functions can be derived accordingly.

Inserting the Lagrangian of QCD into the mother equation Eq. (2), we obtain the gap equation for the quark propagator as $(S(p) = 1/[i\gamma \cdot p A(p^2) + B(p^2)])$

$$S^{-1}(p) = S_0^{-1}(p) + \int \frac{d^4q}{(2\pi)^4} g^2 D_{\mu\nu}(p - q) \frac{\lambda_a}{2} \gamma_\mu S(q) \frac{\lambda_a}{2} \Gamma_\nu(q, p), \tag{3}$$

where $S_0^{-1}(p) = i\gamma \cdot p + m$ is the bare quark propagator; $D_{\mu\nu}$ is the gluon propagator; and Γ_μ is the dressed quark-gluon vertex. In Feynman diagram, the gap equation reads:

$$\tag{4}$$

where gray circular blobs denote dressed propagators or vertices and black dots denote bare propagators or vertices. In the above equation, the loop term is called quark self-energy. Similarly, projecting the quark-antiquark four-point Green function onto a channel with specified J^P quantum number, we obtain that the resulted vertex satisfies the inhomogeneous Bethe–Salpeter (BS) equation:

$$[\Gamma_{J^P}(k, P)]_{\alpha\beta} = [\gamma_{J^P}]_{\alpha\beta} + \int \frac{d^4q}{(2\pi)^4} [K^{(2)}(k, q, P)]_{\alpha\alpha';\beta'\beta} [S(q_+)\Gamma_{J^P}(q, P)S(q_-)]_{\alpha'\beta'}, \tag{5}$$

where $q_\pm = q \pm \eta_\pm P$ with $\eta_+ + \eta_- = 1$; $\gamma_{J^P} = \mathbf{1}, \gamma_5, \gamma_\mu, \gamma_5\gamma_\mu$, and etc., is the inhomogeneous driving term corresponding to the J^P quantum number; and $K^{(2)}$ is the full quark-antiquark scattering kernel kernel. In Feynman diagram, the BS equation reads:

$$\tag{6}$$

A meson state with a given J^P quantum number appears as a simple pole in the corresponding projected vertex. Equating the residues in both sides of the inhomogeneous BS equation (5), we obtain that the BS amplitude of the meson satisfies the homogeneous BS equation:

$$[\tilde{\Gamma}_{J^P}(k, P)]_{\alpha\beta} = \int \frac{d^4q}{(2\pi)^4} [K^{(2)}(k, q, P)]_{\alpha\alpha';\beta'\beta}[S(q_+)\tilde{\Gamma}_{J^P}(q, P)S(q_-)]_{\alpha'\beta'}, \tag{7}$$

where the total momentum P yields the on-shell condition $P^2 = -M^2$.

The coupled system which consists of the gap equation and the BS equation involves three unknown elements: the gluon propagator, the quark-gluon vertex, and the BS kernel. Once these elements are specified, the system can be solved self-consistently. The simplest truncation approximates the quark-gluon vertex and the BS kernel as follows:

$$\Gamma_\nu(q, p) = \gamma_\nu, \tag{8}$$

and

$$[K^{(2)}(k, q, P)]_{\alpha\alpha';\beta'\beta} = -g^2 D_{\mu\nu}(k - q) \left[\frac{\lambda_a}{2}\gamma_\mu\right]_{\alpha\alpha'} \left[\frac{\lambda_a}{2}\gamma_\nu\right]_{\beta'\beta}. \tag{9}$$

Inserting the approximation into Eqs. (3) and (5), the quark propagator S is expressed as a sum of infinitely many rainbow-like Feynman diagrams and the vertex Γ_{J^P} as a sum of infinitely many ladder-like Feynman diagrams. Thus the simplest truncation is also called rainbow-ladder (RL) approximation. As we shall see, the RL approximation preserves the chiral symmetry. Although numerous successful results are obtained in the RL approximation [8,9], drawbacks are also exposed in many aspects [4,12]. Therefore, in the following sections, we will focus on constructing the quark-gluon vertex and the BS kernel beyond the RL approximation.

3 Quark-Gluon Vertex

In order to construct the quark-gluon vertex, one can in principle solve its own Dyson–Schwinger equation. But the equation itself is extremely complicated, which involves four-point and five-point Green functions. One can of course make an approximation to drop those complicated terms. However, here we will follow another way. As we do in classic mechanics, we analyze the symmetries which the theory exhibits. In QFT, the symmetries result in Ward–Green–Takahashi identities (For non-Abelian theories, the identities are called Slavnov–Taylor identities.) which relate Green function of different orders to each other. Typically, WGTIs are simple algebraic equations rather than complicated integral equations. As we shall see, WGTIs can expose structures of the quark-gluon vertex. To avoid complicities at the first step, we limit ourselves in Abelian theory.

First, we consider the gauge symmetry which the Standard Model was built on. The infinitesimal gauge transformations of the fields read

$$\delta\psi(x) = ig\alpha(x)\psi(x). \tag{10}$$

The gauge invariance results in the famous WGTI:

$$iq_\mu\Gamma_\mu(k, p) = S^{-1}(k) - S^{-1}(p), \tag{11}$$

where $q = k - p$ and Γ_μ is the fermion–gauge-boson vertex. This identity is called longitudinal WGTI because it relates the longitudinal projection of the vertex to the fermion propagators. Note that Eq. (11) can only partially constrains the vertex. Thus one can find infinitely many solutions for the identity. Among them, a straightforward one reads

$$\Gamma_\mu^0(k, p) = -\frac{iq_\mu}{q^2}\left[S^{-1}(k) - S^{-1}(p)\right].$$ (12)

Choosing another Lorentz basis, Ball and Chiu (BC) [1] obtained a solution as

$$\Gamma_\mu^{BC}(k, p) = \gamma_\mu \Sigma_A(k^2, p^2) + \frac{1}{2}t_\mu\gamma\cdot t\Delta_A(k^2, p^2) - it_\mu\Delta_B(k^2, p^2).$$ (13)

where $t = k + p$, $\Sigma_F(x, y) = F(x) + F(y)$, and $\Delta_F(x, y) = [F(x) - F(y)]/(x - y)$. Especially, the term $\Delta_B(k^2, p^2)$ is proportional to the magnitude of DCSB. The BC vertex is used as an ansatz of the quark-gluon vertex because it not only has a correct asymptotic limit in the ultraviolet region but also includes a DCSB term in the infrared region.

Now introducing a smooth function $f(x, y)$ with $f(\infty, \infty) = 0$, we can obtain a new solution for the longitudinal WGTI:

$$\Gamma_\mu^f(k, p) = \left[1 - f(k^2, p^2)\right]\Gamma_\mu^{BC}(k, p) + f(k^2, p^2)\Gamma_\mu^0(k, p).$$ (14)

Compared with the BC vertex, this form of the vertex has the same ultraviolet behavior while can have very different infrared behavior. Thus, in order to constrain the fermion–gauge-boson vertex further, we need to consider more symmetries.

Next, we consider a composite transformation of gauge transformation and Lorentz transformation. The infinitesimal composite transformation [6] reads

$$\delta_T\psi(x) = \delta_{\text{Lorentz}}(\delta\psi(x)) = -\frac{i}{2}\epsilon_{\mu\nu}S_{\mu\nu}\delta\psi(x),$$ (15)

where $S_{\mu\nu}$ is the generator of the infinitesimal Lorentz transformation for the field $\delta\psi$. The composite transformation is called transverse gauge transformation. Accordingly, we obtain the transverse WGTI:

$$q_\mu\Gamma_\nu(k, p) - q_\nu\Gamma_\mu(k, p) = S^{-1}(p)\sigma_{\mu\nu} + \sigma_{\mu\nu}S^{-1}(k) + 2im\Gamma_{\mu\nu}(k, p) + t_\lambda\varepsilon_{\lambda\mu\nu\rho}\Gamma_\rho^A(k, p) + A_{\mu\nu}^V(k, p),$$ (16)

where m is the fermion mass; $\text{tr}[\gamma_5\gamma_\mu\gamma_\nu\gamma_\rho\gamma_\sigma] = -4\varepsilon_{\mu\nu\rho\sigma}$; $\sigma_{\mu\nu} = \frac{1}{2}[\gamma_\mu, \gamma_\nu]$; $\Gamma_{\mu\nu}$ is the tensor vertex; Γ_ρ^A is the axial-vector vertex; and $A_{\mu\nu}^V$ stands for contributions from high-order Green functions. Note that the transverse WGTI has two features: (1) it is not closed since it involves contributions of high-order Green functions; (2) it mixes vertices of different channels together. Thus Eq. (16) can give us few useful information. Combining the Lorentz transformation and the chiral transformation, we obtain the transverse WGTI for the axial-vector vertex:

$$q_\mu\Gamma_\nu^A(k, p) - q_\nu\Gamma_\mu^A(k, p) = S^{-1}(p)\sigma_{\mu\nu}^5 - \sigma_{\mu\nu}^5 S^{-1}(k) + t_\lambda\varepsilon_{\lambda\mu\nu\rho}\Gamma_\rho(k, p) + V_{\mu\nu}^A(k, p),$$ (17)

where $\sigma_{\mu\nu}^5 = \gamma_5\sigma_{\mu\nu}$ and $V_{\mu\nu}^A(k, p)$ stands for contributions from high-order Green functions.

In order to extract information useful, we need to decouple Eqs. (16) and (17). For this purpose, we introduce two projection tensors [13,14]:

$$T_{\mu\nu}^1 = \frac{1}{2}\varepsilon_{\alpha\mu\nu\beta}t_\alpha q_\beta\mathbf{I}_D, \quad T_{\mu\nu}^2 = \frac{1}{2}\varepsilon_{\alpha\mu\nu\beta}\gamma_\alpha q_\beta.$$ (18)

Both the left-hand sides of Eqs. (16) and (17) vanish by contracting with the projection tensors. Then, an interesting result comes out, i.e., that the transverse axial-vector (vector) WGTI constrains the vector (axial-vector) vertex. In other words, the chiral symmetry could suggest the structure of the fermion–gauge-boson vertex. Now we collect the decoupled transverse WGTIs for the fermion–gauge-boson vertex as

$$0 = T_{\mu\nu}^1\left[S^{-1}(p)\sigma_{\mu\nu}^5 - \sigma_{\mu\nu}^5 S^{-1}(k)\right] + \left[t^2 q\cdot\Gamma(k, p) - q\cdot t t\cdot\Gamma(k, p)\right] + T_{\mu\nu}^1 V_{\mu\nu}^A(k, p),$$ (19)

$$0 = T_{\mu\nu}^2\left[S^{-1}(p)\sigma_{\mu\nu}^5 - \sigma_{\mu\nu}^5 S^{-1}(k)\right] + \left[\gamma\cdot t q\cdot\Gamma(k, p) - q\cdot t\gamma\cdot\Gamma(k, p)\right] + T_{\mu\nu}^2 V_{\mu\nu}^A(k, p).$$ (20)

These identities combining with the longitudinal one Eq. (11) can determine the fermion–gauge-boson vertex completely and uniquely. Solving the identities, we obtain a solution [13]:

$$\Gamma_\mu^{\text{Full}}(k, p) = \Gamma_\mu^{\text{BC}}(k, p) + \Gamma_\mu^{\text{T}}(k, p) + X_\mu^{\text{T}}(k, p), \tag{21}$$

where the superscript T means 'transverse', i.e., $q \cdot \Gamma^{\text{T}} = q \cdot X^{\text{T}} = 0$; the second term which only depends on the fermion propagator reads

$$\Gamma_\mu^{\text{T}}(k, p) = \gamma_\mu^{\text{T}} q^2 \Delta_A(k^2, p^2) - \sigma_{\mu\nu} q_\nu \Delta_B(k^2, p^2) + \frac{1}{2}[t_\mu^{\text{T}}\gamma \cdot q + i\gamma_\mu^{\text{T}}\sigma_{\nu\rho}t_\nu q_\rho]\Delta_A(k^2, p^2); \tag{22}$$

and X^{T} denotes contributions from high-order Green functions (it can be decomposed as eight terms which only depend on V^A in Eqs. (19) and (20), see Ref. [13] for details). Note that the solution includes two contributions from the fermion propagator and high-order Green functions. However, in contrast to Eq. (14), the longitudinal part has been completely determined, i.e., the BC vertex. For the transverse vertex, it is expect that the term $\sigma_{\mu\nu} q_\nu \Delta_B(k^2, p^2)$ must be significant for a theory with DCSB. Therefore, for QCD which exhibits strong DCSB, this term must play an important role to explain phenomena and thus a realistic ansatz for the quark-gluon vertex should put it into consideration.

4 Quark-Antiquark Scattering Kernel

In QCD, there are two colorless conserved currents: the vector current and the axial-vector current. As a consequence of current conservations, the color-singlet vector and axial-vector vertices satisfy the following identities ($k_\pm = k \pm P/2$),

$$i P_\mu \Gamma_\mu(k, P) = S^{-1}(k_+) - S^{-1}(k_-), \tag{23}$$

$$P_\mu \Gamma_{5\mu}(k, P) + 2im\Gamma_5(k, P) = S^{-1}(k_+)i\gamma_5 + i\gamma_5 S^{-1}(k_-). \tag{24}$$

These identities are called the color-singlet vector and axial-vector WGTIs, respectively. The vector WGTI guarantees vector current conservation, e.g., the pion form factor $F_\pi(Q^2 = 0) = 1$. The axial-vector WGTI is derived by chiral transformations and shows special importance in the chiral limit. Letting $P_\mu \to 0$, the right-hand side of Eq. (24) equals $i\gamma_5 B(k^2)$ which is finite because of DCSB, while the left-hand side vanishes if there is no any singularity in $\Gamma_{5\mu}(k, P)$. In other words, for $P_\mu \to 0$, the axial-vector vertex $\Gamma_{5\mu}(k, P)$ must have a pole [7,14]:

$$\Gamma_{5\mu}(k, P) \sim \frac{2i\gamma_5 P_\mu E_\pi(k^2)}{P^2}. \tag{25}$$

It follows that there must be a massless pseudoscalar bound-state in the chiral limit and its Lorentz covariant amplitude $E_\pi(k^2)$ satisfies [7,14]

$$f_\pi E_\pi(k^2) = B(k^2), \tag{26}$$

where f_π is a constant for the normalization. In fact, the bound-state is nothing else but pion. The massless pion as well as the relation (26) is an expression of the Goldstone theorem for DCSB in terms of Green functions. As we shall see, the massless pion is not a sufficient and necessary condition to guarantee the chiral symmetry since one can always obtain a massless bound-state by tuning the binding energy.

Now we understand that the color-singlet vector and axial-vector WGTIs, i.e., Eqs. (23) and (24), are of great importance for a plausible truncation scheme. Then the remaining question is: how these WGTIs constrain the structure of the scattering kernel so that it is valid when the quark-gluon vertex is fully dressed? Note that the color-singlet vertices satisfy the the inhomogeneous BS equations. Inserting the inhomogeneous BS equations and the quark gap equation into the WGTIs, we can relate the scattering kernel to the quark-gluon vertex as follows [10]:

$$\int_q K^{(2)}_{\alpha\alpha',\beta'\beta}[S(q_-) - S(q_+)]_{\alpha'\beta'} = \int_q D_{\mu\nu}(k-q)\gamma_\mu[S(q_+)\Gamma_\nu(q_+, k_+) - S(q_-)\Gamma_\nu(q_-, k_-)], \tag{27}$$

$$\int_q K^{(2)}_{\alpha\alpha',\beta'\beta}[S(q_+)\gamma_5 + \gamma_5 S(q_-)]_{\alpha'\beta'} = \int_q D_{\mu\nu}(k-q)\gamma_\mu[S(q_+)\Gamma_\nu(q_+, k_+)\gamma_5 - \gamma_5 S(q_-)\Gamma_\nu(q_-, k_-)]. \tag{28}$$

One can easily check that the RL truncation, i.e., Eqs. (8) and (9), satisfies the above equations. Thus, the RL truncation is also called the leading symmetry-preserving truncation.

In the previous section, we have discussed how to construct a plausible ansatz for the quark-gluon vertex beyond the rainbow approximation. Next we consider two straightforward extensions of the ladder approximation for the scattering kernel [3]:

$$K^{(2)}_{\gamma\Gamma,\Gamma\Gamma} = \quad \text{and} \quad .$$

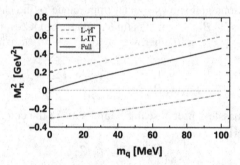

(29)

The first extension, denoted $\gamma\Gamma$, substitutes the bare vertex of one side in the ladder kernel with the dressed one, while the second extension, denoted $\Gamma\Gamma$, substitutes both sides. Inserting these kernels into the integral identities Eqs. (27) and (28), it is found that the identities break down. Thus it is expected that the kernels may give massive pion in the chiral limit. Using the BC vertex, we calculated pion mass with different current quark masses. The result is shown in Fig. 1. For both kernels, pion is indeed massive in the chiral limit $m_q = 0$. However, it is also possible to obtain an accidental massless pion for $m_q = 0$ by choosing models for the quark-gluon vertex and the gluon propagator. For instance, with the kernel $K_{\Gamma\Gamma}$, we obtain a massless pion in the chiral limit by introducing a transverse term $\tau_5 = -it^T_\mu \Delta_B(k^2, p^2)$ to the BC vertex and tuning the coupling strength in the gluon model. As we mentioned before, this is not an evidence that the kernel preserves the chiral symmetry. The identity (26) must be demonstrated as well. The comparison between the amplitude E_π and the mass function B is shown in Fig. 2. It is found that E_π and B are not proportional to each other, i.e., the chiral symmetry is broken in fact.

Next, for any given quark-gluon vertex, we propose a kernel which can reconcile the color-singlet vector and axial-vector WGTIs, i.e., Eqs. (27) and (28), simultaneously. The kernel has a diagrammatic representation as follows (cross circular blobs denotes γ_5):

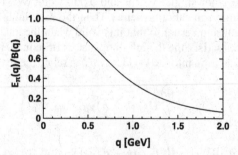

Fig. 1 Pion mass-squared versus current-quark mass, obtained with the BC vertex for Γ_μ in the gap equation: *dashed (blue) curve*, kernel is $K^{(2)}_{\gamma\Gamma}$; *dot-dashed (red) curve*, kernel is $K^{(2)}_{\Gamma\Gamma}$; and *solid (black) curve*, complete BC-vertex-consistent kernel. (Results obtained with the gluon model in Ref. [11]: $D = 0.5\,\text{GeV}^2$, $\omega = 0.5\,\text{GeV}$, and without the ultraviolet tail.) (Color figure online)

Fig. 2 The ratio $E_\pi(q)/B(q)$ with the momenta, where $E_\pi(q)$ and $B(q)$ are scaled at $q = 0$ in order that the ratio would be constantly one if they are proportional to each other. (Results obtained with the same gluon model as Fig. 1 but $D = 0.29\,\text{GeV}^2$ and the quark-gluon vertex ansatz as described in the text.)

$$K_{\text{Full}}^{(2)} = \quad + \quad + \quad ,$$

(30)

which includes three terms: the first term is the ladder-like term whose violation of the symmetries we have already shown; and the other two terms are the symmetry-rescuing terms. Note that it is necessary to introduce two assumed symmetry-rescuing terms, i.e., the solid square and diamond boxes, because of a simple algebraic truth that two equations (i.e., the color-singlet vector and axial-vector WGTIs) can determine two unknowns. With this new kernel, it is found that the plot of $M_\pi^2(m_q)$ always starts from the origin point, i.e., $M_\pi = 0$ for $m_q = 0$ (see the solid line in Fig. 1, for example). Moreover, the amplitude of massless pion is always proportional to the mass function in contrast to the case in Fig. 2. Therefore, the new kernel must be useful in Bethe–Salpeter studies.

5 Epilogue

In this paper, we discussed the coupled system of the gap equation for the quark propagator and the Bethe–Salpeter equation for the meson wavefunction. How to construct the elements of the coupled system, i.e., the quark-gluon vertex and the quark-antiquark scattering kernel, is a long-standing problem. The simplest rainbow-ladder approximation for them has been widely used in phenomenological studies of QCD. However, drawbacks which the RL approximation has exposed in many aspects force us to pursue a sophisticated scheme. Analyzing symmetries which the theory exhibits, we obtained a group of identities called Ward-Green-Takahashi identities. The identities are in fact an expression of the symmetries in terms of Green functions. Solving the longitudinal and transverse WGTIs together, we suggested some useful structures for the quark-gluon vertex (or the fermion–gauge-boson vertex in a general Abelian theory). Based on the color-singlet vector and axial-vector WGTIs, we proposed a form of scattering kernel which can guarantee that the vector current is conserved and the chiral symmetry is preserved. Using the new truncation beyond rainbow-ladder approximation, the spectrum of light-flavor mesons including ground and radially excited states can be well described. Moreover, the scheme can potentially be extended to more applications, e.g., electromagnetic form factors and baryon properties in the diquark picture.

Acknowledgments The author would like to thank C. D. Roberts for helpful discussions. The work was supported by the Office of the Director at Argonne National Laboratory through the Named Postdoctoral Fellowship Program—Maria Goeppert Mayer Fellowship.

References

1. Ball, J.S., Chiu, T.W.: Analytic properties of the vertex function in gauge theories I. Phys. Rev. D **22**, 2542 (1980)
2. Binosi, D., Chang, L., Papavassiliou, J., Roberts, C.D.: Bridging a gap between continuum-QCD and ab initio predictions of hadron observables. Phys. Lett. B **742**, 183–188 (2015)
3. Binosi, D., Chang, L., Papavassiliou, J., Qin, S.-X., Roberts, C.D.: Symmetry preserving truncations of the gap and Bethe–Salpeter equations. Phys. Rev. D **93**, 096010 (2016)
4. Chang, L., Roberts, C.D.: Tracing masses of ground-state light-quark mesons. Phys. Rev. C **85**, 052201 (2012)
5. Durr, S., et al.: Ab-initio determination of light hadron masses. Science **322**, 1224–1227 (2008)
6. He, H.-X.: Transverse symmetry transformations and the quark-gluon vertex function in QCD. Phys. Rev. D **80**, 016004 (2009)
7. Maris, P., Roberts, C.D.: pi- and K meson Bethe–Salpeter amplitudes. Phys. Rev. C **56**, 3369–3383 (1997)
8. Maris, P., Tandy, P.C.: Electromagnetic transition form-factors of light mesons. Phys. Rev. C **65**, 045211 (2002)
9. Maris, P., Roberts, C.D., Tandy, P.C.: Pion mass and decay constant. Phys. Lett. B **420**, 267–273 (1998)
10. Qin, S.-X.: A systematic approach to sketch Bethe-Salpeter equation. EPJ Web Conf. **113**, 05024 (2016)
11. Qin, S.-X., Chang, L., Liu, Y.-X., Roberts, C.D., Wilson, D.J.: Interaction model for the gap equation. Phys. Rev. C **84**, 042202 (2011)
12. Qin, S.-X., Chang, L., Liu, Y.-X., Roberts, C.D., Wilson, D.J.: Investigation of rainbow-ladder truncation for excited and exotic mesons. Phys. Rev. C **85**, 035202 (2012)
13. Qin, S.-X., Chang, L., Liu, Y.-X., Roberts, C.D., Schmidt, S.M.: Practical corollaries of transverse Ward–Green–Takahashi identities. Phys. Lett. B **722**, 384–388 (2013)
14. Qin, S.-X., Roberts, C.D., Schmidt, S.M.: Ward–Green–Takahashi identities and the axial-vector vertex. Phys. Lett. B **733**, 202–208 (2014)
15. Roberts, C.D., Williams, A.G.: Dyson–Schwinger equations and their application to hadronic physics. Prog. Part. Nucl. Phys. **33**, 477–575 (1994)

Few-Body Syst (2016) 57:1019–1026
DOI 10.1007/s00601-016-1143-8

V. M. Braun

Hadron Wave Functions from Lattice QCD

Received: 14 February 2016 / Accepted: 26 July 2016 / Published online: 17 August 2016
© Springer-Verlag Wien 2016

Abstract I give a brief account of the status and perspectives of lattice calculations of the light-front wave functions at small transverse separations, usually referred to as hadron distribution amplitudes (DAs). The existing calculations indicate that the corrections to the asymptotic form of such distributions at large scales are rather small as compared to earlier model estimates. Lattice calculations also suggest an alternating pattern of such corrections for the nucleon resonances with increasing mass. Several recent results are discussed, including precise determination of the second moment of the pion DA, leading-twist DAs of the nucleon and $N^*(1535)$, and the first calculation of the flavor-symmetry breaking corrections in the DAs of the baryon octet.

1 Introduction

The experimental program [1] for the measurement of nucleon elastic form factors and electroproduction of nucleon resonances at large photon virtualities up to $Q^2 = 12$ GeV^2 at the upgraded Jefferson Lab facilities presents a considerable step in the quest to understand hadron structure. The challenge for theory is to understand the expected data from first principles, starting from quarks and gluons as fundamental degrees of freedom.

The question how to transfer a large momentum to a weakly bound partonic system has been discussed for a long time [2–4]. The heuristic picture supported by the study of leading regions in the relevant Feynman diagrams in QCD perturbation theory is that quarks can acquire large transverse momenta when they appear to be at small transverse separations and exchange gluons. Such "hard" gluon exchanges can be separated from "soft" nonperturbative wave functions following the procedure known as collinear factorization. One general consequence of this picture is that only valence Fock states (i.e. states with a minimum number of constituents) can contribute at a large momentum transfer as the probability to "squeeze" a state with many partons to small transverse volume is suppressed by powers of Q, which is just the phase space effect. This selectivity to the lowest Fock states is the new element and the most interesting feature that distinguishes hard exclusive reactions from inclusive processes that involve "global" parton densities. Hence these two types of observations probe different aspects of hadron structure that are to a large extent complementary.

Unfortunately, the leading contribution to baryon form factors involves two hard gluon exchanges and is suppressed by the small factor $(\alpha_s/\pi)^2 \sim 0.01$ compared to the "soft" (endpoint) contributions which are subleading in the power counting in $1/Q^2$ but do not involve small coefficients. Hence the collinear factorization regime is approached very slowly. In this situation the question what exactly do we learn from

Project of the Regensburg Lattice Collaboration (RQCD).
Supported by the German Science Foundation through the Collaborative Research Center Program SFB TR55.

This article belongs to the special issue "Nucleon Resonances".

V. M. Braun (✉)
Institut für Theoretische Physik, Universität Regensburg, 93040 Regensburg, Germany
E-mail: vladimir.braun@ur.de

the studies of form factors is far from trivial. The valence-quark states indeed seem to dominate, however, "squeezing" to small transverse separations occurs very slowly and the helicity selection rules do not work suggesting important role of the quark angular momentum.

The problem has many facets. On the one hand, one needs to go beyond the standard factorization. One direction is to introduce more complicated, transverse momentum dependent (TMD) quark distributions following and extending the work [5] on the pion form factor. Another approach is to calculate the soft contributions to the form factors using dispersion relations and duality [6]. On the other hand, one needs nonperturbative calculations to constrain at least some parameters of hadron wave functions as precisely as possible. Such calculations are absolutely necessary to verify the applicability of the (extended) factorization techniques to the current and expected experimental data. The ultimate goal would be to turn this procedure around and extract the information on the wave functions from the experimental data on elastic and transition form factors in the same way as nucleon parton distributions are extracted from the analysis of inclusive deep-inelastic lepton-hadron scattering.

In this work I give a brief account of the status and perspectives of lattice calculations of the light-front wave functions at small transverse separations, usually referred to as hadron distribution amplitudes (DAs). Several recent results are discussed, including precise determination of the second moment of the pion DA [7], leading-twist DAs of the nucleon and $N^*(1535)$ [8], and the first calculation of the flavor-symmetry breaking corrections in the DAs of the baryon octet [9].

2 Pion Distribution Amplitude

The pion distribution amplitude $\phi_\pi(x)$ determines the momentum sharing in a fast moving pion between its constituents (valence quark and valence antiquark). It is related to the pion Bethe-Salpeter wave function ϕ_π^{BS} by an integral over transverse momenta:

$$\phi_\pi(x, \mu) = Z_2(\mu) \int\limits^{k_\perp < \mu} d^2 k_\perp \, \phi_\pi^{BS}(x, k_\perp). \tag{1}$$

Here x is the quark momentum fraction, Z_2 is the renormalization factor (in the light-cone gauge) for the quark-field operators in the wave function, and μ denotes the renormalisation scale.

The width of the distribution is usually characterized in terms of one of the two parameters

$$\langle \xi^2 \rangle = \int_0^1 dx \, (2x - 1)^2 \phi_\pi(x), \qquad a_2 = \frac{7}{3} \int_0^1 dx \, [1 - 5x(1 - x)] \phi_\pi(x), \tag{2}$$

where the first one is more intuitive and the second one (called second Gegenbauer coefficient) is advantageous for theory as it corresponds to a matrix element of a multiplicatively renormalizable operator (at one-loop level). The summary of the existing lattice calculations is

$$\langle \xi^2 \rangle^{\overline{MS}} = 0.269(39), \qquad a_2^{\overline{MS}} = 0.201(114) \qquad \text{QCDSF 2006 [10]}$$

$$\langle \xi^2 \rangle^{\overline{MS}} = 0.28(1)(2), \qquad a_2^{\overline{MS}} = 0.233(29)(58) \qquad \text{UKQCD 2010 [11]}$$

$$\langle \xi^2 \rangle^{\overline{MS}} = 0.2361(41)(39), \qquad a_2^{\overline{MS}} = 0.1364(154)(145) \qquad \text{Our work [7]}$$

at the scale 2 GeV. The first error combines the statistical uncertainty and the uncertainty of the chiral extrapolation. The second error is the estimated uncertainty contributed by the nonperturbative determination of the renormalization and mixing factors. Our lattice data are collected for the lattice spacing $a = 0.06$–0.08 fm, and this range is not large enough to ensure a reliable continuum extrapolation.

Our result is the most precise determination of the width of the pion DA to date. Since this parameter enters the calculation of physical observables in the combination $1 + a_2(Q^2)$ (in leading order), the achieved accuracy, apart from the continuum extrapolation, corresponds to an uncertainty below 2–3 % in QCD predictions for, e.g., the transition form factor $\gamma^* \to \pi\gamma$ and the weak decay $B \to \pi\ell\nu$. This accuracy is sufficient for phenomenology. Our result also excludes some nontraditional (e.g. "flat") models for the pion DA that have been suggested to explain the BaBar experimental data on the transition form factor $\gamma^* \to \pi\gamma$.

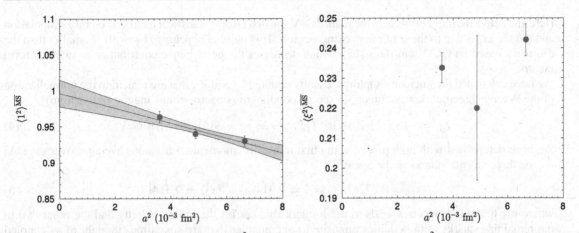

Fig. 1 Continuum extrapolation of the sum $\langle (x_q + x_{\bar{q}})^2 \rangle$ (*left panel*) and the difference $\langle (x_q - x_{\bar{q}})^2 \rangle$ (*right panel*) of the momentum fractions carried by the quark and the antiquark in the pion. The lattice data points correspond to the measurements with pion mass $m_\pi \sim 280$ MeV and three different lattice spacings in the range $a = 0.06$–0.08 fm for (almost) the same lattice volumes. Each lattice point was obtained by averaging over ca. 3000–5000 measurements

The increase of accuracy, roughly by a factor three as compared to the earlier lattice calculations, has been achieved by implementing several methodical improvements. First, our analysis includes new lattice ensembles with much lower pion masses down to the physical point. Since the quantities considered here do not involve any chiral logarithms, the chiral extrapolation is in general unproblematic and rather reliable. Second, we have used the variational approach with two pion interpolators in order to suppress contributions of excited states. This procedure allows us to reduce statistical fluctuations in our data. Third, and most importantly, we have implemented, for the first time, a fully nonperturbative determination of the renormalization factors including the mixing with operators containing total derivatives. It turns out that the nonperturbatively determined mixing coefficient differs from the corresponding result in first order of lattice perturbation theory by up to an order of magnitude. This difference is the main reason for the smaller values of a_2 and $\langle \xi^2 \rangle$ obtained in this work as compared to the previous lattice studies. Fourth, we have made, also for the first time, a detailed study of the discretization errors in lattice derivatives that lead to a violation of the product rule and, for matrix elements, to a violation of the energy conservation. As the result, the relation between a_2 and $\langle \xi^2 \rangle$ at finite lattice spacing is different from the continuum. This difference is significant and must be taken into account in the lattice data analysis.

The difficulty with continuum extrapolation is illustrated in Fig. 1 where we show the renormalized lattice results for the sum (left panel) and the difference (right panel) of the average momentum fractions carried by the quark and the antiquark in the pion. Energy conservation implies that $x_q \equiv x$ and $x_{\bar{q}} = 1 - x$ sum up to unity up to discretization errors, and, indeed, we see that our data extrapolate nicely to the expected continuum limit. Unfortunately, statistical fluctuations for $\langle \xi^2 \rangle = \langle (x_q - x_{\bar{q}})^2 \rangle$ are too large to allow the extrapolation (and for a_2 they are even bigger). There are no reasons to expect much larger discretization errors for $\langle (x_q - x_{\bar{q}})^2 \rangle$ compared to $\langle (x_q + x_{\bar{q}})^2 \rangle$, however, the effect cannot be quantified from our present data. The corresponding uncertainty will be addressed in future.

3 Nucleon Distribution Amplitudes

Nucleon DAs describe the momentum distribution between the three valence quarks, e.g., in the proton. They are more numerous because the valence quarks can have different helicities. Different possibilities to couple the quark helicities to the helicity 1/2 of the proton correspond, roughly speaking, to the S-wave, P-wave, etc., contributions to the proton light-front wave function. A detailed discussion of this connection can be found in [12–14].

Our main results [8] can be summarized as follows:

– We have calculated the nucleon coupling f_N that corresponds to the probability amplitude to find the three valence quarks at one space point (S-wave wave function at the origin)

$$f_N = 2.84(1)(33) \cdot 10^{-3} \text{ GeV}^2. \tag{3}$$

Here and below the numbers refer to the scale 2 GeV, the first error is statistical including chiral extrapolation and the second is due to the continuum extrapolation. This number appears to be ~ 30 % smaller than the estimates based on QCD sum rules [15], which decreases the perturbative contribution to nucleon form factors.

– We have calculated the nucleon couplings, usually denoted λ_1 and λ_2, that are related to the normalization of the P-wave three-quark wave functions (corresponding to nonzero orbital angular momentum)

$$\lambda_1 = -4.13(2)(20) \cdot 10^2 \, \text{GeV}^2, \quad \lambda_2 = 8.19(5)(39) \cdot 10^2 \, \text{GeV}^2. \tag{4}$$

– We have determined with high precision the first moments (momentum fractions averages over the DA) for the three valence quarks in the S-state

$$\langle x_1 \rangle = 0.372(7), \quad \langle x_2 \rangle = 0.314(3), \quad \langle x_2 \rangle = 0.314(7), \tag{5}$$

where the first number corresponds to the u-quark that carries the proton helicity and the other two to the remaining quarks with helicities opposite to one another that are sometimes thought of as coupled in a scalar "diquark". The uncertainty due to continuum extrapolation is not included in the given error bars. The approximate equality $\langle x_2 \rangle \simeq \langle x_3 \rangle$ is unexpected and can be viewed as supporting the "diquark" picture.

– We have calculated the second moments $\langle x_1^2 \rangle$, $\langle x_1 x_2 \rangle$, etc. The statistical fluctuations are rather large in this case and do not allow for a real measurement. The obtained constraints are, however, significant and exclude the possibility of large corrections to asymptotic DAs suggested in early models based on QCD sum rules, see e.g. [15].

The results can be presented in a more systematic way using the expansion of the DA in a set of orthogonal polynomials of three variables

$$\varphi_N(x_i, \mu) = 120 x_1 x_2 x_3 f_N \sum_{n=0}^{\infty} \sum_{k=0}^{n} \phi_{nk}(\mu) \mathcal{P}_{nk}(x_1, x_2, x_3)$$

that are chosen as eigenfunctions of one-loop evolution equations. Here n is the order of the polynomial $\mathcal{P}_{nk}(x_1, x_2, x_3)$ and k enumerates different orthogonal polynomials of order n. We refer to the coefficients in this expansion as shape parameters. Our complete results for the shape parameters are summarized in Fig. 2. They agree well with the parameters extracted from the fits to the nucleon form factors in the light-cone sum rule framework [16] but are mostly much lower as compared to the old QCD sum rule estimates [15].

These calculations are in general much more complicated as for the pion and required a lot of preparatory work. A consistent perturbative renormalization scheme for the baryon DAs based on dimensional regularization was developed in [18]. The chiral perturbation theory extrapolation for the coupling constants and

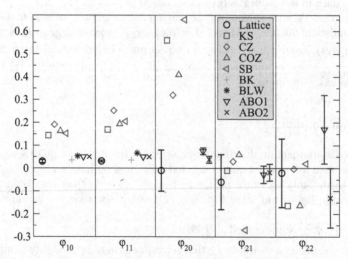

Fig. 2 Comparison of lattice results [8] for the nucleon shape parameters (*black circles*) to several QCD sum rule predictions, e.g. [15] (*red symbols*), light-cone sum rules [16] (*blue symbols*) and the Bolz–Kroll model [17] (BK) (*orange crosses*) (color figure online)

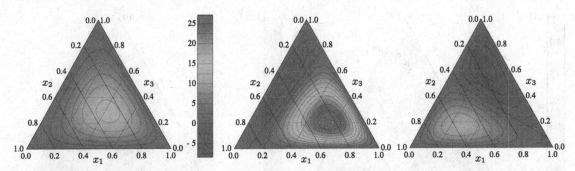

Fig. 3 Barycentric plots of the nucleon (*left*), $N^*(1650?)$ (*center*) and $N^*(1535?)$ (*right*) DAs based on the lattice calculation of the first moments

shape parameters of the DAs for this work was worked out in [19,20]. The construction of irreducible lattice multiplets and nonperturbative renormalization of relevant three quark operators was implemented using the results in [21].

4 Distribution Amplitudes of Negative Parity Nucleon Resonances

Identification of negative parity baryons on the lattice is considerably more difficult than that of the nucleon: In addition to the two lowest-lying $J^P = 1/2^-$ states $N^*(1535)$ and $N^*(1650)$, which only have a small mass difference, there are also contributions of pion-nucleon scattering states.

The spectra of negative parity states were studied before by several authors. In particular in Ref. [22] a variational approach was used involving two different three-quark and one five-quark interpolator. Comparing the eigenvectors of the variational basis for the two- and three-state analyses, the authors suggest that the lower mass state of the two-state analysis splits into the $N\pi$ state and the $N^*(1650)$, while the higher mass state of the two-state analysis becomes the $N^*(1535)$. This is phenomenologically plausible, since the $N^*(1535)$ is not expected to mix strongly with the $N\pi$ continuum as the observed $N^*(1535) \to N\pi$ decay width is rather small.

Our study is the first one that addresses the wave functions. Due to the high cost of five-quark interpolators we have used only the three-quark ones for our analysis. Following the identification suggested in Ref. [22] we label the lower mass state of our two-state variational analysis as $N^*(1650?)$ and the higher mass state as $N^*(1535?)$, where the question marks indicate that this identification requires further study. In the case of $N^*(1650?)$ we expect that there is also considerable contamination by pion-nucleon scattering states.

For the negative parity states, we observe that the leading twist DA of $N^*(1650?)$ is similar to that of the nucleon, whereas $N^*(1535?)$ is qualitatively different as illustrated by the barycentric plots of the DAs in Fig. 3. It can be seen that the DA of $N^*(1650?)$ (in reality, likely a mixture of $N^*(1650)$ and the pion-nucleon nonresonant background) is similar to the nucleon, but with larger deviations from the asymptotic form. The DA of $N^*(1535?)$ appears to be completely different: It is approximately *antisymmetric* under the exchange of the quarks in the diquark. This feature can be related to the observed small decay width of the $N^*(1535)$ to a pion-nucleon final state. It is also interesting that the next-to-leading twist couplings $\lambda_{1,2}$ for the nucleon and both negative parity states are comparable, which is an indication that the quark angular momentum plays a similar role. The consequences of this structure for the electroproduction cross section of the negative parity resonances at large momentum transfer are studied in [14].

5 Distribution Amplitudes of the $\frac{1}{2}^+$ Baryon Octet

This is a very new work [9] where we extend the previous analysis to the full positive parity baryon octet and our first study on a set of $N_f = 2 + 1$ ensembles provided by the coordinated lattice simulations (CLS) effort [23]. These are obtained using the tree-level Symanzik improved gauge action and 2+1 dynamical Wilson (clover) quark flavors. A special feature of CLS program is the use of open boundary conditions in time direction. This will eventually allow for simulations at very fine lattices without topological freezing. We perform a nonperturbative renormalization in a RI′/SMOM scheme, followed by a conversion to $\overline{\text{MS}}$ applying

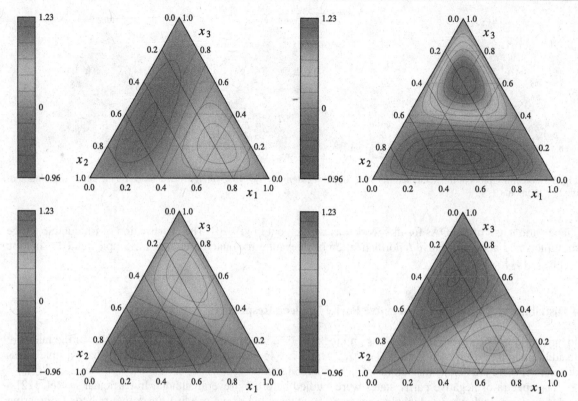

Fig. 4 Deviation from the asymptotic DA for the nucleon (*upper left*), Λ (*upper right*), Σ (*lower left*) and Ξ (*lower right*)

continuum perturbation theory at one-loop accuracy. The results are extrapolated to the physical point using three-flavor BChPT [20]. Full implementation of the CLS strategy lies beyond the scope of this work, where we have focused on the development of the necessary formalism to describe patterns of $SU(3)$ breaking at the wave function level.

In our calculation we start at the flavor symmetric point, where all quark masses are equal, and approach the real world in such a way that u and d quark masses decrease and simultaneously the s quark mass increases so that the average mass is kept (approximately) constant. We find significant SU(3) flavor breaking effects for the leading twist normalization constants

$$\frac{f^\Sigma}{f^N} = 1.41(4), \qquad \frac{f_T^\Sigma}{f^N} = 1.36(4), \qquad \frac{f^\Xi}{f^N} = 1.50(4), \qquad \frac{f_T^\Xi}{f^N} = 1.52(4), \qquad \frac{f^\Lambda}{f^N} = 1.22(4), \qquad (6)$$

and somewhat smaller symmetry breaking for the higher twist couplings

$$\frac{\lambda_1^\Sigma}{\lambda_1^N} = 0.93(2), \qquad \frac{\lambda_1^\Xi}{\lambda_1^N} = 0.98(2), \qquad \frac{\lambda_1^\Lambda}{\lambda_1^N} = 0.81(2), \qquad \frac{\lambda_T^\Lambda}{\lambda_1^N} = 1.05(3), \qquad (7)$$

where the number in parentheses gives a combined statistical and chiral extrapolation error, while the uncertainty from the renormalization procedure is negligible for these ratios. It is likely that these ratios are less sensitive to discretization effects than the couplings themselves.

Deviations from the asymptotic DAs are quantified by the values of shape parameters. They are small for all baryons in the octet, in agreement with the findings of Ref. [8] for the nucleon, and much smaller than results of old QCD sum rule calculations [24]. The SU(3) breaking in the shape parameters is, however, very large. For the isospin-nonsinglet baryons one can identify two competing patterns: First, the strange quarks carry, in general, a larger fraction of the momentum. Second, the quark that carries baryon helicity also carries a larger momentum. These rules do not apply to the Λ baryon due to its reversed symmetry properties. The interplay of these two patterns leads to the rather elaborate structure shown in Fig. 4.

To first order in the SU(3) symmetry breaking the following relation holds between the DAs of the Σ and Ξ hyperons [20]:

$$\Phi_+^{\Sigma}(x_1, x_2, x_3) - \Pi^{\Sigma}(x_1, x_2, x_3) = \Pi^{\Xi}(x_1, x_2, x_3) - \Phi_+^{\Xi}(x_1, x_2, x_3). \qquad (8)$$

This relation has the same theory status as the renowned Gell-Mann–Okubo relation for the masses, and is satisfied with similarly high accuracy $\sim 1\,\%$ in our data.

The analysis in [9] will in future be complemented by a second set of lattices using the trajectory at fixed physical strange quark mass. The extrapolation to the physical point along this second path can be described using chiral perturbation theory with only two flavors, while any other path requires a full $SU(3)$ treatment. The combination of these two methods will provide one with an additional tool to estimate systematic errors.

6 Outlook

Our future studies will have access to a rich landscape of CLS ensembles along both trajectories, including ensembles at (nearly) physical quark masses and various lattice spacings down to $a \sim 0.04$ fm, thus allowing for a reliable continuum extrapolation.

The work is planned along the following lines:

- We will elaborate on our calculation [9] of the DAs for the positive parity octet, aiming at a precision of about 5 % for the coupling constants and 1–2 % for the momentum fractions in Λ^0, $\Sigma^{\pm,0}$ and $\Xi^{\pm}$ baryons. To this end we need to include the ensembles with lower pion mass in order to ensure reliable chiral extrapolation. The statistics for the existing ensembles will be increased.

- Repeat our analysis for the nucleon DAs and DAs of negative parity resonances including CLS ensembles with very small lattice spacing $a \sim 0.04$–0.045 fm. Such gauge configurations are very expensive and their generation will take considerable time. Using CLS ensembles will allow us to get a reliable continuum extrapolation of the results, which is presently the only uncertainty still not under control. Using $N_f = 2+1$ configurations instead of $N_f = 2$ in our published work [8] is not expected to have a large effect on the nucleon, but can have interesting consequences for $N^*(1535)$ since this resonance has a large branching fraction for the decay into $N\eta$.

- Calculate the coupling constants and quark momentum fractions for the $\Delta(1232)$. We have made a feasibility study for such a calculation using a limited set of CLS configurations and the results are quite encouraging: The quality of the signal appears to be very good and statistical fluctuations are small, at least for the coupling constants. From this (very preliminary) analysis we see that the coupling for Δ with helicity $\frac{1}{2}$ appears to be in the ballpark of the existing QCD sum rules estimates, whereas for helicity $\frac{3}{2}$ the raw lattice data indicate a much lower coupling.

- Our calculation of the second moment of the pion DA in Ref. [7] is, at present, state-of the art, and the continuum extrapolation presents the only remaining significant uncertainty. This is difficult to quantify because of generically large statistical errors in the matrix elements of the relevant operators with two derivatives. A larger lever-arm is needed in the lattice spacing, and the corresponding gauge configurations will be produced within the next years by the CLS collaboration. We plan a reanalysis of the pion DA and a similar analysis for the K-meson.

- We will consider flavor-singlet mesons, such as η and η'. This is a major new development from the technical point of view, because of the necessity to take into account disconnected contributions that require using "all-to-all" propagators. Our goal is to calculate η and η' decay constants to at least 5 % precision and estimate the second moments of the DAs. This calculation is very challenging because of the necessity to take into account disconnected contributions and, at the same time, make simulations at physical (or almost physical) quark masses in order to obtain reliable estimates of the $SU(3)$-flavor symmetry breaking. The necessary techniques are currently being developed.

- We will calculate the first and the second moment of the vector meson DAs, ρ, K^* and ϕ. This project is motivated primarily by the studies of rare weak decays $B \to (\rho, K^*)\gamma$ and $B \to (\rho, K^*)\ell^+\ell^-$ at LHCb and BELLEII. The continuum extrapolation is nontrivial because in the real world $\rho(770)$ is a broad resonance decaying mostly to two pions and it is not obvious how the short-distance structure encoded in the DAs can be influenced. the same and we will accept this hypothesis as the working assumption. We are currently checking feasibility of such calculations using our "old" $N_f = 2$ ensembles with $m_\pi \sim 150$ MeV and the results are encouraging.

– . Several supporting projects are on the way, e.g. to calculate nonperturbatively the renormalization factors and to obtain a better control over chiral and finite volume extrapolation using low-energy effective field theory.

Acknowledgments The results described in this report are drawn from work by the Regensburg Lattice Collaboration. In particular the contribution of G. Bali, S. Collins, B. Gläßle, M. Gruber, M. Göckeler, F. Hutzler, P. Pérez-Rubio, A. Schäfer, R. Schiel, J. Simeth, W. Söldner, A. Sternbeck, P. Wein is gratefully acknowledged.

References

1. Aznauryan, I.G., et al.: Studies of nucleon resonance structure in exclusive meson electroproduction. Int. J. Mod. Phys. E **22**, 1330015 (2013)
2. Efremov, A.V., Radyushkin, A.V.: Factorization and asymptotical behavior of pion form-factor in QCD. Phys. Lett. B **94**, 245 (1980)
3. Lepage, G.P., Brodsky, S.J.: Exclusive processes in perturbative quantum chromodynamics. Phys. Rev. D **22**, 2157 (1980)
4. Chernyak, V.L., Zhitnitsky, A.R.: Asymptotics of hadronic form-factors in the quantum chromodynamics. Sov. J. Nucl. Phys. **31**, 544 (1980)
5. Li, H.n, Sterman, G.F.: The Perturbative pion form-factor with Sudakov suppression. Nucl. Phys. B. **381**, 129 (1992)
6. Offen, N.: Light-cone sum rule approach for Baryon form factors. arXiv:1607.01227 [hep-ph]
7. Braun, V.M., et al.: Second moment of the pion light-cone distribution amplitude from lattice QCD. Phys. Rev. D **92**(1), 014504 (2015)
8. Braun, V.M., et al.: Light-cone distribution amplitudes of the nucleon and negative parity nucleon resonances from lattice QCD. Phys. Rev. D **89**, 094511 (2014)
9. Bali, G.S., et al.: Light-cone distribution amplitudes of the baryon octet. JHEP **02**, 070 (2006)
10. Braun, V.M., et al.: Moments of pseudoscalar meson distribution amplitudes from the lattice. Phys. Rev. D **74**, 074501 (2006)
11. Arthur, R., et al.: Lattice results for low moments of light meson distribution amplitudes. Phys. Rev. D **83**, 074505 (2011)
12. Ji, X.d, Ma, J.P., Yuan, F.: Three quark light cone amplitudes of the proton and quark orbital motion dependent observables. Nucl. Phys. B **652**, 383 (2003)
13. Belitsky, A.V., Radyushkin, A.V.: Unraveling hadron structure with generalized parton distributions. Phys. Rept. **418**, 1 (2005)
14. Anikin, I.V., Braun, V.M., Offen, N.: Electroproduction of the $N^*(1535)$ nucleon resonance in QCD. Phys. Rev. D **92**(1), 014018 (2015)
15. Chernyak, V.L., Zhitnitsky, A.R.: Asymptotic behavior of exclusive processes in QCD. Phys. Rept. **112**, 173 (1984)
16. Anikin, I.V., Braun, V.M., Offen, N.: Nucleon form factors and distribution amplitudes in QCD. Phys. Rev. D **88**, 114021 (2013)
17. Bolz, J., Kroll, P.: Modeling the nucleon wave function from soft and hard processes. Z. Phys. A **356**, 327 (1996)
18. Krankl, S., Manashov, A.: Two-loop renormalization of three-quark operators in QCD. Phys. Lett. B **703**, 519 (2011)
19. Wein, P., Bruns, P.C., Hemmert, T.R., Schäfer, A.: Chiral extrapolation of nucleon wave function normalization constants. Eur. Phys. J. A **47**, 149 (2011)
20. Wein, P., Schäfer, A.: Model-independent calculation of $SU(3)_f$ violation in baryon octet light-cone distribution amplitudes. JHEP **1505**, 073 (2015)
21. Göckeler, M., et al.: Non-perturbative renormalization of three-quark operators. Nucl. Phys. B **812**, 205 (2009)
22. Lang, C.B., Verduci, V.: Scattering in the πN negative parity channel in lattice QCD. Phys. Rev. D **87**(5), 054502 (2013)
23. Bruno, M., et al.: Simulation of QCD with $N_f = 2 + 1$ flavors of non-perturbatively improved Wilson fermions. JHEP **1502**, 043 (2015)
24. Chernyak, V.L., Ogloblin, A.A., Zhitnitsky, I.R.: Wave functions of octet baryons. Z. Phys. C **42**, 569 (1989)

Few-Body Syst (2016) 57:893–900
DOI 10.1007/s00601-016-1124-y

Raúl A. Briceño

Meson Electro-/Photo-Production from QCD

Received: 10 February 2016 / Accepted: 25 April 2016 / Published online: 2 July 2016
© Springer-Verlag Wien 2016

Abstract I present the calculation of the $\pi^+\gamma^\star \to \pi^+\pi^0$ transition amplitude from quantum chromodynamics performed by the Hadron Spectrum Collaboration. The amplitude is determined for a range of values of the photon virtuality and the final state energy. One observes a clear dynamical enhancement due to the presence of the ρ resonance. By fitting the transition amplitude and analytically continuing it onto the ρ-pole, the $\rho \to \pi\gamma^\star$ form factor is obtained. This exploratory calculation, performed using lattice quantum chromodynamics, constitutes the very first determination of an electroweak decay of a hadronic resonance directly from the fundamental theory of quarks and gluons. In this talk, I highlight some of the necessary steps that made this calculation possible, placing emphasis on recently developed formalism. Finally, I discuss the status and outlook of the field for the study of $N\gamma^\star \to N^\star \to N\pi$ transitions.

1 Introduction

The study of hadronic resonances is entering an exciting era. For the first time since the identification of quantum chromodynamics (QCD) as the fundamental theory of the strong interaction, one can hope to study hadronic resonances and their properties in a systematically controlled fashion. This is in part due to the tremendous progress made by the lattice QCD community.

The need for rigorous determinations of resonant properties directly from the standard model expands a wide range of phenomenology. These include the field of hadron spectroscopy (e.g., exotic hadrons [1–3]), hadronic structure (e.g., $N \to N^\star$ transitions) and heavy meson decays (e.g., $B \to K^\star \ell^+ \ell^-$ weak decays [4]), among others. The theoretical progress towards the study of these processes has been historically limited by the fact that QCD is non-perturbative at low to medium energies. Presently, there is only one tool at our disposal that allows for the reliable study of QCD in this kinematic regime, this is the aforementioned lattice QCD.

By definition, lattice QCD requires one to place the theory in a finite, discretized Euclidean spacetime. Discretizing the spacetime introduces a UV cutoff, normally referred to as the lattice spacing. The truncation of the spacetime introduces an IR cutoff. For concreteness, I will only consider cubic volumes in the spacial extent with length L and a temporal extent of length T.

The fact that the volume is finite leads to a drastic alteration of the theory's analytic structure. To illustrate this, it is sufficient to consider a generic quantum field theory. If we assume a theory that in the infinite volume has a bound state, followed by multiple thresholds, narrow and broad resonances, its spectrum can be

This article belongs to the special issue "Nucleon Resonances".

R. A. Briceño
Thomas Jefferson National Accelerator Facility, 12000 Jefferson Avenue, Newport News, VA 23606, USA

R. A. Briceño (✉)
Department of Physics, Old Dominion University, Norfolk, VA 23529, USA
E-mail: rbriceno@jlab.org

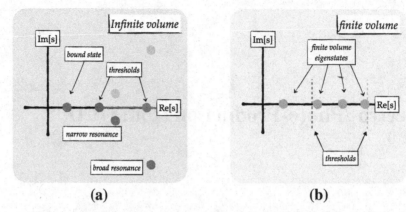

(a) **(b)**

Fig. 1 Shown is the analytic structure of the spectrum of a generic quantum field theory in **a** the infinite-volume limit and **b** a finite volume. On the *left*, the *red dots* are meant to poles, while the *blue* and *magenta* dots denote branch-points associated with the opening of thresholds. Along the real axis above the thresholds, there is a continuum of states. Poles on the real axis are bound states, while poles with non-zero imaginary components are resonances. In a finite volume, there is a discrete number of states and they lie on the real axis. In the text I describe the interpretation of finite volume states that lie above two-body thresholds (color figure online)

qualitatively depicted by Fig. 1a. The bound state would appear as an S-matrix pole on the real axis below all open thresholds, the thresholds emerge as branch-points whose cuts are commonly aligned along the positive Re[s]-axis, and resonances appear as poles off the Re[s]-axis. For resonances that lie above just one open thresholds, there will be two poles associated with the resonances appearing in the second Riemann sheet. These two poles correspond to complex conjugates of each other. Poles associated with a narrow resonance will be comparatively close to the real axis than the broader resonances. Experimentally, one only has access to quantities along the Re[s]-axis above thresholds, and resonant poles must deduced by analytically continuing fits of the S-matrix.

There are two important consequences of placing a theory in a finite volume. The first is that, as is illustrated in Fig. 1b, the spectrum becomes discretized and all poles reside on the Re[s]-axis. This emphasizes the fact that *there are no hadronic resonances in a finite volume*. Second, in a finite volume there is no means of defining asymptotic states, and one cannot "*perform scattering*" in a finite volume.

The only means to circumvent this limitation is to find non-perturbative relations between finite-volume quantities and infinite-volume observables. For the spectrum, this is typically attributed to Lüscher in the literature [5,6], and as a result I will refer to this formalism and its extensions [7–11] as *Lüscher formalism*. The most general extension of this formalism was presented in Ref. [12], which accommodates for any number of two-body coupled channels. For electromagnetic transitions in the presence of an external current one needs another formalism to related finite-volume matrix elements of the current to infinite-volume amplitudes. This was first addressed by Lellouch and Lüscher [13] for $K \to \pi\pi$ weak decays. This formalism has been since extended to accommodate increasingly complex systems [8,9,11,14,15]. The most general formalism for $1 \to 2$ [16,17] and $2 \to 2$ elastic/inelastic reactions [18] has been recently derived.[1]

In Sect. 2, I give some details about the necessary formalism for the analysis of the spectrum and matrix elements, and in Sect. 3 I review its implementation in the calculation of the $\pi^+\gamma^\star \to \pi^+\pi^0$ transition amplitude, presented in Ref. [22,23] by the Hadron Spectrum Collaboration. Finally, in Sect. 4 I give a biased outlook for similar calculations of $N \to N^\star$ transitions from lattice QCD.

2 Finite-Volume Formalism

Here I give a *bird's-eye view* of the steps needed at arriving at a resonant amplitude or a form factor of a resonance from lattice QCD.[2] To supplement the discussion, it is useful to look at Fig. 2, where I give a schematic flowchart for the steps needs to go from lattice QCD quantities to physical observables. For

---

[1] Reference [16] also presented the most general result for $0 \to 2$ transitions, which was first presented in Ref. [19] and later implemented in the study of $\gamma^\star \to \pi\pi$ in Refs. [20,21].

[2] For more detailed discussions on the topic, I point the reader to recent reviews on the topic [24–28].

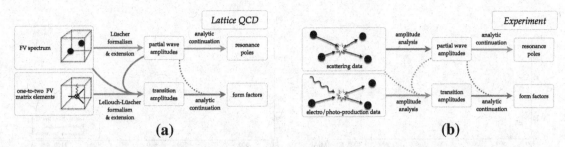

Fig. 2 a Shown is a flow chart depicting the procedure for obtaining resonance pole and form factors from lattice QCD. This is compared to the procedure for obtaining these same observables from experiment (**b**). Each step is further explained in the text

comparison, I also present a similar flowchart for experiments. I will refer to these figures periodically through the text.

I begin by reviewing the necessary formalism to obtain $2 \to 2$ scattering amplitudes from lattice QCD calculations. One can show that a finite-volume two-particle energy, E_L, satisfies the following relation [5–12]

$$\det[F^{-1}(E_L, L) + \mathcal{M}(E_L)] = 0, \tag{1}$$

where $\mathcal{M}$ is the infinite-volume scattering amplitude, $F(E_L, L)$ is a known function of the volume, and the determinant acts on the space of partial waves and open channels, of which can be an infinite number of both. In general, due to the reduction of rotational symmetry, different partial waves mix.

This equation qualitatively can be understood as follows. If one has access to the spectrum, one can constrain the scattering amplitude. Equivalently one can constrain the spectrum if one has the scattering amplitude. This is what is being depicted by the light green arrow on the upper panel of Fig. 2a. Obtaining partial wave amplitude from the lattice QCD finite-volume spectrum is analogous to the determination of partial wave amplitudes from experimental scattering data (depicted by the red arrow of the upper panel of Fig. 2b).

This formalism has been extensively implemented in the literature in the study of elastic [29–32] and inelastic [33–35] resonant scattering amplitudes. In Fig. 3a I highlight a recent calculation of the $\pi\pi$ scattering phase shift by the Hadron Spectrum Collaboration using two different values of the quark masses corresponding to $m_\pi \approx 240, 400$ MeV [29,33]. Having a physical scattering amplitude, one can proceed to compare it to the experimental one. As experimentalist are constrained to use a quark masses corresponding to $m_\pi \approx 140$ MeV, we must first extrapolate the lattice QCD results obtained using heavy quark masses to the physical ones. This was carried out in Ref. [36]. The resulting phase shift is plotted and compared with the experiment in Fig. 3b.

Having determined the resonant scattering amplitude, one can obtain the resonance's mass and width by analytically continuing the amplitude onto the complex plane and finding its pole. This procedure, depicted in Fig. 2, is the same for lattice QCD calculations as experiment. Following this procedure, Ref. [36] found the ρ pole to be $E_\rho = \left[755(2)\binom{20}{02} - \frac{i}{2} 129(3)\binom{7}{2}\right]$ MeV. The first uncertainty corresponds to the statistical and the second is a combination of various systematic errors, which include scale setting and uncertainties in the input masses. Its mass and width are given by the real and imaginary components of the pole position: $m_\rho = \text{Re}(E_\rho) = 755(2)\binom{20}{02}$ MeV and $\Gamma_\rho = -2\text{Im}(E_\rho) = 129(3)\binom{2}{7}$ MeV.

For sometime now it has been well understood how one can study strong decays of hadronic resonances. Until recently it was not evident how to study electroweak decays of these, for example $\rho \to \pi\gamma^\star$ transitions. Presently, the only means to do this is to determine $\pi \to \pi\pi$ infinite-volume transition amplitude, $\mathcal{H}^{\text{out}}$, from finite-volume matrix element, $\langle \pi\pi | \mathcal{J} | \pi \rangle_L$, of an external current $\mathcal{J}$. It is important to recognize that the π state is composed of a single stable hadron while the $\pi\pi$ state is composed of two. With this in mind, we can frame the $\pi \to \pi\pi$ transition as a subset of a general class of $1 \to 2$ transitions, where the numbers denote the number of QCD-stable hadrons present in the initial and final state. Since the ρ is unstable and couples to $\pi\pi$, it must be labeled as a state with "2" hadrons.

One can find a model-independent relation between finite-volume matrix elements and infinite-volume transitions amplitudes [16,17]

$$|\langle 2 | \mathcal{J} | 1 \rangle_L| = \sqrt{\frac{1}{2E_1}} \sqrt{\mathcal{H}^{\text{in}} \, \mathcal{R} \, \mathcal{H}^{\text{out}}}, \tag{2}$$

where $\mathcal{R}$ is a known function that depends on the spectrum, volume and the scattering amplitude. This equation is the generalization of Lellouch's and Lüscher's original relation [13] and it holds for any local current. In this

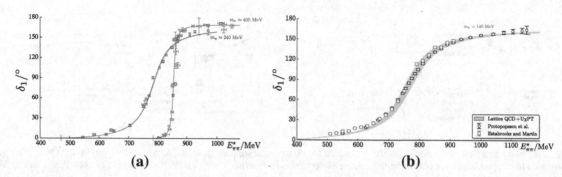

Fig. 3 a Shown is the determination of the $\ell = 1$ $\pi\pi$ scattering phase shift determined by the Hadron Spectrum Collaboration using two different values of the pion mass, $m_\pi \approx 240$ MeV, 400 MeV [29,33]. $E_{\pi\pi}^\star$ denotes the c.m. energy of the $\pi\pi$ system. **b** Depicted by the *blue band* is the extrapolation performed in Ref. [36] to the physical quark masses of the $\pi\pi$ scattering phase shift determined in Ref. [33]. The extrapolated result is compared with the experimental phase shifts, depicted here in *black circles* [37] and *green squares* [38] (color figure online)

equation $\mathcal{R}$ is a matrix in the space of partial waves and open channels, while $\mathcal{H}^{\text{in}}$ and $\mathcal{H}^{\text{out}}$ are row and column vectors in this space. Given the finite-volume spectrum, the $2 \to 2$ scattering amplitude extracted from it, and the matrix elements calculated in a finite volume, one can constrain infinite-volume transition amplitudes. This procedure is qualitatively depicted by the three blue arrows in the bottom panel of Fig. 2a. This parallels the amplitude analysis needed to obtain transition amplitudes from experimental data for electro/photo-production processes. In the following section, I describe in further details of the implementation of this formalism by considering a specific example: $\pi^+\gamma^\star \to \pi^+\pi^0$.

3 $\pi^+\gamma^\star \to \pi^+\pi^0$ and the Unstable $\rho \to \pi\gamma^\star$ Form-Factor

The formalism presented in the previous section, Eq. 2, holds for generic $1 \to 2$ processes. Eventually it would be desirable to perform calculations of transition amplitude involving baryonic resonances, but for now processes involving mesonic degrees of freedom are simpler and serve as a natural testing ground for more these complex systems. In Sect. 4 I review some of the challenges for calculations involving baryons.

As a stepping stone towards the study of $N \to N^\star$ transitions, the Hadron Spectrum Collaboration recently performed an exploratory calculation of the $\pi^+\gamma^\star \to \pi^+\pi^0$ transition amplitude [23]. This was done, following the prescription outlined in the previous section. First, the finite-volume matrix element, $\langle \pi; L | \mathcal{J}_{x=0}^\mu | \pi\pi; L \rangle$, are calculated for a range of discrete $\pi\pi$ center of mass (c.m.) energies, $E_{\pi\pi}^\star$, and a virtuality of the photon, $Q^2 = -(P_{\pi\pi} - P_\pi)^2$. $\mathcal{J}^\mu = (2\bar{u}\gamma^\mu u - \bar{d}\gamma^\mu d)/3$ is the electromagnetic current and u and d are the up- and down-quark fields. This exploratory calculation is performed using quark masses corresponding to $m_\pi \approx 400$ MeV and it was made possible by the technology developed and tested in Ref. [39]. Having determined the matrix elements and the scattering amplitude, one can then use Eq. 2 to obtain the infinite-volume transition amplitude.

Given that the amplitude, $\mathcal{H}_{\pi\pi,\pi\gamma^\star}^\mu$, is a Lorentz vector, it is convenient to define a Lorentz scalar amplitude, $\mathcal{A}_{\pi\pi,\pi\gamma^\star}$. One choice to parametrize the amplitude is

$$\mathcal{H}_{\pi\pi,\pi\gamma^\star}^\mu = \epsilon^{\mu\nu\alpha\beta} P_{\pi,\nu} P_{\pi\pi,\alpha} \epsilon_\beta(\lambda_{\pi\pi}, \mathbf{P}_{\pi\pi}) \frac{2\,\mathcal{A}_{\pi\pi,\pi\gamma^\star}}{m_\pi}, \tag{3}$$

where ϵ_β is the polarization of the $\pi\pi$ channel and $\lambda_{\pi\pi}$ is its helicity. The calculated transition amplitude is shown in Fig. 4a for two values of the virtuality of the photon. This amplitude is compared to the elastic scattering amplitude. One important observation is that these amplitudes exhibit the same resonant behavior.

Another convenient illustration of the result is to plot the $\pi^+\gamma \to \pi^+\pi^0$ cross section for a real photon, $\sigma(\pi^+\gamma \to \pi^+\pi^0)$. In Fig. 4b, this cross section is plotted in physical units and is contrasted with the elastic cross section. One observes that, although the resonant behavior of these two are quite similar, the latter is nearly three orders of magnitude larger. This is to be expected, as the $\pi^+\gamma \to \pi^+\pi^0$ cross section is proportional to the electromagnetic fine structure constant.

Having constrained the transition amplitude, one can proceed to determine the $\rho \to \pi\gamma^\star$ form factor. This can only be rigorously defined as the residue of the transition amplitude at the ρ-pole. Given that the amplitude

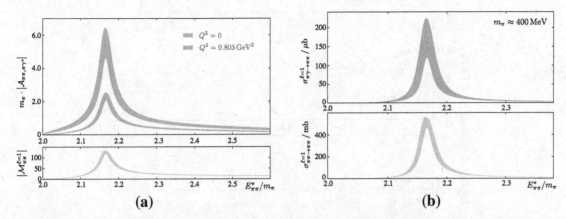

Fig. 4 **a** *Above* is shown the absolute value of the transition amplitude $\mathcal{A}_{\pi\pi,\pi\gamma^\star}$ as a function of the $E_{\pi\pi}^\star$ for two different values of Q^2 from Ref. [23]. This is compared with the elastic $\ell = 1$ $\pi\pi$ amplitude shown below [29]. **b** *Above* is shown the $\pi^+\gamma \to \pi^+\pi^0$ cross section, and it is compared to the elastic $\ell = 1$ scattering cross section shown below

is only constrained on the real axis, one can expect the form factor to be dependent upon the parametrization chosen for the $E_{\pi\pi}^\star$ dependence. By trying a range of choices, a systematic uncertainty can be estimated, which we find to be small for the narrow ρ resonance. This is the non-trivial step depicted by the magenta arrows in the lower panel of Fig. 2a. The dashed lines connecting the partial wave amplitude emphasize that one may use information obtained from the partial wave amplitude in order to perform the analytic continuation. This is not a necessity, since both the transition amplitudes and elastic scattering amplitudes should have the same pole structure.

The definition used for the form factor, $F(E_{\pi\pi}^\star, Q^2)$, is the following,

$$\mathcal{A}_{\pi\pi,\pi\gamma^\star} = F_{\pi\rho}(E_{\pi\pi}^\star, Q^2)\sin\delta_1 \sqrt{\frac{16\pi}{q_{\pi\pi}^\star \Gamma_P(E_{\pi\pi}^\star)}}\, e^{i\delta_1}, \tag{4}$$

where $\Gamma_P(E_{\pi\pi}^\star)$ is the energy-dependent width of the ρ which can be constrained from the elastic $\pi\pi$ scattering amplitude. Using this relation, Hadron Spectrum Collaboration found the $\rho \to \pi\gamma^\star$ form factor show in Fig. 5. This was done using two values of the quark masses. The first corresponds to $m_\pi \approx 700$ MeV [39] where the ρ is stable, and the second for $m_\pi \approx 400$ where the ρ decays to $\pi\pi$ [23]. Since the ρ is stable for the $m_\pi \approx 700$ MeV calculation, the finite formalism was not necessary in its analysis.

One can observe two important facts concerning the behavior of the form factor as a function of the quark masses. First, it appears to follow a natural trend towards the physical point, as the $m_\pi \approx 400$ MeV ensemble is closer to the experimental point than the $m_\pi \approx 700$ MeV ensemble. The second important observation is that the $m_\pi \approx 400$ MeV ρ-pole is located off the Re[s]-axis. As a result, the form factor, as defined in Eq. 4, can in general be complex. The imaginary component of the form factor is found to be approximately two orders of magnitude smaller than its real component. This can be understood by the fact that the ρ is rather narrow for these quark masses.

4 Outlook to $N \to N^\star$ Transitions from QCD

Having performed the first calculation of a resonant radiative transition process, one could naturally ask "*why not perform a similar calculation for $N\gamma^\star \to N^\star$?*".[3] Here I briefly discuss the obtasbles associated with these calculations and give an outlook for the future. There are several challenges, some of which I list: more open thresholds, three-particle thresholds might be important, larger number of contractions, deterioration of signal, larger number of partial waves, and partial-wave mixing. To clarify what is meant by each one of these, I describe them below.

The first two of these point are important since the $N\gamma^\star \to N^\star$ form factor is only determinable from the residue of the transition amplitude for $N\gamma^\star \to N\pi, N\pi\pi, N\eta, \ldots$. The formalism presented in Refs. [16,17]

---

[3] For a complimentary discussion on the present status of $N^\star$ studies from lattice QCD, I point the reader to Ref. [40]

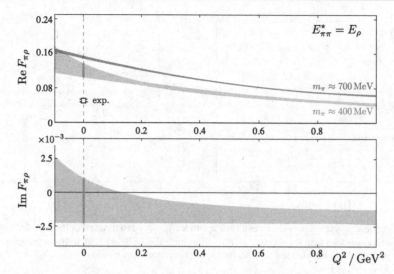

Fig. 5 *Above* is shown the real component of the $\rho \to \pi\gamma^\star$ form factor determined in using $m_\pi \approx 400$ MeV [23] [*orange band*]. This is evaluated at the ρ pole. For comparison are shown the form factor determined in Ref. [39] [*green band*] using $m_\pi \approx 700$ MeV, where the ρ resonance is QCD stable, and the experimentally determined $\rho\pi$ photocoupling [41,42]. The *lower panel* shows the imaginary component of the form factor obtained for $m_\pi \approx 400$ MeV (color figure online)

accommodates energies above any number of two-body thresholds but below three-particle thresholds. This is of particular importance for resonances like the *Roper* $N(1440)$, which experimentally couples approximately $30 - 40\%$ of the time to $N\pi\pi$. A universal formalism for three-particles in a finite volume is not yet in place, but important steps have been taken in this direction [43–46].

Given that baryonic systems have a larger number of valence quarks, when computing correlations functions this leads to a larger number of Wick contractions than in the mesonic sector. This is a challenge but certainly not a limitation. For instance, Ref. [47] has already presented an exploratory calculation of resonant $N\pi$ scattering in the negative parity sector, and recently Ref. [48] presented a calculation of meson-baryon phase-shifts in various non-resonant channels. As is clear from these and all other studies in the baryonic sector, the signal deteriorates quicker for baryonic systems than mesonic ones. Reducing the statistical noise of these calculations is possible by lengthening their computational running time.

Lastly, given that baryons carry nonzero intrinsic spin, there is a larger number of partial waves for baryonic systems than in systems of spin-0. Due to the reduction of rotational symmetry these mix, as is dictated by Eqs. 1 and 2. Unlike $\pi\pi$, for the $N\pi$ system positive- and negative-parity partial waves will mix when the system is boosted. This would imply that, for example, in a rigorous lattice QCD calculation one would have to simultaneously study the $N(1440)$, $N(1520)$, $N(1535)$, ... resonances. This is a challenge, but it has been previously addressed in, for example, the πK, ηK channels by the Hadron Spectrum Collaboration [34,35]. In this work, the authors performed a simultaneously determination of the κ, $K_0^\star$, $K_1^\star$, and $K_2^\star$ resonance poles.

In summary, the historical limitations associated with the study of resonant hadrons are currently being overcome. There is still more development needed, but this is currently being addressed by the lattice QCD community. Therefore, it is not unrealistic to expect calculation of resonant elastic/inelastic scattering and transitions processes in the baryonic sector in the near future. As in the mesonic sector, the first calculations will be of scattering processes, followed by transition processes involving electroweak external currents.

5 Final Remarks

Lattice QCD has proven to be a remarkable tool for the determination of low-lying, QCD-stable states. The determination of properties of resonances via lattice QCD has been historically limited. In the past few years we have witnessed a tremendous amount of progress that has opened the doors towards increasingly complex and important observables. It is worth reiterating that we are entering an exciting time for the study of few-body systems, and we should expect many more studies of phenomenologically interesting processes directly from QCD. In this talk I have highlighted some important developments towards this goal.

Acknowledgments I would like to thank my colleagues within the Hadron Spectrum Collaboration, in particular Jozef Dudek, Robert Edwards, Christian Shultz, Christopher Thomas and David Wilson, for granting me permission to share the results of their hard work. Also, I would like to thank my other collaborators, Daniel Bolton and Maxwell Hansen, whose work I have also presented in this talk.

References

1. Briceño, R.A. et al.: Issues and Opportunities in Exotic Hadrons. arxiv:1511.06779 (2015)
2. Swanson, E.S.: XYZ states: theory overview. In: Proceedings, 16th International Conference on Hadron Spectroscopy (Hadron 2015). arxiv:1512.04853 (2015)
3. Prelovsek, S., Leskovec, L.: Evidence for X(3872) from DD* scattering on the lattice. Phys. Rev. Lett. **111**, 192001 (2013). arxiv:1307.5172
4. Horgan, R.R., Liu, Z., Meinel, S., Wingate, M.: Calculation of $B^0 \to K^{*0}\mu^+\mu^-$ and $B_s^0 \to \phi\mu^+\mu^-$ observables using form factors from lattice QCD. Phys. Rev. Lett. **112**, 212003 (2014). arxiv:1310.3887
5. Luscher, M.: Volume dependence of the energy spectrum in massive quantum field theories. 2. Scattering states. Commun. Math. Phys. **105**, 153 (1986)
6. Luscher, M.: Two particle states on a torus and their relation to the scattering matrix. Nucl. Phys. B **354**, 531 (1991)
7. Rummukainen, K., Gottlieb, S.A.: Resonance scattering phase shifts on a nonrest frame lattice. Nucl. Phys. B **450**, 397 (1995). arxiv:hep-lat/9503028
8. Kim, C., Sachrajda, C., Sharpe, S.R.: Finite-volume effects for two-hadron states in moving frames. Nucl. Phys. B **727**, 218 (2005). arxiv:hep-lat/0507006
9. Christ, N.H., Kim, C., Yamazaki, T.: Finite volume corrections to the two-particle decay of states with non-zero momentum. Phys. Rev. D **72**, 114506 (2005). arxiv:hep-lat/0507009
10. Briceño, R.A., Davoudi, Z.: Moving multichannel systems in a finite volume with application to proton-proton fusion. Phys. Rev. D **88**, 094507 (2013). arxiv:1204.1110
11. Hansen, M.T., Sharpe, S.R.: Multiple-channel generalization of Lellouch-Luscher formula. Phys. Rev. D **86**, 016007 (2012). arxiv:1204.0826
12. Briceño, R.A.: Two-particle multichannel systems in a finite volume with arbitrary spin. Phys. Rev. D **89**, 074507 (2014). arxiv:1401.3312
13. Lellouch, L., Luscher, M.: Weak transition matrix elements from finite volume correlation functions. Commun. Math. Phys. **219**, 31 (2001). arxiv:hep-lat/0003023
14. Lin, C.D., Martinelli, G., Sachrajda, C.T., Testa, M.: $K \to \pi\pi$ decays in a finite volume. Nucl. Phys. B **619**, 467 (2001). arxiv:hep-lat/0104006
15. Meyer, H.B.: Photodisintegration of a bound state on the torus. arxiv:1202.6675 (2012)
16. Briceño, R.A., Hansen, M.T.: Multichannel $0 \to 2$ and $1 \to 2$ transition amplitudes for arbitrary spin particles in a finite volume. Phys. Rev. D **92**, 074509 (2015). arxiv:1502.04314
17. Briceño, R.A., Hansen, M.T., Walker-Loud, A.: Multichannel $1 \to 2$ transition amplitudes in a finite volume. Phys. Rev. D **91**, 034501 (2015). arxiv:1406.5965
18. Briceño, R.A., Hansen, M.T.: Relativistic, model-independent, multichannel $2 \to 2$ transition amplitudes in a finite volume. arxiv:1509.08507 (2015)
19. Meyer, H.B.: Lattice QCD and the timelike pion form factor. Phys. Rev. Lett. **107**, 072002 (2011). arxiv:1105.1892
20. Feng, X., Aoki, S., Hashimoto, S., Kaneko, T.: Timelike pion form factor in lattice QCD. arxiv:1412.6319 (2014)
21. Bulava, J., Horz, B., Fahy, B., Juge, K.J., Morningstar, C., Wong, C.H.: Pion-pion scattering and the timelike pion form factor from $N_f = 2 + 1$ lattice QCD simulations using the stochastic LapH method. In: Proceedings, 33rd International Symposium on Lattice Field Theory (Lattice 2015). arxiv:1511.02351 (2015)
22. Briceño, R.A., Dudek, J.J., Edwards, E.G., Shultz, C.J., Thomas, C.E., Wilson, D.J.: The $\pi\pi \to \pi\gamma^*$ amplitude and the resonant $\rho \to \pi\gamma^*$ transition from lattice QCD. arXiv:1604.03530 (2016)
23. Briceño, R.A., Dudek, J.J., Edwards, R.G., Shultz, C.J., Thomas, C.E., Wilson, D.J.: The resonant $\pi^+\gamma \to \pi^+\pi^0$ amplitude from quantum chromodynamics. Phys. Rev. Lett. **115**, 242001 (2015). arxiv:1507.06622
24. Hansen, M.T.: Extracting three-body observables from finite-volume quantities. In: Proceedings, 33rd International Symposium on Lattice Field Theory (Lattice 2015). arxiv:1511.04737 (2015)
25. Briceño, R.A.: Few-body physics. PoS **LATTICE2014**, 008 (2015). arxiv:1411.6944
26. Briceño, R.A., Davoudi, Z., Luu, T.C.: Nuclear reactions from lattice QCD. J. Phys. G **42**, 023101 (2015d). arxiv:1406.5673
27. Prelovsek, S.: Hadron Spectroscopy. PoS **LATTICE2014**, 015 (2014). arxiv:1411.0405
28. Mohler, D.: Review of lattice studies of resonances. PoS **LATTICE2012**, 003 (2012). arxiv:1211.6163
29. Dudek, J.J., Edwards, R.G., Thomas, C.E.: Energy dependence of the ρ resonance in $\pi\pi$ elastic scattering from lattice QCD. Phys. Rev. D **87**, 034505 (2013). arxiv:1212.0830
30. Lang, C., Mohler, D., Prelovsek, S., Vidmar, M.: Coupled channel analysis of the rho meson decay in lattice QCD. Phys. Rev. D **84**, 054503 (2011). arxiv:1105.5636
31. Lang, C.B., Leskovec, L., Mohler, D., Prelovsek, S., Woloshyn, R.M.: Ds mesons with DK and D*K scattering near threshold. Phys. Rev. D **90**, 034510 (2014). arxiv:1403.8103
32. Feng, X., Jansen, K., Renner, D.B.: Resonance parameters of the rho-Meson from lattice QCD. Phys. Rev. D **83**, 094505 (2011). arxiv:1011.5288
33. Wilson, D.J., Briceño, R.A., Dudek, J.J., Edwards, R.G., Thomas, C.E.: Coupled $\pi\pi$, $K\bar{K}$ scattering in P-wave and the ρ resonance from lattice QCD. Phys. Rev. D **92**, 094502 (2015). arxiv:1507.02599
34. Wilson, D.J., Dudek, J.J., Edwards, R.G., Thomas, C.E.: Resonances in coupled πK, ηK scattering from lattice QCD. Phys. Rev. D **91**, 054008 (2015). arxiv:1411.2004

35. Dudek, J.J., Edwards, R.G., Thomas, C.E., Wilson, D.J.: Resonances in coupled $\pi K - \eta K$ scattering from quantum chromodynamics. Hadron Spectr. Phys. Rev. Lett. **113**, 182001 (2014). arxiv:1406.4158
36. Bolton, D.R., Briceño, R.A., Wilson, D.J.: Connecting physical resonant amplitudes and lattice QCD. arxiv:1507.07928 (2015)
37. Protopopescu, S., Alston-Garnjost, M., Barbaro-Galtieri, A., Flatte, S.M., Friedman, J., et al.: Pi pi partial wave analysis from reactions pi+ p —> pi+ pi- Delta++ and pi+ p —> K+ K- Delta++ at 7.1-GeV/c. Phys. Rev. D **7**, 1279 (1973)
38. Estabrooks, P., Martin, A.D.: pi pi phase shift analysis below the K anti-K threshold. Nucl. Phys. B **79**, 301 (1974)
39. Shultz, C.J., Dudek, J.J., Edwards, R.G.: Excited meson radiative transitions from lattice QCD using variationally optimized operators. Phys. Rev. D **91**, 114501 (2015). arxiv:1501.07457
40. Richards, D.: N^* resonances in lattice QCD from (mostly) low to (sometimes) high virtualities. In: ECT* Workshop on Nucleon Resonances: from Photoproduction to High Photon Virtualities, Trento, Italy, 12–16 October 2015 (2016). https://inspirehep.net/record/1444997/files/arXiv:1604.02988.pdf
41. Huston, J., Berg, D., Chandlee, C., Cihangir, S., Collick, B., et al.: Measurement of the resonance parameters and radiative width of the ρ^+. Phys. Rev. D **33**, 3199 (1986)
42. Capraro, L., Levy, P., Querrou, M., Van Hecke, B., Verbeken, M., et al.: The ρ radiative decay width: a measurement at 200-GeV. Nucl. Phys. B **288**, 659 (1987)
43. Hansen, M.T., Sharpe, S.R.: Relativistic, model-independent, three-particle quantization condition. Phys. Rev. D **90**, 116003 (2014). arxiv:1408.5933
44. Hansen, M.T., Sharpe, S.R.: Expressing the three-particle finite-volume spectrum in terms of the three-to-three scattering amplitude. Phys. Rev. D **92**, 114509 (2015). arxiv:1504.04248
45. Polejaeva, K., Rusetsky, A.: Three particles in a finite volume. Eur. Phys. J. A **48**, 67 (2012). arxiv:1203.1241
46. Briceño, R.A., Davoudi, Z.: Three-particle scattering amplitudes from a finite volume formalism. Phys. Rev. D **87**, 094507 (2012). arxiv:1212.3398
47. Lang, C.B., Verduci, V.: Scattering in the πN negative parity channel in lattice QCD. Phys. Rev. D **87**, 054502 (2013). arxiv:1212.5055
48. Detmold, W., Nicholson, A.: Low energy scattering phase shifts for meson-baryon systems. arxiv:1511.02275 (2015)

Few-Body Syst (2016) 57:975–983
DOI 10.1007/s00601-016-1135-8

Nils Offen

Light-Cone Sum Rule Approach for Baryon Form Factors

Received: 15 February 2016 / Accepted: 1 June 2016 / Published online: 4 August 2016
© Springer-Verlag Wien 2016

Abstract We present the state-of-the-art of the light-cone sum rule approach to Baryon form factors. The essence of this approach is that soft Feynman contributions are calculated in terms of small transverse distance quantities using dispersion relations and duality. The form factors are thus expressed in terms of nucleon wave functions at small transverse separations, called distribution amplitudes, without any additional parameters. The distribution amplitudes, therefore, can be extracted from the comparison with the experimental data on form factors and compared to the results of lattice QCD simulations.

1 Introduction

Understanding the characteristics of Hadrons in terms of QCD degrees fo freedom, namely quarks and gluons, is one of the central challenges of particle physics. It is understood that form factors at large momentum transfer Q^2 can be described in terms of distribution amplitudes, i.e. light-cone wave functions at small light-like separation [1–9]. In this way experimental measurements of form factors can be connected to the momentum distribution of quarks inside the involved hadrons. For mesons the gold plated modes are the so called transition form factors $\pi, \eta^{(\prime)} \to \gamma\gamma*$ where the hard formally leading contribution in $\frac{1}{Q^2}$ is not suppressed by powers of $\frac{\alpha_s}{\pi}$. But even there power corrections can reach up to $\sim 20\%$ at large $Q^2 \sim 40\,\text{GeV}^2$ [10–14].

For electromagnetic Baryon form factors the hard contributions are suppressed by $\left(\frac{\alpha_s}{\pi}\right)^2 \sim 0.01$ compared to the so called Feynman (soft) terms where one quark carries almost all of the momentum of the parent hadron and interacts solely via soft gluons. Therefore the asymptotic regime where the perturbative description in terms of distribution amplitudes is correct is postponed to very high Q^2 far out of reach of current experiments.

Under these circumstances additional model assumptions have to be made to interpret experimental data. One possibility is to model transverse momentum dependent (TMD) light-cone wave function and use Sudakov suppression of large transverse distances as initially suggested by Li and Sterman [15].

The possibility we advocate is called light-cone sum rules [16–18]. It is based on an light-cone expansion in baryon distribution amplitudes of increasing twist using dispersion relations and duality. Soft- and hard contributions are calculated on the same footing and there is no double counting [19]. This method gives up to now the most direct connection between form factors and distribution amplitudes and has already been succesfully applied to several meson decays.

For baryons the case is more complicated and it shall be discussed in some detail in the next section.

This article belongs to the special issue "Nucleon Resonances".

N. Offen (✉)
Fakultät Physik, Universität Regensburg, 93040 Regensburg, Germany
E-mail: nils.offen@ur.de
Tel:+49-941-9432003

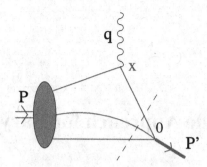

Fig. 1 Schematic structure of the light-cone sum rule for baryon form factors

2 Light-Cone Sum Rules for N and N^* Form Factors

The basic object of the LCSR approach to baryon form factors [20–22] is the correlation function

$$\Pi_\mu(P, q) = \int dx\, e^{-iqx} \langle 0|T\{\eta(0) j_\mu(x)\}|P\rangle \tag{1}$$

where j_μ represents the electromagnetic probe and η is a suitable operator with nucleon quantum numbers. The other (in this example, initial state) nucleon is explicitly represented by its state vector $|P\rangle$, see a schematic representation in Fig. 1. LCSRs are then derived by matching two different representations of the correlation function: If both the momentum flowing through the η-vertex $P'^2 = (P - q)^2$ and the momentum transfer Q^2 are large and negative it can be shown that the main contribution to the integral in (1) comes from the region $x^2 \approx 0$. Hence it can be studied using the operator product expansion of the time ordered product of the two currents around the light-cone. The light-cone divergence of the coefficient function is governed by the twist, i.e. dimension minus spin, of the respective operator. The matrix element of the operator is related to the baryon distribution amplitude. The resulting expression is then analytically continued to positive P'^2 by dispersion relations.

On the other hand the correlation function can be represented as a complete sum over intermediate hadron states and can be written as a dispersion integral in P'^2 with the nucleon or N^* contribution separated explicitly from the higher states.

Quark-Hadron duality allows to equate both representations from a certain duality threshold s_0 on giving an expression for the form factor

$$\frac{\lambda_1 F_1(Q^2)}{m_N^2 - P'^2} = \int_0^{s_0} \frac{ds}{s - P'^2} \operatorname{Im}_s \Pi(s, Q^2) + \text{subtractions.}$$

A Borel-transformation gets rid of subtraction terms needed to render the dispersion integrals finite and suppresses higher order states (contributions of large s) on the cost of introducing an additional parameter M^2. The dependence on this parameter is artificial in a similar way as the dependence on the factorization scale μ in perturbation theory.

2.1 Form Factors

The electromagnetic or electroproduction form factors we are considering are conventionally defined as the matrix element of the electromagnetic current

$$j_\mu^{\text{em}}(x) = e_u \bar{u}(x)\gamma_\mu u(x) + e_d \bar{d}(x)\gamma_\mu d(x) \tag{2}$$

taken either between nucleon states or between the negative parity spin $\frac{1}{2}$ partner N^* and a nucleon:

$$\langle N(P')|j_\mu^{\text{em}}|N(P)\rangle = \bar{u}_N(P')\left[\gamma_\mu F_1(Q^2) - i\frac{\sigma_{\mu\nu}q^\nu}{2m_N}F_2(Q^2)\right]u_N(P),$$

$$\langle N^*(P')|j_\mu^{\text{em}}|N(P)\rangle = \bar{u}_{N^*}(P')\left[\gamma_5\frac{G_1(q^2)}{m_N^2}(\slashed{q}q_\mu - q^2\gamma_\mu) - i\frac{G_2(q^2)}{m_N}\sigma_{\mu\rho}q^\rho\right]u_N(P). \tag{3}$$

Roughly speaking they are a measure for the probability of a nucleon being hit by an energetic photon to form a nucleon or a N^*. The second form factor is needed to describe the effect of the anomalous magnetic moment of the respective hadron. Possible third and fourth form factors are not necessary due to current conservation and parity invariance of the electromagnetic interaction. For the nucleon these form factors are called Dirac F_1 and Pauli F_2 form factor. For experimental measurements it is more convenient to consider the so called electric and magnetic Sachs form factors

$$G_M(Q^2) = F_1(Q^2) + F_2(Q^2), \quad G_E(Q^2) = F_1(Q^2) - \frac{Q^2}{4m_N^2} F_2(Q^2). \tag{4}$$

They lead to a separation of the form factors in the famous Rosenbluth scattering cross-section.

The helicity amplitudes $A_{1/2}(Q^2)$ and $S_{1/2}(Q^2)$ for the electroproduction of $N^*(1535)$ can be expressed in terms of the form factors [23] via:

$$A_{1/2} = eB \Big[Q^2 G_1(Q^2) + m_N(m_{N^*} - m_N)G_2(Q^2) \Big], \quad S_{1/2} = \frac{eBC}{\sqrt{2}} \Big[(m_N - m_{N^*})G_1(Q^2) + m_N G_2(Q^2) \Big], \tag{5}$$

where $e = \sqrt{4\pi\alpha}$ is the elementary charge and B, C are kinematic factors defined as

$$B = \sqrt{\frac{Q^2 + (m_{N^*} + m_N)^2}{2m_N^5(m_{N^*}^2 - m_N^2)}}, \quad C = \sqrt{1 + \frac{(Q^2 - m_{N^*}^2 + m_N^2)^2}{4Q^2 m_{N^*}^2}}. \tag{6}$$

2.2 Distribution Amplitudes

One of the attractive features about light-cone sum rules is that one can calculate form factors in terms of distribution amplitudes which correspond to light-cone wave functions at small transverse distances and are fundamental process independent functions that describe the longitudinal momentum distribution of the partons inside the hadron.

They are defined as matrix elements of non-local light-ray operators. The leading twist distribution amplitude of the nucleon is given by [24,25]

$$\langle 0|\epsilon^{ijk} \left(u_i^\uparrow(a_1 n)C \,\not{h}u_j^\downarrow(a_2 n) \right) \not{h}d_k^\uparrow(a_3 n)|P\rangle = -\frac{1}{2} f_N \, Pn \not{h} N^\uparrow(P) \int [dx] e^{-iPn \sum x_i a_i} \varphi_N(x_i), \tag{7}$$

where n is a light-like vector $n^2 = 0$ and f_N is the decay constant of the nucleon. The distribution amplitudes can be expanded into a set of orthogonal polynomials which are eigenfunctions of the corresponding one-loop evolution kernel [26–29]. The coefficients are matrix elements of local conformal operators which can be calculated on the lattice [30–33] to constrain the shape of the respective distribution amplitude, see also Tables 1 and 2. In the following we will call these coefficients, shape parameters. Higher twist distribution amplitudes

Table 1 LCSR (ABO1) and LCSR (ABO2) refer to the two models ABO1 and ABO2 extracted in [34]

Method	f_N/λ_1	φ_{10}	φ_{11}	φ_{20}	φ_{21}	φ_{22}	η_{10}	η_{11}	Refs.
LCSR(ABO1)	−0.17	0.05	0.05	0.075(15)	−0.027(38)	0.17(15)	−0.039(5)	0.140(16)	[34]
LCSR(ABO2)	−0.17	0.05	0.05	0.038(15)	−0.018(37)	−0.13(13)	−0.027(5)	0.092(15)	[34]
LATTICE	−0.083(6)	0.043(15)	0.041(14)	0.038(100)	−0.14(15)	−0.47(33)	–	–	[30]
LATTICE	−0.075(5)	0.038(3)	0.039(6)	−0.050(80)	−0.19(12)	−0.19(14)	–	–	[31]
QCDSR (NLO)	−0.15	–	–	–	–	–	–	–	[35]

The values of the normalization of the leading and next-to-leading twist distribution amplitudes f_N/λ_1 and the first order shape parameter of the leading distribution amplitude, φ_{10} and φ_{11} have been fixed before fitting to the experimental data. The comparatively large value of f_N was needed to get the normalization of the experimental data right and is in agreement with a recent NLO sum rule determination [35]. φ_{20}, φ_{21} and φ_{22} refer to the second order shape parameters of the leading twist distribution amplitude. η_{10} and η_{11} are the first order shape parameter of the twist 4 distribution amplitudes. All values are given at a scale $\mu^2 = 2 \, \text{GeV}^2$

Table 2 Similar to Table 1

Method	$\lambda_1^{N*}/\lambda_1^N$	f_{N*}/λ_1^{N*}	φ_{10}	φ_{11}	φ_{20}	φ_{21}	φ_{22}	η_{10}	η_{11}	Refs.
LCSR (1)	0.633	0.027	0.36	−0.95	0	0	0	0.00	0.94	[36]
LCSR (2)	0.633	0.027	0.37	−0.96	0	0	0	−0.29	0.23	[36]
LATTICE	0.633 (43)	0.027 (2)	0.28 (12)	−0.86 (10)	1.7 (14)	−2.0 (18)	1.7 (26)	–	–	[32]

LCSR (1) corresponds to a fit to the form factors $G_1(Q^2)$ and $G_2(Q^2)$ extracted from the measurements of helicity amplitudes in [37]. The uncertainties of the extracted form factors were added in quadrature. LCSR (2) is obtained from a fit to helicity amplitudes including all available data at $Q^2 > 1.7$ GeV^2 [37–40] $\lambda_1^{N*}/\lambda_1^N$ and f_{N*}/λ_1^{N*} were fixed to the lattice results

either describe Fock-states with additional partons, e.g. $qqqG$-states, or with relative angular momentum or both [25]. For the N^* there is some freedom in defining the distribution amplitudes by choosing different positions of γ_5. We have defined them in such a way that all the relations for the nucleon case stay intact and that the coefficient functions in the light-cone sum rules are exactly the same. Since distribution amplitudes with additional partons are up to now very poorly known we don't consider them in our calculation.

3 Results

The results presented here needed several prerequisites which were derived in the last several years.

1. a consistent and practical renormalization scheme for three-quark operators was developed [41]
2. expressions for matrix elements of operators with non light-like distance in terms of distribution amplitudes were derived [34]
3. next to leading order corrections both for twist 3 and 4 were calculated [34,42]
4. the kinematic contributions to higher twist distribution amplitudes, the so called Wandzura–Wilczek contributions, were taken into account [43]
5. off light-cone corrections (x^2-corrections) to leading twist distribution amplitudes were recalculated [34]
6. the leading twist distribution amplitude was expanded up to second order [34]

These advances allowed for the first time to make quantitative statements on the shape of the nucleon and N^* distribution amplitude based on experimental data. The extracted shape parameters for the nucleon and N^* with lattice results for comparison are given in Tables 1 and 2. The shape of the resulting distribution amplitudes is plotted in Fig. 7.

3.1 Nucleon Electromagnetic Form Factors

We did two separate fits fixing the normalization f_N/λ_1, and the lowest order shape parameters φ_{10}, φ_{11} to the values given in Table 1 for two Borel-prameters to the proton data on the magnetic form factor $G_M^p(Q^2)$ and the ratio $G_E^p(Q^2)/G_M^p(Q^2)$ in the interval $1 < Q^2 < 8.5$ GeV^2. The fitted values of the shape parameters are given in Table 1 and the corresponding form factors are shown in Fig. 2. Several noteworthy points are seen in the result:

First, the experimental data prefers larger values for the ratio of f_N/λ_1 compatible with NLO sum rule calculations but a factor two larger than the lattice result. A fit with all parameters free gets unstable but we see that for different fixed parameters it is a rather robust feature that a large normalization and small first order shape parameters are favored.

Second, the neutron magnetic form factor $G_M^n(Q^2)$ which is not fitted comes out about 20 to 30 % too low. This feature is pretty robust. A fit to both proton and neutron data simultaneously leads to very large values of η_{10}, $\eta_{11} \sim \mathcal{O}(1)$ and leads to a worse description of proton data. We think this is an artefact of missing information on even higher twist distribution amplitudes and of the more involved OPE of the form factor F_2. Part of this can be understood with the help of Fig. 4 where the experimental data is separated into u- and d-quark contribution. It is seen that the Dirac form factors $F_1^{u,d}(Q^2)$ are described rather well while there are considerable deviations in the Pauli form factors $F_2^{u,d}(Q^2)$ at low Q^2. This does not come unexpected. At low Q^2 one would expect F_2 to get sizeable corrections from very high twist, e.g. factorizable five quark distribution amplitudes and the structure of the correlation function is so, that to get the same accuracy in

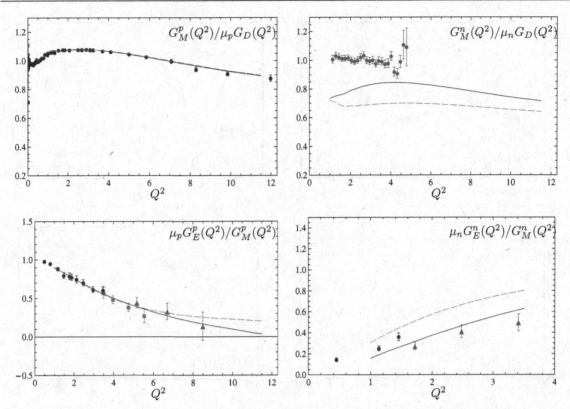

Fig. 2 Nucleon electromagnetic form factors from LCSRs compared to the experimental data [44–50]. Parameters of the nucleon DAs correspond to the sets ABO1 and ABO2 in Table 1 for the solid and dashed curves, respectively. The fits were done for different Borel parameters, i.e. $M^2 = 1.5\,\text{GeV}^2$ for ABO1 and $M^2 = 2\,\text{GeV}^2$ for ABO2

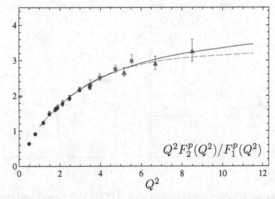

Fig. 3 The ratio of Pauli and Dirac electromagnetic proton form factors from LCSRs compared to the experimental data [46–48]. Parameters of the nucleon DAs correspond to the sets ABO1 and ABO2 in Table 1 for the *solid* and *dashed curves*, respectively. Borel parameter $M^2 = 1.5\,\text{GeV}^2$ for ABO1 and $M^2 = 2\,\text{GeV}^2$ for ABO2

the NLO contributions for F_2 one would need to take into account second order corrections in the deviation from the light-cone which are reserved for a future project. Additionally it is seen that the NLO corrections to the d-quark contribution are generally very large, probably a feature of the spin-flavor structure of the Ioffe-current, which means they are generally less precise and potentially stronger affected by higher QCD corrections. Since in the neutron the role of the d-quark is taken by the u-quark, the larger charge factor leads to an enhancement of aforementioned problems and therefore to lesser accuracy in the neutron form factors.

Third, we did not take into account the uncertainty due to the Borel-parameter separately but rather did two fits with different Borel-parameters. The difference in the shape parameters between the two fits can be seen as a measure for the induced deviation. We have illustrated the separate variation of the Borel-parameter

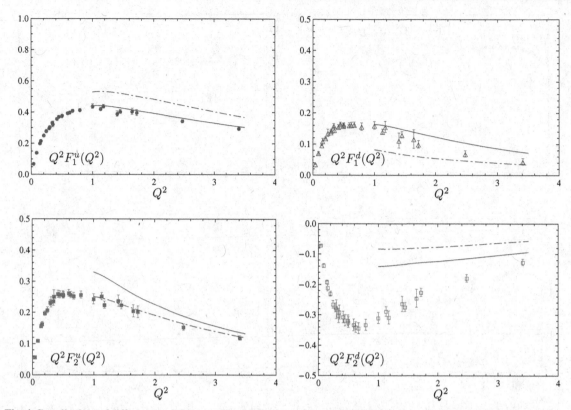

Fig. 4 Contributions of different quark flavors to the proton electromagnetic form factors compared to the compilation of experimental data in Ref. [51]. The corresponding leading-order results are shown by the *dash-dotted curves* for comparison. Parameters of the nucleon DAs correspond to the set ABO1 in Table 1

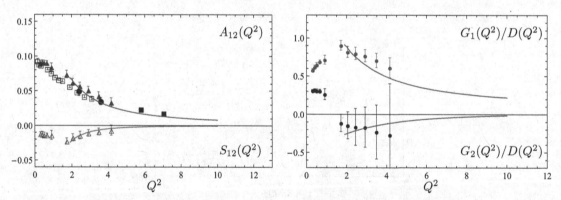

Fig. 5 Helicity amplitudes A_{12} and S_{12} for electroproduction of $N^*(1535)$ (*left panel*) and the form factors $G_1(Q^2)$, $G_2(Q^2)$, normalized to the dipole formula (*right panel*). Experimental data on the left panel are taken from [38] (empty squares) [39] (*filled squares*) [40] (*filled circles*) and [37] (*triangles*). The form factors on the right panel are calculated from the data [37] on helicity amplitudes adding the errors in quadrature. The *curves* show the results of the NLO LCSR fit to the form factors $G_1(Q^2)$ and $G_2(Q^2)$ for $Q^2 \geq 1.7$ GeV^2 with parameters of the $N^*(1535)$ DAs specified in the first line in Table 1

in figure 6 of [34]. Fourth, the factorization scale dependence increases with increasing momentum transfer Q^2. This might at first glance seem counterintuitive but it is consistent with the expected dominant role of the hard scattering corrections which start at next-next-to-leading order (NNLO).

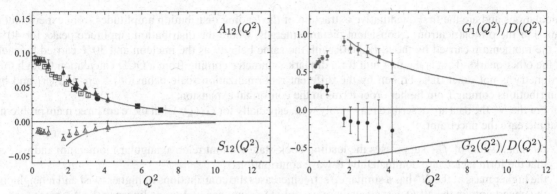

Fig. 6 The same as in Fig. 2 but for the fit to helicity amplitudes A_{12}, S_{12} including all available data at $Q^2 \geq 1.7$ GeV^2. The fitted parameters of the $N^*(1535)$ DAs are specified in the *second line* in Table 1

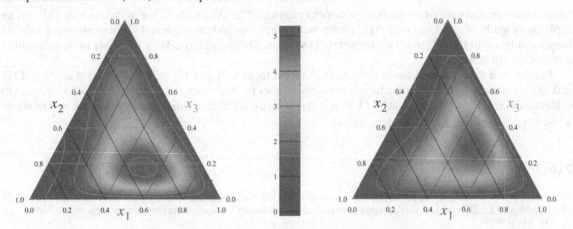

Fig. 7 Leading twist distribution amplitude of the proton $\varphi(x_i)$ for the parameter sets ABO1 (*left*) and ABO2 (*right*) in Table 1. *Central values* are used for the second order parameters

3.2 N^* Electroproduction Form Factors

Due to the larger mass of the N^* the light-cone sum rules get unstable for $Q^2 < 2$ GeV^2 in this case. Since data for $Q^2 > 2$ GeV^2 is relatively scarce we set φ_{20}, φ_{21} and φ_{22} to zero and fix λ_1^*, f_{N^*}, φ_{10} and φ_{11} to the lattice values. In this way we are left only with the twist 4 parameters η_{10} and η_{11}. We did two separate fits. One where we extracted the form factors from the helicity amplitudes A_{12} and S_{12} from [37] adding the uncertainties in quadrature and then fit the shape parameters to the form factors.

And one where we fitted directly to all data on the helicity amplitudes. The latter fit is driven by the data from [38–40] on the helicity amplitude A_{12} which is not entirely consistent with [37]. Therefore a worse description of the extracted form factors is expected.

In general the sum rules have dominant contributions from P-wave states that is states with one unit of angular momentum. Especially the helicity amplitude A_{12} and the Dirac-like form factor G_1 are nearly insensitive to the shape of the leading twist distribution amplitude mainly due to the very small normalization constant f_{N^*}. Even the sensitivity on η_{10} and η_{11} is rather mild. They are predominantly affected by the ratio $\lambda_1^{N^*}/\lambda_1^N$ which comes out rather robust in the range of the lattice result. S_{12} and the Pauli-like form factor G_2 on the other hand are far more sensitive to η_{10} and η_{11} and due to cancellations of higher twist contributions even to the leading twist distribution amplitude but to a lesser degree. More data will be needed to make this extraction more robust.

4 Conclusions

We have presented the results of the first consistent NLO light-cone sum rules description of the nucleon electromagnetic form factor and of the N^* electroproduction form factors. The results are consistent with lattice calculations and are the first quantitative extraction of the leading distribution amplitudes from experimental data. For the proton(neutron) a consistent picture emerges, where the distribution amplitude peaks for 40 % of the momentum carried by the u(d)-quark with the same helicity as the nucleon and 30 % carried by each of the other quarks. This is the first hint for a diquark-symmetry coming from a QCD calculation though this symmetry is not exact: It is broken by the different renormalization scale behavior of φ_{10} and φ_{11} and by contributions coming from higher order terms in the conformal expansion.

For the N^* the data are described reasonably well, especially for $G_1(Q^2)$, but there are three main problems that increase the uncertainty:

1. the small value of f_{N^*} suppresses the leading Fock-state without relative angular momentum and we see that the form factors are dominated by P-wave contributions
2. the higher mass of the N^* has a similar effect. It increases the contributions of higher twist and it heightens the uncertainty in the NLO part since there we only took into account terms linear in the N^* mass
3. strong cancellations of higher twist contributions for the Pauli-like form factor $G_2(Q^2)$

Several more projects are either planned or work in progress. The axial form factor of the nucleon [52] and an exploratory study of the $\Lambda_{c(b)} \to N^*$ [53] form factors are close to being finished. An extension towards the Roper-resonance or the $N^*(1650)$ is planned. In both cases a better understanding of higher twist distribution amplitudes will be needed.

Finally on a longer time scale to bring both the form factors F_2 and G_2 on the same level as F_1 and G_1 and to lessen the uncertainty for the higher mass resonances in view of expected new data [54,55] we plan to calculate the $m_{N(N^*)}^2$ -corrections at NLO. This will require a dedicated calculation where several new relations at the twist 5 level will have to be derived.

References

1. Chernyak, V.L., Zhitnitsky, A.R.: Asymptotic behavior of Hadron form-factors in Quark model. JETP Lett. **25**, 510 (1977)
2. Chernyak, V.L., Zhitnitsky, A.R.: Asymptotics of hadronic form-factors in the quantum chromodynamics. Sov. J. Nucl. Phys. **31**, 544 (1980)
3. Chernyak, V.L., Zhitnitsky, A.R., Serbo, V.G.: Asymptotic hadronic form-factors in quantum chromodynamics. JETP Lett. **26**, 594 (1977)
4. Chernyak, V.L., Zhitnitsky, A.R., Serbo, V.G.: Calculation of asymptotics of the pion electromagnetic form-factor in the QCD perturbation theory. Sov. J. Nucl. Phys. **31**, 552 (1980)
5. Radyushkin, A.V.: JINR report R2-10717 (1977), Deep elastic processes of composite particles in field theory and asymptotic freedom. arXiv:hep-ph/0410276 (English translation)
6. Efremov, A.V., Radyushkin, A.V.: Electromagnetic form-factor In QCD. Theor. Math. Phys. **42**, 97 (1980)
7. Efremov, A.V., Radyushkin, A.V.: Factorization and asymptotical behavior of pion form-factor in QCD. Phys. Lett. B **94**, 245 (1980)
8. Lepage, G.P., Brodsky, S.J.: Exclusive processes in quantum chromodynamics: evolution equations for hadronic wave functions and the form-factors of mesons. Phys. Lett. B **87**, 359 (1979)
9. Lepage, G.P., Brodsky, S.J.: Exclusive processes in perturbative quantum chromodynamics. Phys. Rev. D **22**, 2157 (1980)
10. Agaev, S.S., Braun, V.M., Offen, N., Porkert, F.A.: Light cone sum rules for the pi0-gamma*-gamma form factor revisited. Phys. Rev. D **83**, 054020 (2011)
11. Agaev, S.S., Braun, V.M., Offen, N., Porkert, F.A.: BELLE Data on the $\pi^0\gamma * \gamma$ Form Factor: A Game Changer?. Phys. Rev. D **86**, 077504 (2012)
12. Agaev, S.S., Braun, V.M., Offen, N., Porkert, F.A., Schäfer, A.: Transition form factors $\gamma^*\gamma \to \eta$ and $\gamma^*\gamma \to \eta'$ in QCD. Phys. Rev. D **90**(7), 074019 (2014)
13. Mikhailov, S.V., Stefanis, N.G.: Transition form factors of the pion in light-cone QCD sum rules with next-to-next-to-leading order contributions. Nucl. Phys. B **821**, 291 (2009). doi:10.1016/j.nuclphysb.2009.06.027. arXiv:0905.4004 [hep-ph]
14. Bakulev, A.P., Mikhailov, S.V., Pimikov, A.V., Stefanis, N.G.: Pion-photon transition: the new QCD frontier. Phys. Rev. D **84**, 034014 (2011). doi:10.1103/PhysRevD.84.034014. arXiv:1105.2753 [hep-ph]
15. Li, H., Sterman, G.F.: Pion-photon transition: the perturbative pion form-factor with Sudakov suppression. Nucl. Phys. B **381**, 129 (1992)
16. Balitsky, I.I., Braun, V.M., Kolesnichenko, A.V.: Sigma-P gamma decay in QCD. Sov. J. Nucl. Phys. **44**, 1028 (1986). (IN RUSSIAN)
17. Balitsky, I.I., Braun, V.M., Kolesnichenko, A.V.: Radiative decay sigma-p gamma in quantum chromodynamics. Nucl. Phys. B **312**, 509 (1989)
18. Chernyak, V.L., Zhitnitsky, I.R.: B meson exclusive decays into baryons. Nucl. Phys. B **345**, 137 (1990)
19. Braun, V.M., Khodjamirian, A., Maul, M.: Pion form-factor in QCD at intermediate momentum transfers. Phys. Rev. D **61**, 073004 (2000)
20. Braun, V.M., Lenz, A., Mahnke, N., Stein, E.: Light cone sum rules for the nucleon form-factors. Phys. Rev. D **65**, 074011 (2002)
21. Braun, V.M., Lenz, A., Wittmann, M.: Nucleon form factors in QCD. Phys. Rev. D **73**, 094019 (2006)

22. Lenz, A., Wittmann, M., Stein, E.: Improved light cone sum rules for the electromagnetic form-factors of the nucleon. Phys. Lett. B **581**, 199 (2004)
23. Aznauryan, I.G., Burkert, V.D., Lee, T.S.H.: Asymptotic behavior of exclusive processes in QCD. arXiv:0810.0997 [nucl-th]
24. Chernyak, V.L., Zhitnitsky, A.R.: Asymptotic behavior of exclusive processes in QCD. Phys. Rep. **112**, 173 (1984)
25. Braun, V., Fries, R.J., Mahnke, N., Stein, E.: Higher twist distribution amplitudes of the nucleon in QCD. Nucl. Phys. B **589**, 381 (2000) [Erratum-ibid. B **607**, 433 (2001)]
26. Braun, V.M., Korchemsky, G.P., Mueller, D.: The uses of conformal symmetry in QCD. Prog. Part. Nucl. Phys. **51**, 311 (2003). doi:10.1016/S0146-6410(03)90004-4. arXiv:hep-ph/0306057
27. Bukhvostov, A.P., Frolov, G.V., Lipatov, L.N., Kuraev, E.A.: Evolution equations for quasi-partonic operators. Nucl. Phys. B **258**, 601 (1985). doi:10.1016/0550-3213(85)90628-5
28. Braun, V.M., Manashov, A.N., Rohrwild, J.: Baryon operators of higher twist in QCD and nucleon distribution amplitudes. Nucl. Phys. B **807**, 89 (2009). doi:10.1016/j.nuclphysb.2008.08.012. arXiv:0806.2531 [hep-ph]
29. Braun, V.M., Manashov, A.N., Rohrwild, J.: Renormalization of twist-four operators in QCD. Nucl. Phys. B **826**, 235 (2010). doi:10.1016/j.nuclphysb.2009.10.005. arXiv:0908.1684 [hep-ph]
30. Braun, V.M., et al.: QCDSF collab.: nucleon distribution amplitudes and proton decay matrix elements on the lattice. Phys. Rev. D **79**, 034504 (2009)
31. Schiel, R., et al.: Wave functions of the nucleon and the $N^*(1535)$, invited talk at the 31st International Symposium on Lattice Gauge Theory, July 29–August 03 (2013), Mainz
32. Braun, V.M., et al.: Light-cone distribution amplitudes of the nucleon and negative parity nucleon resonances from lattice QCD. Phys. Rev. D **89**, 094511 (2014)
33. Braun, V.M., et al.: Electroproduction of the N*(1535) resonance at large momentum transfer. Phys. Rev. Lett. **103**, 072001 (2009)
34. Anikin, I.V., Braun, V.M., Offen, N.: Nucleon form factors and distribution amplitudes in QCD. Phys. Rev. D **88**, 114021 (2013)
35. Gruber, M.: The nucleon wave function at the origin. Phys. Lett. B **699**, 169 (2011)
36. Anikin, I.V., Braun, V.M., Offen, N.: Electroproduction of the N*(1535) nucleon resonance in QCD. Phys. Rev. D **92**(1), 014018 (2015). doi:10.1103/PhysRevD.92.014018. arXiv:1505.05759 [hep-ph]
37. Aznauryan, I.G., et al.: Electroexcitation of nucleon resonances from CLAS data on single pion electroproduction. [CLAS Collaboration]. Phys. Rev. C **80**, 055203 (2009)
38. Denizli, H., et al.: [CLAS Collaboration]. Q*2 dependence of the S(11)(1535) photocoupling and evidence for a P-wave resonance in eta electroproduction. Phys. Rev. C **76**, 015204 (2007)
39. Dalton, M.M., Adams, G.S., Ahmidouch, A., Angelescu, T., Arrington, J., Asaturyan, R., Baker, O.K., Benmouna, N., et al.: Electroproduction of eta mesons in the S(11)(1535) resonance region at high momentum transfer. Phys. Rev. C **80**, 015205 (2009)
40. Armstrong, C.S., et al.: [Jefferson Lab E94014 Collaboration]: electroproduction of the S(11)(1535) resonance at high momentum transfer. Phys. Rev. D **60**, 052004 (1999)
41. Krankl, S., Manashov, A.: Two-loop renormalization of three-quark operators in QCD. Phys. Lett. B **703**, 519 (2011). doi:10.1016/j.physletb.2011.08.028. arXiv:1107.3718 [hep-ph]
42. Passek-Kumericki, K., Peters, G.: Nucleon form factors to next-to-leading order with light-cone sum rules. Phys. Rev. D **78**, 033009 (2008)
43. Anikin, I.V., Manashov, A.N.: Higher twist nucleon distribution amplitudes in Wandzura-Wilczek approximation. Phys. Rev. D **89**(1), 014011 (2014). doi:10.1103/PhysRevD.89.014011. arXiv:1311.3584 [hep-ph]
44. Arrington, J., Melnitchouk, W., Tjon, J.A.: Global analysis of proton elastic form factor data with two-photon exchange corrections. Phys. Rev. C **76**, 035205 (2007)
45. Lachniet, J., et al.: CLAS collab.: a precise measurement of the neutron magnetic form factor G**n(M)in the few-GeV**2 region. Phys. Rev. Lett. **102**, 192001 (2009)
46. Gayou, O., et al.: [Jefferson Lab Hall A Collaboration]: measurement of G(Ep) / G(Mp) in polarized-e p - e polarized-p to $Q**2 = 5.6 - GeV**2$. Phys. Rev. Lett. **88**, 092301 (2002)
47. Punjabi V., et al.: Proton elastic form-factor ratios to $Q**2 = 3.5 - GeV**2$ by polarization transfer. Phys. Rev. C **71**, 055202 (2005) [Erratum-ibid. C **71**, 069902 (2005)]
48. Puckett, A.J.R., et al.: Recoil polarization measurements of the proton electromagnetic form factor ratio to $Q2 = 8 : 5$ GeV2. Phys. Rev. Lett. **104**, 242301 (2010)
49. Plaster, B., et al.: [Jefferson Laboratory E93-038 Collaboration]: Measurements of the neutron electric to magnetic form-factor ratio G(En) / G(Mn) via the H-2 (polarized-e, e-prime, polarized-n) H-1 reaction to Q**2=1.45-(GeV/c)**2. Phys. Rev. C **73**, 025205 (2006)
50. Riordan, S., et al.: Measurements of the Electric Form Factor of the Neutron up to $Q^2 = 3.4 GeV^2$ using the Reaction $^3He^{->}(e^{->}, e'n)pp$. Phys. Rev. Lett. **105**, 262302 (2010)
51. Diehl, M., Kroll, P.: Nucleon form factors, generalized parton distributions and quark angular momentum. Eur. Phys. J. C **73**, 2397 (2013)
52. Anikin, I.V., Braun, V.M., Offen, N.: Axial form factor of the nucleon at large momentum transfers. arxiv:1607.01504 [hep-ph]
53. Emmerich, M., Offen, N., Schäfer, A.: The decays $\Lambda_{b,c} \to N^*l\nu$ in QCD. arXiv:1604.06595 [hep-ph]
54. Aznauryan, I.G., et al.: Studies of nucleon resonance structure in exclusive meson electroproduction. Int. J. Mod. Phys. E **22**, 1330015 (2013)
55. Gothe, R.W., et al.: Nucleon Resonance Studies with CLAS12, Experiment E12-09-003

Few-Body Syst (2016) 57:1051–1058
DOI 10.1007/s00601-016-1147-4

David G. Richards

N^* Resonances in Lattice QCD from (Mostly) Low to (Sometimes) High Virtualities

Received: 11 April 2016 / Accepted: 4 August 2016 / Published online: 20 September 2016
© Springer-Verlag Wien (Outside the USA) 2016

Abstract I present a survey of calculations of the excited N^* spectrum in lattice QCD. I then describe recent advances aimed at extracting the momentum-dependent phase shifts from lattice calculations, notably in the meson sector, and the potential for their application to baryons. I conclude with a discussion of calculations of the electromagnetic transition form factors to excited nucleons, including calculations at high Q^2.

1 Introduction

Lattice gauge theory provides a means of *solving* QCD in the low-energy regime, and thereby has a key role in hadronic and nuclear physics. The calculation of the low-lying spectrum, the masses of hadrons stable under the strong interactions, has long been a benchmark calculation for lattice QCD since it provides a confrontation between lattice QCD and precisely determined quantities in experiment. Such calculations are demanding, since they require a high degree of control over the systematic uncertainties inherent to lattice calculations, namely those arising from the finite volume in which they are performed, non-zero lattice spacing, and, finally, the need until recently to extrapolate from unphysical u and d quark masses to the physical light quark masses.

The focus of this talk is the spectrum of excited-state nucleons, and on the determination of their properties and photoproduction mechanisms. In fact, my emphasis will very much be on the former since it is a prerequisite for the latter; we will hear elsewhere in this workshop of recent theoretical developments aimed at a rigorous determination of hadronic matrix elements that involve multi-hadron states [16]. I will begin by reminding us briefly of the formalism of lattice QCD, and note some of the computational challenges. I will point out some recent successes at looking at the properties of stable hadrons such as the nucleon, before proceeding to describe how the spectrum of excited states can be determined on the lattice. In particular, I will focus on the use of the variational method, and how that can provide important insights both into the masses of their states, and their quark and gluon structure. I will then describe how resonances are treated in a lattice calculation, and the determination of the momentum-dependent phase shifts. I will conclude with a discussion of lattice calculations of transition form factors, and a summary.

2 Lattice QCD and the Properties of Ground-state Hadrons

Lattice QCD is Quantum Chromodynamics formulated on a Euclidean space-time lattice, with the "lattice spacing" fulfilling the rôle of a hard, ultraviolet cut-off. An "observable" in lattice QCD can be calculated

This article belongs to the special issue "Nucleon Resonances".

D. G. Richards (✉)
Jefferson Lab, 12000 Jefferson Avenue, Newport News, VA 23606, USA
E-mail: dgr@jlab.org
Tel.: +1-757-269-7736

through the evaluation of the discretized path integral

$$\langle \mathcal{O} \rangle = \frac{1}{\mathcal{Z}} \prod_{x,\mu} dU_\mu(x) \prod_x d\psi(x) \prod_x d\bar{\psi}(x) \mathcal{O}(U, \psi, \bar{\psi}) e^{-S(U, \psi, \bar{\psi})}, \tag{1}$$

where $U_\mu(x)$ are $SU(3)$ matrices representing the gluonic degrees of freedom, and $\psi, \bar{\psi}$ are Grassmann variables representing the quarks. Writing the QCD action in terms of the pure gauge and fermion components, we have

$$S(U, \psi, \bar{\psi}) = S_g(U) + \bar{\psi} M(U) \psi,$$

whence we can integrate the fermion degrees of freedom to yield

$$\langle \mathcal{O} \rangle = \frac{1}{\mathcal{Z}} \prod_{x,\mu} dU_\mu(x) \mathcal{O}(U) \det M(U) e^{-S_g(U)}. \tag{2}$$

The need to formulate the theory in Euclidean space is now clear: Eq. 2 can be estimated using the *importance sampling* methods familiar to statistical physics, since the action $S_g(U)$, and indeed the determinant, are real. A lattice gauge calculation therefore consists, firstly, of generating an ensemble of equilibrated gauge configuration $\{U_\mu^n(x) : n = 1, \ldots, N_{cfg}\}$ distributed according to

$$P(U) \propto \det M(U) e^{-S_g(U)} \tag{3}$$

and then calculating the expectation value of the operator $\mathcal{O}$ on

$$\langle \mathcal{O} \rangle = \frac{1}{N_{cfg}} \sum_{n=1}^{N_{cfg}} \mathcal{O}(U^n, G[U^n]) \tag{4}$$

where $G[U^n]$ represent quark propagators computed on the gauge field $U_\mu^n(x)$. The generation of the gauge fields in Eq. 3 is a *capability* computing task in which the whole, or at least a sizable proportion, of a leadership-class computer has to be devoted to a single sequence of Monte Carlo calculations to yield a thermalized distribution. The evaluation of $\det M(U)$ is dominant, since any local change to gauge linke $U_\mu(x)$ requires the evaluation of the determinant over the whole lattice, in contrast to a pure-gauge calculation where the change in $S_g(U)$ can be evaluated locally. The evaluation of Eq. 4 is inherently a *capacity* computing task, since the calculation of $\mathcal{O}(U^n)$ can be performed on each configuration independently. For many of the calculations described below, the integrated cost of these computations can far exceed those of Eq. 3, but these can exploit cost-optimized clusters and GPU-accelerated clusters.

The spectrum of QCD is determined through the exponential decay of two-point correlation functions

$$C(t) = \sum_{\mathbf{x}} \langle 0 \mid \mathcal{O}(\mathbf{x}, t) \bar{\mathcal{O}}(0) \mid 0 \rangle \longrightarrow \sum_n A_n e^{-E_n t}, \tag{5}$$

where $\mathcal{O}$ is an interpolating operator, say for the nucleon, and the E_n and A_n are the energies and residues of all states n with $\langle n \mid \bar{\mathcal{O}} \mid 0 \rangle \neq 0$; the energies E_n are real, reflecting the formulation in Euclidean space. The masses of the low-lying states of the spectrum, that is states stable under the strong interaction, are determined from the leading exponential in Eq. 5.

As noted earlier, calculations that can directly confront experiment require control over a variety of systematic uncertainties. A prominent example of such a calculation, for states composed of the u, d and s quarks, was that of the BMW collaboration [31], and they have recently extended their work through the inclusion of QED computations to provide a calculation of the proton-neutron mass difference, as well as that for other hadrons [13]. Beyond the low-lying spectrum, calculations are also being performed of the matrix elements of the nucleon, including those for the (space-like) electric and magnetic form factors, and for the moments of parton distributions and Generalized Parton Distributions, initially for the isovector properties, but now extended to include the effects of the sea quarks; for a recent review, see ref. [26]. For each of these quantities, calculations in Euclidean space can be directly related to quantities in Minkowski space. None-the-less, even for ground-state hadrons, there are limitations to properties that can be straightforwardly calculated, notably the matrix elements of the quark bilinears separated along the light cone that give rise to the (generalized) parton distribution functions and transverse-momentum-dependent distributions, and methods have been proposed [37,45,55] and indeed applied, to circumvent these issues in Euclidean-space calculations.

3 The excited-State Spectrum in Lattice QCD

The first challenge in the study of the excited-state spectrum is to effectively determine the subleading exponentials in Eq. 5. A robust way of doing so is by means of the variational method, whereby we compute a matrix of correlation functions

$$C_{ij}(t) = \sum_{\mathbf{x}} \langle 0 \mid \mathcal{O}_i^{J^P}(\mathbf{x}, t) \bar{\mathcal{O}}_j^{J^P}(\mathbf{0}, 0) \mid 0 \rangle \longrightarrow \sum_n A_{ij}^n e^{-E_n t}, \tag{6}$$

where $\{\mathcal{O}_i^{J^P} : i = 1, \ldots, N\}$ is a basis of operators, each having common quantum numbers. We now solve the generalized eigenvalue equation

$$C(t)u(t, t_0) = \lambda(t, t_0)C(t_0)u(t, t_0) \tag{7}$$

yielding a set of real eigenvalues $\{\lambda_n(t, t_0) : n = 1, \ldots, N\}$ with corresponding eigenvectors $\{u^n(t, t_0) : n = 1, \ldots, N\}$, where, for sufficiently large t, we have $\lambda_0 \geq \lambda_1 \geq \ldots$. These delineate the different states

$$\lambda_n(t, t_0) \to (1 - A)e^{-E_n(t-t_0)} + A e^{-(E_n + \Delta E_n)(t-t_0)}. \tag{8}$$

The application of the variational method replies on the construction of a suitable basis of interpolating operators. In the case of the nucleon, there are only three local interpolating operators that can be constructed from three quark fields, such that at most three energies can be determined:

$$\mathcal{O}^{1/2} = \begin{cases} (uC\gamma_5 d)u \\ (uCd)\gamma_5 u \\ (uC\gamma_4\gamma_5 d)u. \end{cases} \tag{9}$$

The basis of operators can be extended to include quasi-local quark operators through the use of gauge-invariant smearing whereby a quark field $\psi(\mathbf{x}, t) \longrightarrow \tilde{\psi}(\mathbf{x}, t) = \sum_{\mathbf{y}} L(\mathbf{x}, \mathbf{y})\psi(\mathbf{y})$ where $L(\mathbf{x}, \mathbf{y})$ is a gauge-covariant operator, such as the inverse of a three-dimensional Laplacian; such a construction does not alter the angular-momentum structure of a nucleon operator. There have been many studies in the last few years that have aimed at extracting the excited-state nucleon spectrum of both parities through the use of the variational method [7,24,25,46–49], as well as through techniques such as the sequential-Bayes method aimed at delineating subleading terms from only a single correlator [50]. A basis of operators with different smearing radii for the quark fields can capture the radial structure of the nucleons, and indeed allow for nodes in the wave function. However, it appears incomplete in that it does not capture the orbital structure. To do so, and indeed to study states of spin higher than 3/2, requires operators that are non-local, either by displacing one or more of the quarks, or equivalently, through the use of covariant derivatives acting on the quark fields. Such constructions, in which the bases of operators were designed in the first instance to satisfy the symmetries of the lattice, were introduced in refs. [10] and [11], and applied in refs. [12,22,23], providing, for the first time, access to states of spin 5/2 and higher.

The lack of rotational symmetry introduced through the discretisation onto a finite space-time lattice has the consequence that angular momentum is no longer a good quantum number at any finite spacing. We find in the meson sector that a remarkable degree of rotational symmetry in operator overlaps is realized at the hadronic scale, enabling the "single-particle" spectrum to be classified according to the total angular momentum of the states [28,29]. We exploit this observation in constructing a basis of interpolating operators for baryons. We begin by expressing continuum baryon interpolating operators of definite J^P as [32,33]

$$\mathcal{O}^{J^P} \sim \left(F_{\Sigma_F} \otimes (S^{P_s})_{\Sigma_S}^n \otimes D_{L,\Sigma_D}^{[d]} \right)^{J^P}, \tag{10}$$

where F, S and D are the flavor, Dirac spin and orbital angular momentum parts of the wave function, and the Σ's express the corresponding permutation symmetry: Symmetric (S), mixed-symmetric (MS), mixed anti-symmetric (MA), and anti-symmetric (A).

Non-zero orbital angular momentum is introduced through the use of gauge-covariant derivatives, written in a circular basis and acting on the quark fields. In the notation above, $D_{L,\Sigma_D}^{[d]}$ corresponds to an orbital wave function constructed from d derivatives, and projected onto orbital angular momentum L. In the calculations described here, up to two covariant derivatives are employed, enabling orbital angular momentum up to $L = 2$

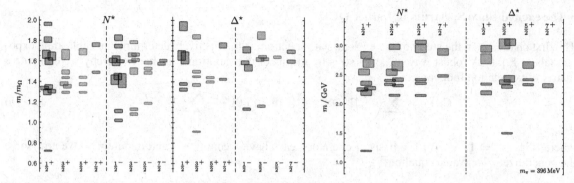

Fig. 1 The *left-hand plot* shows the spin-identified excited nucleon and Δ spectra in units of the Ω mass obtained on a $16^3 \times 64$ anisotropic clover lattice with the strange quark mass at its physical value, and the light-quark masses corresponding to a pion mass of 396 MeV [32]. The *right-hand plot* shows the positive-parity spectrum for both the nucleon and Δ in physical units using an operator basis with the "hybrid" operators included [27]. The additional states that couple predominantly to the hybrid-type operators are shown in *blue* (colour figure online)

to be accessed, and therefore states up to spin 7/2 to be studied; the spin-identified spectrum obtained using such a basis of operators is shown as the left-hand plot in Fig. 1. The spectrum reveals a counting consistent with that of a qqq non-relativistic quark model, and richer than certain quark-diquark pictures.

A faithful extraction of the spectrum is contingent on having a sufficiently complete basis of operators. In the analysis described above, one particular operator is absent, namely the mixed-symmetric combination $D_{L=1,M}^{[2]}$, the commutator of two covariant derivatives projected to $L = 1$, that corresponds to a chromo-magnetic field that would vanish for trivial gauge field configuration; operators with this construction we identify as *hybrid* operators, associated with a manifest gluon content [27]. The right-hand plot of Fig. 1 is the positive-parity nucleon and Δ spectrum on the same ensemble as that used in the left-hand plot, showing the additional states appearing when this more complete basis of operators is employed. The conclusion is that, with the addition of these operators, the counting of states is still richer than the quark model, with the additional states attributable to a coupling to a colour-octet gluonic excitation.

One feature of the experimentally observed nucleon spectrum that is absent in Fig. 1 is a low-lying Roper resonance, a positive-parity excitation *below* the lowest-lying negative-parity state. This feature of a quark-model type ordering of the low-lying positive- and negative-parity excitations is a feature of most of the lattice calculations discussed above. The exception to this is the calculation of the χQCD group, using overlap fermions on an ensemble generated using domain-wall fermions [42]; their calculation, together with the results of other groups, is shown in Fig. 2. It has been argued that the lattice calculations of the Roper are incomplete, lacking the multi-hadron operators as we discuss below, but not inconsistent with the observed spectrum [39].

3.1 Multi-Hadron Spectrum and the Nature of Resonances

Lattice QCD is formulated in Euclidean space, and the energies entering into the spectral decomposition of Eq. 6 are real. On a finite, discretized Euclidean lattice, the spatial momenta are quantized, and the calculated discrete spectrum, that is the energies of Eq. 5, should include two- and higher-body scattering states. Indeed, even at the unphysically large quark masses used in Fig. 1, many of the higher excitations are above decay thresholds. For non-interacting particles, the energies are given by the symmetries of the volume in which we are working and the allowed spatial momenta. The finite, periodic spatial volume forces those hadrons to interact thereby shifting the energies from their non-interacting values. For the case of elastic scattering, the so-called Lüscher method enables the shift in energies at a finite volume to be related to the infinite volume phase shift [43,44], shortly thereafter extended to states with non-zero total momentum [52]. In the meson sector, the precise calculation of the momentum-dependent phase shifts for states such as the ρ meson in $I = 1 \pi\pi$ scattering has now been accomplished [30], and the formalism has been extended to the extraction of the momentum-dependent amplitudes for inelastic scattering [17,34–36].

Analogous calculations in the baryon sector are less advanced. For most of the calculations discussed above, the interpolating operators were constructed from three quarks, albeit with quite elaborate orbital structures. The two-hadron energy levels that should be seen in the spectrum are in general absent, and their non-observation is attributed to the volume suppression of single-hadron operators to multi-hadron states.

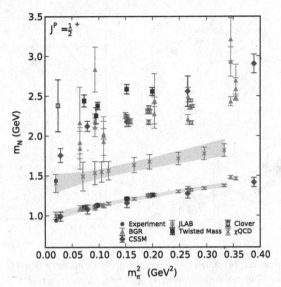

Fig. 2 Calculations of the lowest-lying positive parity excitation of the nucleon, taken from reference [42]. The χQCD results are obtained using overlap fermions on an ensemble generated using domain-wall fermions

Several groups have included multi-hadron or five-quark operators in their basis [38,54]. Notably, one group has included interpolators of the form $\mathcal{O}_{N\pi}(\mathbf{p}) = N^i(\mathbf{p})\pi(\mathbf{0})$, where $N^i, i = 1, 2, 3$, are the three local interpolators of Eq. 9, and observe $N\pi$ states as expected, but also the presence of an additional energy level in the spectrum. A different approach, applicable to transitions near threshold [51], has been applied to the decays of decuplet baryon resonances to an octet baryon-pion final state [5,6]. However, the application of the full panoply of the Lüscher method is currently limited.

4 Electromagnetic Transitions in N^* Resonances

The formalism to extract infinite-volume transition matrix elements between states containing two or more hadrons from Euclidean space lattice calculations has only recently been developed [19–21], and applied to meson transitions [18]; it will be presented workshop [16]. Instead, I will backtrack somewhat and talk about calculations of electromagnetic properties in which both the incoming and outgoing particles are treated as single-particle states.

Hadronic matrix elements in lattice QCD are computed through the calculation of three-point functions:

$$C_{3\mathrm{pt}}(t_f, t; \mathbf{p}, \mathbf{q}) = \sum_{\mathbf{x}} \sum_{\mathbf{y}} \langle 0 \mid \mathcal{O}_1(\mathbf{x}, t_f) J(\mathbf{y}, t) \bar{\mathcal{O}}_2(0) \mid 0 \rangle e^{-i\mathbf{p}\cdot\mathbf{x}} e^{-i\mathbf{q}\cdot\mathbf{y}}$$

$$\rightarrow \langle 0 \mid \mathcal{O}_1 \mid N_1, \mathbf{p} \rangle \langle N_1, \mathbf{p} \mid \bar{\mathcal{O}}_2 \mid 0 \rangle \langle N_1, \mathbf{p} \mid J \mid N_2, \mathbf{p} + \mathbf{q} \rangle$$

$$\times e^{-E_1(\mathbf{p})(t_f - t)} e^{-E_2(\mathbf{p}+\mathbf{q})t}, \tag{11}$$

with $Q^2 = (E_2(\mathbf{p}) - E_2(\mathbf{p}+\mathbf{q}))^2 - \mathbf{q}^2$, and where we treat $\mathcal{O}_1$ and $\mathcal{O}_2$ as ideal interpolating operators for the incoming and outgoing states, obtained, for example, from the eigenvectors obtained in the variational method [53].

The most widely studied transition is that to the Δ resonance [3,4], though there was an earlier investigation of the P_{11} Roper transtion [40]. The former is of particular interest both because there has been considerable experimental effort, and because a non-zero quadrupole moment can provide important information about the structure nucleon inaccessible from the proton form factor. A recent example of such a calculation is shown in Fig. 3, for both the magnetic dipole transition form factor and for the ratio of the quadrupole to dipole form factors. As noted earlier, it must be emphasised that the interpretation of infinite-volume matrix elements for multi-hadron states requires considerable theoretical work [1,2], and the application is computationally demanding.

As the momenta in Eq. 11 increase, lattice calculations of the form factors become increasingly demanding, both because of decreasing signal-to-noise ratios and because of increasing discretisation errors. There have

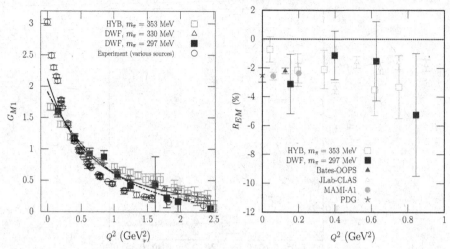

Fig. 3 The *left-hand plot*, taken from ref. [4] shows the magnetic dipole form factor $G_{M1}(Q^2)$ for the N to Δ transition. The *right-hand plot*, taken from the same paper, shows the ratio $R_{EM} = -G_{E2}(Q^2)/G_{M1}(Q^2)$

been several ideas that aim to overcome these issues [8,41] to enable the direct calculation of the Q^2 dependence of the form factor, but another approach, applicable to hard, exclusive processes, is through the calculation of the hadronic wave functions for both the nucleon and for its excitations [9,15]; that approach is described elsewhere at this workshop [14].

5 Summary

There has been far-reaching theoretical and computational progress aimed at understanding the excited state spectrum of QCD through lattice calculations. Much of the computational work has been centered on the meson sector, where the computational are considerably reduced. However, with the increasing computational resources, and our increasing theoretical understanding, these ideas will be applied to the N^* spectrum, enabling lattice calculations to truly confront experiment.

Acknowledgments The author would like to thank his colleagues in the *HadSpec Collaboration* for discussions and collaboration on some of the work. I am grateful to the authors of refs. [42] and [4] for use of their figures. This material is based upon work supported by the U.S. Department of Energy, Office of Science, Office of Nuclear Physics under contract DE-AC05-06OR23177.

References

1. Agadjanov, A., Bernard, V., Meißner, U.G., Rusetsky, A.: A framework for the calculation of the $N\Delta\gamma^*$ transition form factors on the lattice. Nucl. Phys. **B886**, 1199–1222 (2014). doi:10.1016/j.nuclphysb.2014.07.023. arXiv:1405.3476
2. Agadjanov, A., Bernard, V., Meißner, U.G., Rusetsky, A.: Resonance matrix elements on the lattice. EPJ Web Conf. **112**, 01,001 (2016). doi:10.1051/epjconf/201611201001
3. Alexandrou, C., Leontiou, T., Negele, J.W., Tsapalis, A.: The Axial N to Delta transition form factors from Lattice QCD. Phys. Rev. Lett. **98**, 052,003 (2007). doi:10.1103/PhysRevLett.98.052003. arXiv:hep-lat/0607030
4. Alexandrou, C., Koutsou, G., Negele, J.W., Proestos, Y., Tsapalis, A.: Nucleon to Delta transition form factors with $N_F = 2+1$ domain wall fermions. Phys. Rev. D **83**, 014,501 (2011). doi:10.1103/PhysRevD.83.014501. arXiv:1011.3233
5. Alexandrou, C., Negele, J.W., Petschlies, M., Pochinsky, A.V., Syritsyn, S.N.: Study of decuplet baryon resonances from lattice QCD (2015). arXiv:1507.02724
6. Alexandrou, C., Negele, J.W., Petschlies, M., Pochinsky, A.V., Syritsyn, S.S.: Calculation of the decay width of decuplet baryons. In: Proceedings, 33rd international symposium on lattice field theory (Lattice 2015). https://inspirehep.net/record/1403565/files/arXiv:1511.02752.pdf, arXiv:1511.02752
7. Allton, C.R., et al.: Gauge invariant smearing and matrix correlators using Wilson fermions at Beta = 6.2. Phys. Rev. D **47**, 5128–5137 (1993). doi:10.1103/PhysRevD.47.5128. arXiv:hep-lat/9303009
8. Bali, G.S., Lang, B., Musch, B.U., Schäfer, A.: A novel quark smearing for hadrons with high momenta in lattice QCD (2016). arXiv:1602.05525
9. Bali, G.S., et al.: Light-cone distribution amplitudes of the baryon octet. JHEP **02**, 070 (2016). doi:10.1007/JHEP02(2016)070. arXiv:1512.02050

10. Basak, S., Edwards, R., Fleming, G.T., Heller, U.M., Morningstar, C., Richards, D., Sato, I., Wallace, S.J.: Clebsch–Gordan construction of lattice interpolating fields for excited baryons. Phys. Rev. D **72**, 074,501 (2005a). doi:10.1103/PhysRevD.72.074501. arXiv:hep-lat/0508018

11. Basak, S., Edwards, R.G., Fleming, G.T., Heller, U.M., Morningstar, C., Richards, D., Sato, I., Wallace, S.: Group-theoretical construction of extended baryon operators in lattice QCD. Phys. Rev. D **72**, 094,506 (2005b). doi:10.1103/PhysRevD.72.094506. arXiv:hep-lat/0506029

12. Basak, S., Edwards, R.G., Fleming, G.T., Juge, K.J., Lichtl, A., Morningstar, C., Richards, D.G., Sato, I., Wallace, S.J.: Lattice QCD determination of patterns of excited baryon states. Phys. Rev. D **76**, 074,504 (2007). doi:10.1103/PhysRevD.76.074504. arXiv:0709.0008

13. Borsanyi, S., et al.: Ab initio calculation of the neutron-proton mass difference. Science **347**, 1452–1455 (2015). doi:10.1126/science.1257050. arXiv:1406.4088

14. Braun, V.: Hadron Wave Functions from Lattice QCD. Few-Body Syst. (2016). doi:10.1007/s00601-016-1143-8

15. Braun, V.M., Collins, S., Gläßle, B., Göckeler, M., Schäfer, A., Schiel, R.W., Söldner, W., Sternbeck, A., Wein, P.: Light-cone distribution amplitudes of the nucleon and negative parity nucleon resonances from lattice QCD. Phys. Rev. D **89**, 094,511 (2014). doi:10.1103/PhysRevD.89.094511. arXiv:1403.4189

16. Briceno, R.: Meson electro-/photo-production from QCD. Few-Body Syst. (2016). doi:10.1007/s00601-016-1124-y

17. Briceno, R.A., Davoudi, Z.: Moving multichannel systems in a finite volume with application to proton-proton fusion. Phys. Rev. D **88**(9), 094,507 (2013). doi:10.1103/PhysRevD.88.094507. arXiv:1204.1110

18. Briceno, R.A., Dudek, J.J., Edwards, R.G., Shultz, C.J., Thomas, C.E., Wilson, D.J.: The resonant $\pi^+\gamma \rightarrow \pi^+\pi^0$ amplitude from quantum chromodynamics. Phys. Rev. Lett. **115**, 242,001 (2015). doi:10.1103/PhysRevLett.115.242001. arXiv:1507.06622

19. Briceño, R.A., Hansen, M.T.: Multichannel $0 \rightarrow 2$ and $1 \rightarrow 2$ transition amplitudes for arbitrary spin particles in a finite volume. Phys. Rev. **D92**(7), 074,509 (2015a). doi:10.1103/PhysRevD.92.074509. arXiv:1502.04314

20. Briceño, R.A., Hansen, M.T.: Relativistic, model-independent, multichannel $2 \rightarrow 2$ transition amplitudes in a finite volume (2015b). arXiv:1509.08507

21. Briceño, R.A., Hansen, M.T., Walker-Loud, A.: Multichannel $1 \rightarrow 2$ transition amplitudes in a finite volume. Phys. Rev. **D91**(3), 034,501 (2015). doi:10.1103/PhysRevD.91.034501

22. Bulava, J., Edwards, R.G., Engelson, E., Joo, B., Lin, H.W., Morningstar, C., Richards, D.G., Wallace, S.J.: Nucleon, Δ and Ω excited states in $N_f = 2 + 1$ lattice QCD. Phys. Rev. D **82**, 014,507 (2010). doi:10.1103/PhysRevD.82.014507. arXiv:1004.5072

23. Bulava, J.M., et al.: Excited state nucleon spectrum with two flavors of dynamical fermions. Phys. Rev. D **79**, 034,505 (2009). doi:10.1103/PhysRevD.79.034505. arXiv:0901.0027

24. Burch, T., Gattringer, C., Glozman, L.Y., Kleindl, R., Lang, C.B., Schaefer, A.: Spatially improved operators for excited hadrons on the lattice. Phys. Rev. D **70**, 054,502 (2004). doi:10.1103/PhysRevD.70.054502. arXiv:hep-lat/0405006

25. Burch, T., Gattringer, C., Glozman, L.Y., Hagen, C., Lang, C.B., Schafer, A.: Excited hadrons on the lattice: Mesons. Phys. Rev. D **73**, 094,505 (2006). doi:10.1103/PhysRevD.73.094505. arXiv:hep-lat/0601026

26. Constantinou, M.: Recent progress in hadron structure from Lattice QCD. In: 8th international workshop on chiral dynamics (CD 2015) Pisa, Italy, June 29–July 3, 2015. https://inspirehep.net/record/1402363/files/arXiv:1511.00214.pdf, arXiv:1511.00214

27. Dudek, J.J., Edwards, R.G.: Hybrid baryons in QCD. Phys. Rev. D **85**, 054,016 (2012). doi:10.1103/PhysRevD.85.054016. arXiv:1201.2349

28. Dudek, J.J., Edwards, R.G., Peardon, M.J., Richards, D.G., Thomas, C.E.: Highly excited and exotic meson spectrum from dynamical lattice QCD. Phys. Rev. Lett. **103**, 262,001 (2009). doi:10.1103/PhysRevLett.103.262001. arXiv:0909.0200

29. Dudek, J.J., Edwards, R.G., Peardon, M.J., Richards, D.G., Thomas, C.E.: Toward the excited meson spectrum of dynamical QCD. Phys. Rev. D **82**, 034,508 (2010). doi:10.1103/PhysRevD.82.034508. arXiv:1004.4930

30. Dudek, J.J., Edwards, R.G., Thomas, C.E.: Energy dependence of the ρ resonance in $\pi\pi$ elastic scattering from lattice QCD. Phys. Rev. D **87**(3), 034,505 (2013). doi:10.1103/PhysRevD.87.034505, doi:10.1103/PhysRevD.90.099902. [Erratum: Phys. Rev.D90,no.9,099902(2014)], arXiv:1212.0830

31. Durr, S., et al.: Ab-initio determination of light hadron masses. Science **322**, 1224–1227 (2008). doi:10.1126/science.1163233. arXiv:0906.3599

32. Edwards, R.G., Dudek, J.J., Richards, D.G., Wallace, S.J.: Excited state baryon spectroscopy from lattice QCD. Phys. Rev. D **84**, 074,508 (2011). doi:10.1103/PhysRevD.84.074508. arXiv:1104.5152

33. Edwards, R.G., Mathur, N., Richards, D.G., Wallace, S.J.: Flavor structure of the excited baryon spectra from lattice QCD. Phys. Rev. D **87**(5), 054,506 (2013). doi:10.1103/PhysRevD.87.054506

34. Guo, P., Dudek, J., Edwards, R., Szczepaniak, A.P.: Coupled-channel scattering on a torus. Phys. Rev. D **88**(1), 014,501 (2013). doi:10.1103/PhysRevD.88.014501. arXiv:1211.0929

35. Hansen, M.T., Sharpe, S.R.: Multiple-channel generalization of Lellouch-Luscher formula. Phys. Rev. D **86**(016), 007 (2012). doi:10.1103/PhysRevD.86.016007. arXiv:1204.0826

36. He, S., Feng, X., Liu, C.: Two particle states and the S-matrix elements in multi-channel scattering. JHEP **07**, 011 (2005). doi:10.1088/1126-6708/2005/07/011. arXiv:hep-lat/0504019

37. Ji, X.: Parton physics on a euclidean lattice. Phys. Rev. Lett. **110**(262), 002 (2013). doi:10.1103/PhysRevLett.110.262002. arXiv:1305.1539

38. Kiratidis, A.L., Kamleh, W., Leinweber, D.B., Owen, B.J.: Lattice baryon spectroscopy with multi-particle interpolators. Phys. Rev. D **91**, 094,509 (2015). doi:10.1103/PhysRevD.91.094509. arXiv:1501.07667

39. Leinweber, D., Kamleh, W., Kiratidis, A., Liu, Z.W., Mahbub, S., Roberts, D., Stokes, F., Thomas, A.W., Wu, J.: N* Spectroscopy from lattice QCD: the roper explained. In: 10th international workshop on the physics of excited nucleons (NSTAR 2015) Osaka, Japan, May 25–28, 2015. https://inspirehep.net/record/1407159/files/arXiv:1511.09146.pdf, arXiv:1511.09146

40. Lin, H.W., Cohen, S.D., Edwards, R.G., Richards, D.G.: First lattice study of the N-P(11)(1440) transition form factors. Phys. Rev. D **78**, 114,508 (2008). doi:10.1103/PhysRevD.78.114508. arXiv:0803.3020

41. Lin, H.W., Cohen, S.D., Edwards, R.G., Orginos, K., Richards, D.G.: Lattice calculations of nucleon electromagnetic form factors at large momentum transfer (2010). arXiv:1005.0799

42. Liu, K.F., Chen, Y., Gong, M., Sufian, R., Sun, M., Li, A.: The roper puzzle. PoS LATTICE **2013**, 507 (2014). arXiv:1403.6847

43. Luscher, M.: Volume dependence of the energy spectrum in massive quantum field theories. 2. scattering states. Commun. Math. Phys. **105**, 153–188 (1986). doi:10.1007/BF01211097

44. Luscher, M.: Two particle states on a torus and their relation to the scattering matrix. Nucl. Phys. **B354**, 531–578 (1991). doi:10.1016/0550-3213(91)90366-6

45. Ma, Y.Q., Qiu, J.W.: Extracting parton distribution functions from lattice QCD calculations (2014). arXiv:1404.6860

46. Mahbub, M.S., Cais, A.O., Kamleh, W., Lasscock, B.G., Leinweber, D.B., Williams, A.G.: Isolating excited states of the nucleon in lattice QCD. Phys. Rev. D **80**, 054,507 (2009). doi:10.1103/PhysRevD.80.054507. arXiv:0905.3616

47. Mahbub, M.S., Kamleh, W., Leinweber, D.B., Cais, A.O., Williams, A.G.: Ordering of spin-$\frac{1}{2}$ excitations of the nucleon in lattice QCD. Phys. Lett. B **693**, 351–357 (2010). doi:10.1016/j.physletb.2010.08.049. arXiv:1007.4871

48. Mahbub, M.S., Kamleh, W., Leinweber, D.B., Moran, P.J., Williams, A.G.: Roper resonance in 2+1 flavor QCD. Phys. Lett. B **707**, 389–393 (2012). doi:10.1016/j.physletb.2011.12.048. arXiv:1011.5724

49. Mahbub, M.S., Kamleh, W., Leinweber, D.B., Moran, P.J., Williams, A.G.: Low-lying odd-parity states of the nucleon in lattice QCD. Phys. Rev. D **87**(1), 011,501 (2013). doi:10.1103/PhysRevD.87.011501. arXiv:1209.0240

50. Mathur, N., Chen, Y., Dong, S.J., Draper, T., Horvath, I., Lee, F.X., Liu, K.F., Zhang, J.B.: Roper resonance and S(11)(1535) from lattice QCD. Phys. Lett. B **605**, 137–143 (2005). doi:10.1016/j.physletb.2004.11.010. arXiv:hep-ph/0306199

51. McNeile, C., Michael, C.: Hadronic decay of a vector meson from the lattice. Phys. Lett. B **556**, 177–184 (2003). doi:10.1016/S0370-2693(03)00130-8. arXiv:hep-lat/0212020

52. Rummukainen, K., Gottlieb, S.A.: Resonance scattering phase shifts on a nonrest frame lattice. Nucl. Phys. **B450**, 397–436 (1995). doi:10.1016/0550-3213(95)00313-H. arXiv:hep-lat/9503028

53. Shultz, C.J., Dudek, J.J., Edwards, R.G.: Excited meson radiative transitions from lattice QCD using variationally optimized operators. Phys. Rev. D **91**(11), 114,501 (2015). doi:10.1103/PhysRevD.91.114501. arXiv:1501.07457

54. Verduci, V., Lang, C.B.: Baryon resonances coupled to pion-nucleon states in lattice QCD. PoS LATTICE **2014**, 121 (2014). arXiv:1412.0701

55. Xiong, X., Ji, X., Zhang, J.H., Zhao, Y.: One-loop matching for parton distributions: nonsinglet case. Phys. Rev. D **90**(1), 0140,51 (2014). doi:10.1103/PhysRevD.90.014051. arXiv:1310.7471

Part VII
Nucleon Resonance Studies in Quark Models

Few-Body Syst (2016) 57:985–991
DOI 10.1007/s00601-016-1137-6

H. García-Tecocoatzi · R. Bijker · J. Ferretti · G. Galatà ·
E. Santopinto

Open Flavor Strong Decays

Received: 29 April 2016 / Accepted: 22 June 2016 / Published online: 22 July 2016
© Springer-Verlag Wien 2016

Abstract In this contribution, we discuss the results of a QM calculation of the open-flavor strong decays of
**** light nucleon resonances. These are the results of a recent calculation, where we used a modified $^3P_0$
model for the amplitudes and the U(7) algebraic model and the hypercentral quark model to predict the baryon
spectrum. The decay amplitudes are compared with the existing experimental data.

1 Introduction

The study of the strong decay processes of baryons is still considered a challenge within theoretical and
experimental hadronic physics. At the moment, the number of known light-quark mesons is much larger than
the number of known baryon resonances [1]. However, it is known that the baryon spectrum is much more
complex than the meson one. The difficulty lies in identifying those high-lying baryon resonances that are only
weakly coupled to the $N\pi$ channel [2–5] and thus cannot be seen in elastic $N\pi$ scattering experiments.

Regarding the strong decays of baryon resonances, no satisfactory description has yet been achieved. We
could list several problems as, for example, the QCD mechanism behind the OZI-allowed strong decays [6,7],
which is still not clear. Theoretical calculations of baryon strong, electromagnetic and weak decays still help
the experimentalists in their search of those resonances that are still unknown, even if interesting results were
provided by CB-ELSA [8,9], TAPS [10–12], GRAAL [13,14], SAPHIR [15,16] and CLAS [17–19].

Invited talk presented at NSTAR:Nucleon Resonances: From Photoproduction to High Photon Virtualities.

This article belongs to the special issue "Nucleon Resonances".

H. García-Tecocoatzi · R. Bijker · G. Galatà
Instituto de Ciencias Nucleares, Universidad Nacional Autónoma de México, 04510 Mexico, DF, Mexico

R. Bijker
E-mail: biker@nucleares.unam.mx

H. García-Tecocoatzi · E. Santopinto
INFN, Sezione di Genova, Via Dodecaneso 33, 16146 Genoa, Italy
E-mail: hgarcia@ge.infn.it

E. Santopinto
E-mail: santopinto@ge.infn.it

J. Ferretti (✉)
Dipartimento di Fisica, INFN, 'Sapienza' Università di Roma, P.le Aldo Moro 5, 00185 Rome, Italy
E-mail: jak.ferretti@gmail.com

Several phenomenological models have been developed in order to carry out strong decay studies, including pair-creation models [20–33], elementary meson emission models [34–44] and effective Lagrangian approaches (for example, see Ref. [45]). A few years after the introduction of the $^3P_0$ model [20], Le Yaouanc et al. used it to compute meson and baryon open flavor strong decays [21,22] and also evaluated the strong decay widths of charmonium states [46,47]. The $^3P_0$ model, extensively applied to the decays of light mesons and baryons [48], has been recently applied to heavy meson strong decays, in the charmonium [49–51], bottomonium [51,52] and open charm [53,54] sectors. In the 90s, Capstick and Roberts calculated the $N\pi$ and the strange decays of nonstrange baryons [3–5], using relativized wave functions for the baryons and mesons.

Recently, we have computed the decay widths of baryon resonances into baryon-pseudoscalar meson pairs [55] within a modified $^3P_0$ model, using two different models for the mass spectrum: the $U(7)$ algebraic model [56,57], by Bijker, Iachello and Leviatan, and the hypercentral model (hQM) [58–60], developed by Giannini and Santopinto. The widths have been computed with harmonic oscillator wave functions. In this contribution, we discuss our main results for the two-body strong decay widths of **** nucleon resonances.

2 U(7) Model for Baryons

The baryon spectrum is computed by means of algebraic methods introduced by Bijker et al. [56,57]. The algebraic structure of the model consists in combining the symmetry of the internal spin–flavor–color part, $SU_{sf}(6) \otimes SU_c(3)$, with that of the spatial part, $U(7)$ into

$$U(7) \otimes SU_{sf}(6) \otimes SU_c(3) . \tag{1}$$

The $U(7)$ model was introduced [56] to describe the relative motion of the three constituent parts of the baryon. The general idea is to introduce a so-called spectrum generating algebra $U(k + 1)$ for quantum systems characterized by k degrees of freedom. For baryons there are the $k = 6$ relevant degrees of freedom of the two relative Jacobi vectors $\boldsymbol{\rho} = \frac{1}{\sqrt{2}}(\mathbf{r}_1 - \mathbf{r}_2)$ and $\boldsymbol{\lambda} = \frac{1}{\sqrt{6}}(\mathbf{r}_1 + \mathbf{r}_2 - 2\mathbf{r}_3)$ and their canonically conjugate momenta, $\mathbf{P}_\rho = \frac{1}{\sqrt{2}}(\mathbf{p}_1 - \mathbf{p}_2)$ and $\mathbf{P}_\lambda = \frac{1}{\sqrt{6}}(\mathbf{p}_1 + \mathbf{p}_2 - 2\mathbf{p}_3)$. The $U(7)$ model is based on a bosonic quantization which consists in introducing two vector boson operators $b_\rho^\dagger$ and $b_\lambda^\dagger$ associated to the Jacobi vectors, and an additional auxiliary scalar boson, $s^\dagger$. The scalar boson does not represent an independent degree of freedom, but is added under the restriction that the total number of bosons N is conserved. The model space consists of harmonic oscillator shells with $n = 0, 1, \ldots, N$.

The baryon mass formula is written as the sum of three terms

$$\hat{M}^2 = M_0^2 + \hat{M}_{space}^2 + \hat{M}_{sf}^2 , \tag{2}$$

where M_0^2 is a constant, $\hat{M}_{space}^2$ is a function of the spatial degrees of freedom and $\hat{M}_{sf}^2$ depends on the internal ones. The spin–flavor part is treated in the same way as in Ref. [57] in terms of a generalized Gürsey–Radicati formula [61], which in turn is a generalization of the Gell-Mann–Okubo mass formula [62,63].

Since the space–spin–flavor wave function is symmetric under the permutation group S_3 of three identical constituents, the permutation symmetry of the spatial wave function has to be the same as that of the spin–flavor part. Thus, the spatial part of the mass operator $\hat{M}_{space}^2$ has to be invariant under the S_3 permutation symmetry. The dependence of the mass spectrum on the spatial degrees of freedom is given by:

$$\hat{M}_{space}^2 = \hat{M}_{vib}^2 + \hat{M}_{rot}^2 . \tag{3}$$

The baryon wave functions are denoted in the standard form as

$$\left| ^{2S+1}\dim\{SU_f(3)\}_J [\dim\{SU_{sf}(6)\}, L_i^P] \right\rangle , \tag{4}$$

where S and J are the spin and total angular momentum $\mathbf{J} = \mathbf{L} + \mathbf{S}$. As an example, in this notation the nucleon and delta wave functions are given by $\left| ^2 8_{1/2} [56, 0_1^+] \right\rangle$ and $\left| ^4 10_{3/2} [56, 0_1^+] \right\rangle$, respectively.

3 HQM for Baryons

In the hQM is supposed that the quark interaction is hypercentral, namely it only depends on the hyperradius [58–60,64],

$$V_{3q}(\boldsymbol{\rho}, \boldsymbol{\lambda}) = V(x), \tag{5}$$

where $x = \sqrt{\rho^2 + \lambda^2}$ is the hyperradius [65]. Thus, the space part of the three quark wave function, ψ_{space}, is factorized as

$$\psi_{space} = \psi_{3q}(\boldsymbol{\rho}, \boldsymbol{\lambda}) = \psi_{\gamma\nu}(x) Y_{[\gamma]l_\rho l_\lambda}(\Omega_\rho, \Omega_\lambda, \xi), \tag{6}$$

where the hyperradial wave function, $\psi_{\gamma\nu}(x)$, is labeled by the grand angular quantum number γ and the number of nodes ν. $Y_{[\gamma]l_\rho l_\lambda}(\Omega_\rho, \Omega_\lambda, \xi)$ are the hyperspherical harmonics, with angles $\Omega_\rho = (\theta_\rho, \phi_\rho)$, $\Omega_\lambda = (\theta_\lambda, \phi_\lambda)$ and hyperangle, $\xi = \arctan\frac{\rho}{\lambda}$ [65]. The dynamics is contained in $\psi_{\gamma\nu}(x)$, which is a solution of the hyperradial equation

$$\begin{aligned} [\frac{d^2}{dx^2} + \frac{5}{x}\frac{d}{dx} - \frac{\gamma(\gamma+4)}{x^2}]\psi_{\gamma\nu}(x) \\ = -2m[E - V_{3q}(x)]\,\psi_{\gamma\nu}(x). \end{aligned} \tag{7}$$

In the hQM, the quark interaction has the form [58–60,64]

$$V(x) = -\frac{\tau}{x} + \alpha x, \tag{8}$$

where τ and α are free parameters, fitted to the reproduction of the experimental data. Equation (8) can be seen as the hypercentral approximation of a Cornell-type quark interaction [23–27], whose form can be reproduced by Lattice QCD calculations [66–69]. Now, to introduce splittings within the $SU(6)$ multiplets, an SU(6)-breaking term must be added. In the case of the hQM, such violation of the $SU(6)$ symmetry is provided by the hyperfine interaction [70–73]. The complete hQM hamiltonian is then [58–60,64]

$$H_{\text{hQM}} = 3m + \frac{\mathbf{p}_\rho^2}{2m} + \frac{\mathbf{p}_\lambda^2}{2m} - \frac{\tau}{x} + \alpha x + H_{\text{hyp}}, \tag{9}$$

where $\mathbf{p}_\rho$ and $\mathbf{p}_\lambda$ are the momenta conjugated to the Jacobi coordinates $\boldsymbol{\rho}$ and $\boldsymbol{\lambda}$. In addition to τ and α, there are two more free parameters in the hQM, the constituent quark mass, m, and the strength of the hyperfine interaction. The former is taken, as usual, as $1/3$ of the nucleon mass. The latter, as in the case of τ and α, is fitted in [58–60] to the reproduction of the *** and **** resonances reported in the PDG [1].

4 Two-Body Strong Decays of Light Nucleon Resonances in the $^3P_0$ Pair-Creation Model

Here, we present some our results for the two-body strong decay widths of nucleon resonances in the $^3P_0$ pair-creation model. The decay widths are computed as [20–22,32,33,49,50,52,55,74]

$$\Gamma_{A \to BC} = \Phi_{A \to BC}(q_0) \sum_{\ell, J} |\langle BC\mathbf{q}_0 \ell J| T^\dagger |A\rangle|^2, \tag{10}$$

where, $\Phi_{A \to BC}(q_0)$ is the relativistic phase space factor:

$$\Phi_{A \to BC}(q_0) = 2\pi q_0 \frac{E_b(q_0) E_c(q_0)}{M_a}, \tag{11}$$

depending on q_0 and on the energies of the two intermediate state hadrons, $E_b = \sqrt{M_b^2 + q_0^2}$ and $E_c = \sqrt{M_c^2 + q_0^2}$. We assumed harmonic oscillator wave functions, depending on a single oscillator parameter α_b for the baryons and α_m for the mesons. The coupling between the final state hadrons $|B\rangle$ and $|C\rangle$ is described in terms of a spherical basis [55]. Specifically, the final state $|BC\mathbf{q}_0 \ell J\rangle$ can be written as

$$\begin{aligned} |BC\mathbf{q}_0 \ell J\rangle = \sum_{m, M_b, M_c} \langle J_b M_b J_c M_c| J_{bc} M_{bc}\rangle \\ \langle J_{bc} M_{bc} \ell m |J M\rangle \frac{Y_{\ell m}(\hat{q})}{q^2}\delta(q - q_0) \\ |(S_b, L_b) J_b M_b\rangle |(S_c, L_c) J_c M_c\rangle, \end{aligned} \tag{12}$$

Table 1 The select strong decay widths of **** nucleon resonances (in MeV) from Ref. [55]

Resonance	Status	M (MeV)	$N\pi$	$N\eta$	ΣK	ΛK	$\Delta\pi$	
$N(1440)P_{11}$	****	1430–1470	110–338	0–5			22–101	Exp.
$^2 8_{1/2}[56, 0_2^+]$		1444	85	–	–	–	13	U(7)
$^2 8_{1/2}[56, 0_2^+]$		1550	105	–	–	–	12	hQM
$N(1520)D_{13}$	****	1515–1530	102	0			342	Exp.
$^2 8_{3/2}[70, 1_1^-]$		1563	134	0	–	–	207	U(7)
$^2 8_{3/2}[70, 1_1^-]$		1525	111	0	–	–	206	hQM
$N(1535)S_{11}$	****	1520–1555	44–96	40–91			<2	Exp.
$^2 8_{1/2}[70, 1_1^-]$		1563	63	75	–	–	16	U(7)
$^2 8_{1/2}[70, 1_1^-]$		1525	84	50	–	–	6	hQM
$N(1650)S_{11}$	****	1640–1680	60–162	6–27	4–20		0–45	Exp.
$^4 8_{1/2}[70, 1_1^-]$		1683	41	72	–	0	18	U(7)
$^2 8_{1/2}[70, 1_2^-]$		1574	51	29	–	0	4	hQM
$N(1675)D_{15}$	****	1670–1685	46–74	0–2		<2	65–99	Exp.
$^4 8_{5/2}[70, 1_1^-]$		1683	47	11	–	0	108	U(7)
$^4 8_{5/2}[70, 1_1^-]$		1579	41	9	–	–	85	hQM
$N(1680)F_{15}$	****	1675–1690	78–98	0–1			6–21	Exp.
$^2 8_{5/2}[56, 2_1^+]$		1737	121	1	–	0	100	U(7)
$^2 8_{5/2}[56, 2_1^+]$		1798	91	0	0	0	92	hQM

The spectrum is computed using the U(7) Model of Sect. 2 and Refs. [56,57] and Hypercentral QM of Sect. 3 and Refs. [58–60,64], in combination with the relativistic phase space factor of Eq. (11) and the values of the model parameters of Table 2 (second column). The experimental values are taken from Ref. [1]. Decay channels labeled by—are below threshold. The symbols (S) and (D) stand for S and D-wave decays, respectively

where the ket $|BC\mathbf{q}_0\,\ell J\rangle$ is characterized by a relative orbital angular momentum ℓ between B and C and a total angular momentum $\mathbf{J} = \mathbf{J}_b + \mathbf{J}_c + \boldsymbol{\ell}$.

The transition operator of the $^3P_0$ model is given by [50,52,55,74]:

$$T^\dagger = -3\,\gamma_0^{\text{eff}} \int d\mathbf{p}_4\,d\mathbf{p}_5\,\delta(\mathbf{p}_4 + \mathbf{p}_5)\,C_{45}\,F_{45}\,e^{-r_q^2(\mathbf{p}_4 - \mathbf{p}_5)^2/6}$$
$$\left[\chi_{45} \times \mathcal{Y}_1(\mathbf{p}_4 - \mathbf{p}_5)\right]_0^{(0)}\,b_4^\dagger(\mathbf{p}_4)\,d_5^\dagger(\mathbf{p}_5). \tag{13}$$

Here, $b_4^\dagger(\mathbf{p}_4)$ and $d_5^\dagger(\mathbf{p}_5)$ are the creation operators for a quark and an antiquark with momenta $\mathbf{p}_4$ and $\mathbf{p}_5$, respectively. The $q\bar{q}$ pair is characterized by a color singlet wave function C_{45}, a flavor singlet wave function F_{45}, a spin triplet wave function χ_{45} with spin $S = 1$ and a solid spherical harmonic $\mathcal{Y}_1(\mathbf{p}_4 - \mathbf{p}_5)$, since the quark and antiquark are in a relative P wave. The operator γ_0^{eff} of Eq. (13) is an effective pair-creation strength [50,52,55,74,75], defined as

$$\gamma_0^{\text{eff}} = \frac{m_n}{m_i}\,\gamma_0, \tag{14}$$

with $i = n$ (i.e. u or d) or s (see Table 2).

Finally, the select results from our study of Ref. [55], obtained with the values of the model parameters of Table 2 (second column) and the relativistic phase space factor of Eq. (11), are reported in Table 1.

5 Discussion and Conclusions

In this contribution, we discussed some recent results for the open-flavor strong decay widths of **** nucleon resonances within a modified $^3P_0$ pair-creation model [55]. The baryon spectrum, we needed in our calculation, was predicted within the U(7) algebraic model [57] and the hQM [58–60], developed by Giannini and Santopinto.

One can observe that the results of Table 1 for **** nucleon resonances from Ref. [55] are quite similar for $N\pi$ and $\Delta\pi$ channels, in both the fits we did for the hQM and U(7) model cases. But it is worthwhile

Table 2 Pair-creation model parameters used in the calculations [55]

Parameter	Value $U(7)$	Value hQM
γ_0	14.3	13.319
α_b	2.99 GeV^{-1}	2.758 GeV^{-1}
α_m	2.38 GeV^{-1}	2.454 GeV^{-1}
α_d	0.52 GeV^{-1}	0
m_n	0.33 GeV	
m_s	0.55 GeV	

In the second column are given the parameters used in the calculations with the relativistic phase space factor of Eq. (11) for U(7) model, while in the third column those for hQM. The values of the constituent quark masses m_n ($n = u, d$) and m_s are taken from Refs. [50,52,74]

noticing that the parameters of the $^3P_0$ model are quite different, see Table 2. On the contrary, in the case of the $\eta\pi$ channels the predictions are different in the hQM and U(7) model.

The possibility of using different models to extract the baryon spectrum helps to understand differences between different types of quark models. Another step towards a deeper understanding of this type of processes could be an extension of the quark model to include the continuum components in the baryon wave function. Thus, in a subsequent paper we will focus on threshold effects and the decays of states close to open and hidden-flavor decay thresholds. This procedure will not only have an effect on the widths, but also on the mass values [76].

Acknowledgments This work was supported in part by INFN, EPOS, Italy and by PAPIIT-DGAPA, Mexico (Grant No. IN107314).

References

1. Olive, K.A., et al.: [Particle data group collaboration], Review of particle physics. Chin. Phys. C **38**, 090001 (2014)
2. Capstick, S.: Photoproduction and electroproduction of nonstrange baryon resonances in the relativized quark model. Phys. Rev. D **46**, 2864 (1992)
3. Capstick, S., Roberts, W.: N pi decays of baryons in a relativized model. Phys. Rev. D **47**, 1994 (1993)
4. Capstick, S., Roberts, W.: Quasi two-body decays of nonstrange baryons. Phys. Rev. D **49**, 4570 (1994)
5. Capstick, S., Roberts, W.: Strange decays of nonstrange baryons. Phys. Rev. D **58**, 074011 (1998)
6. Okubo, S.: Phi meson and unitary symmetry model. Phys. Lett. **5**, 165 (1963)
7. Iizuka, J.: Systematics and phenomenology of meson family. Prog. Theor. Phys. Suppl. **37**, 21 (1966)
8. Crede, V., et al.: [CB-ELSA collaboration], Photoproduction of eta mesons off protons for 0.75-GeV > E(gamma) < 3-GeV. Phys. Rev. Lett. **94**, 012004 (2005)
9. Trnka, D., et al.: [CBELSA/TAPS collaboration], First observation of in-medium modifications of the omega meson. Phys. Rev. Lett. **94**, 192303 (2005)
10. Krusche, B., et al.: New threshold photoproduction of eta mesons off the proton. Phys. Rev. Lett. **74**, 3736 (1995)
11. Harter, F., et al.: Two neutral pion photoproduction off the proton between threshold and 800-MeV. Phys. Lett. B **401**, 229 (1997)
12. Wolf, M., et al.: Photoproduction of neutral pion pairs from the proton. Eur. Phys. J. A **9**, 5 (2000)
13. Renard, F., et al.: [GRAAL collaboration], Differential cross-section measurement of eta photoproduction on the proton from threshold to 1100-MeV. Phys. Lett. B **528**, 215 (2002)
14. Assafiri, Y., et al.: Double pi0 photoproduction on the proton at GRAAL. Phys. Rev. Lett. **90**, 222001 (2003)
15. Tran, M.Q., et al.: [SAPHIR collaboration], Measurement of gamma p -> K+ Lambda and gamma p -> K+ Sigma0 at photon energies up to 2-GeV. Phys. Lett. B **445**, 20 (1998)
16. Glander, K.H., et al.: Measurement of gamma p -> K+ Lambda and gamma p -> K+ Sigma0 at photon energies up to 2.6-GeV. Eur. Phys. J. A **19**, 251 (2004)
17. Dugger, M., et al.: [CLAS collaboration], Eta photoproduction on the proton for photon energies from 0.75-GeV to 1.95-GeV. Phys. Rev. Lett. **89**, 222002 (2002)
18. Dugger, M., et al.: Eta-prime photoproduction on the proton for photon energies from 1.527-GeV to 2.227-GeV. Phys. Rev. Lett. **96**, 062001 (2006)
19. Ripani, M., et al.: [CLAS collaboration], Measurement of e p -> e-prime p pi+ pi- and baryon resonance analysis. Phys. Rev. Lett. **91**, 022002 (2003)
20. Micu, L.: Decay rates of meson resonances in a quark model. Nucl. Phys. B **10**, 521 (1969)
21. Le Yaouanc, A., Oliver, L., Pene, O., Raynal, J.-C.: Naive quark pair creation model of strong interaction vertices. Phys. Rev. D **8**, 2223 (1973)
22. Le Yaouanc, A., Oliver, L., Pene, O., Raynal, J.-C.: Naive quark pair creation model and baryon decays. Phys. Rev. D **9**, 1415 (1974)

23. Eichten, E., Gottfried, K., Kinoshita, T., Kogut, J.B., Lane, K.D., Yan, T.-M.: The spectrum of charmonium. Phys. Rev. Lett. **34**, 369 (1975)
24. Eichten, E., Gottfried, K., Kinoshita, T., Kogut, J.B., Lane, K.D., Yan, T.-M.: The spectrum of charmonium. Phys. Rev. Lett. **36**, 1276 (1976)
25. Eichten, E., Gottfried, K., Kinoshita, T., Lane, K.D., Yan, T.-M.: Charmonium: the model. Phys. Rev. D **17**, 3090 (1978)
26. Eichten, E., Gottfried, K., Kinoshita, T., Lane, K.D., Yan, T.-M.: Charmonium: the model. Phys. Rev. D **21**, 313 (1980)
27. Eichten, E., Gottfried, K., Kinoshita, T., Lane, K.D., Yan, T.-M.: Charmonium: comparison with experiment. Phys. Rev. D **21**, 203 (1980)
28. Alcock, J.W., Burfitt, M.J., Cottingham, W.N.: A string breaking model of heavy meson decays. Z. Phys. C **25**, 161 (1984)
29. Dosch, H.G., Gromes, D.: Theoretical foundation for treating decays allowed by the Okubo-Zweig-iizuka rule and related phenomena. Phys. Rev. D **33**, 1378 (1986)
30. Kokoski, R., Isgur, N.: Meson decays by flux tube breaking. Phys. Rev. D **35**, 907 (1987)
31. Roberts, W., Silvestre-Brac, B.: General method of calculation of any hadronic decay in the $^3P_0$ model. Few Body Syst. **11**, 171 (1992)
32. Ackleh, E.S., Barnes, T., Swanson, E.S.: On the mechanism of open flavor strong decays. Phys. Rev. D **54**, 6811 (1996)
33. Barnes, T., Close, F.E., Page, P.R., Swanson, E.S.: Higher quarkonia. Phys. Rev. D **55**, 4157 (1997)
34. Becchi, C., Morpurgo, G.: Vanishing of the E2 part of the $N_{33}^* \to N\gamma$ amplitude in the non-relativistic quark model of "elementary" particles. Phys. Lett. **17**, 352 (1965)
35. Becchi, C., Morpurgo, G.: Test of the nonrelativistic quark model for 'elementary' particles: radiative decays of vector mesons. Phys. Rev. **140**, B687 (1965)
36. Becchi, C., Morpurgo, G.: Connection between BBP and VPP vertices in the quark model. Phys. Rev. **149**, 1284 (1966)
37. Faiman, D., Hendry, A.W.: Harmonic oscillator model for baryons. Phys. Rev. **173**, 1720 (1968)
38. Faiman, D., Hendry, A.W.: Harmonic-oscillator model for baryons. Phys. Rev. **180**, 1609 (1969)
39. Koniuk, R., Isgur, N.: Baryon Decays in a quark model with chromodynamics. Phys. Rev. D **21**, 1868 (1980)
40. Koniuk, R., Isgur, N.: Baryon decays in a quark model with chromodynamics. Phys. Rev. D **23**, 818 (1981)
41. Koniuk, R., Isgur, N.: Where have all the resonances gone? An analysis of baryon couplings in a quark model with chromodynamics. Phys. Rev. Lett. **44**, 845 (1980)
42. Godfrey, S., Isgur, N.: Mesons in a relativized quark model with chromodynamics. Phys. Rev. D **32**, 189 (1985)
43. Sartor, R., Stancu, F.: Strong decay of hadrons in a semirelativistic quark model. Phys. Rev. D **34**, 3405 (1986)
44. Bijker, R., Iachello, F., Leviatan, A.: Strong decays of nonstrange q**3 baryons. Phys. Rev. D **55**, 2862 (1997)
45. Colangelo, P., De Fazio, F., Giannuzzi, F., Nicotri, S.: New meson spectroscopy with open charm and beauty. Phys. Rev. D **86**, 054024 (2012)
46. Le Yaouanc, A., Oliver, L., Pene, O., Raynal, J.-C.: Strong decays of psi–prime–prime (4.028) as a radial excitation of charmonium. Phys. Lett. B **71**, 397 (1977)
47. Le Yaouanc, A., Oliver, L., Pene, O., Raynal, J.-C.: Why is psi–prime–prime–prime (4.414) SO narrow? Phys. Lett. B **72**, 57 (1977)
48. Blundell, H.G., Godfrey, S.: The Xi (2220) revisited: strong decays of the 1(3) F2 1(3) F4 s anti-s mesons. Phys. Rev. D **53**, 3700 (1996)
49. Barnes, T., Godfrey, S., Swanson, E.S.: Higher charmonia. Phys. Rev. D **72**, 054026 (2005)
50. Ferretti, J., Galatà, G., Santopinto, E.: Interpretation of the X(3872) as a charmonium state plus an extra component due to the coupling to the meson–meson continuum. Phys. Rev. C **88**, 015207 (2013)
51. Ferretti, J., Galatà, G., Santopinto, E.: Quark structure of the $X(3872)$ and $\chi_b(3P)$ resonances. Phys. Rev. D **90**, 054010 (2014)
52. Ferretti, J., Santopinto, E.: Higher mass bottomonia. Phys. Rev. D **90**, 094022 (2014)
53. Close, F.E., Swanson, E.S.: Dynamics and decay of heavy-light hadrons. Phys. Rev. D **72**, 094004 (2005)
54. Segovia, J., Entem, D.R., Fernandez, F.: Scaling of the 3P0 strength in heavy meson strong decays. Phys. Lett. B **715**, 322 (2012)
55. Bijker, R., Ferretti, J., Galatà, G., García-Tecocoatzi, H., Santopinto, E.: Strong decays of baryons and missing resonances. arXiv:1506.07469
56. Bijker, R., Iachello, F., Leviatan, A.: Algebraic models of hadron structure. 1. Nonstrange baryons. Ann. Phys. **236**, 69 (1994)
57. Bijker, R., Iachello, F., Leviatan, A.: Algebraic models of hadron structure. 2. Strange baryons. Ann. Phys. **284**, 89 (2000)
58. Ferraris, M., Giannini, M.M., Pizzo, M., Santopinto, E., Tiator, L.: A three body force model for the baryon spectrum. Phys. Lett. B **364**, 231 (1995)
59. Santopinto, E., Iachello, F., Giannini, M.M.: Exactly solvable models of baryon spectroscopy. Nucl. Phys. A **623**, 100c (1997)
60. Santopinto, E., Iachello, F., Giannini, M.M.: Nucleon form-factors in a simple three-body quark model. Eur. Phys. J. A **1**, 307 (1998)
61. Gürsey, F., Radicati, L.A.: Spin and unitary spin independence of strong interactions. Phys. Rev. Lett. **13**, 173 (1964)
62. Ne'eman, Y.: Derivation of strong interactions from a gauge invariance. Nucl. Phys. **26**, 222 (1961)
63. Gell-Mann, M.: Symmetries of baryons and mesons. Phys. Rev. **125**, 1067 (1962)
64. Giannini, M.M., Santopinto, E.: The hypercentral constituent quark model and its application to baryon properties. Chin. J. Phys. **53**, 1 (2015)
65. Ballot, J., de la Ripelle, M.Fabre: Application of the hyperspherical formalism to the trinucleon bound state problems. Ann. Phys. N.Y. **127**, 62 (1980)
66. Bali, G.S., et al.: Static potentials and glueball masses from QCD simulations with Wilson sea quarks. Phys. Rev. D **62**, 054503 (2000)
67. Bali, G.S.: QCD forces and heavy quark bound states. Phys. Rep. **343**, 1 (2001)
68. Alexandrou, C., de Forcrand, P., Jahn, O.: The ground state of three quarks. Nucl. Phys. Proc. Suppl. **119**, 667 (2003)

69. Suganuma, H., Takahashi, T.T., Okiharu, F., Ichie, H.: Study of quark confinement in baryons with lattice QCD. Nucl. Phys. Proc. Suppl. **141**, 92 (2005)
70. De Rújula, A., Georgi, H., Glashow, S.L.: Hadron masses in a gauge theory. Phys. Rev. D **12**, 147 (1975)
71. Isgur, N., Karl, G.: P-Wave baryons in the quark model. Phys. Rev. D **18**, 4187 (1978)
72. Isgur, N., Karl, G.: Positive parity excited baryons in a quark model with hyperfine interactions. Phys. Rev. D **19**, 2653 (1979)
73. Isgur, N., Karl, G.: Ground state baryons in a quark model with hyperfine interactions. Phys. Rev. D **20**, 1191 (1979)
74. Ferretti, J., Galatà, G., Santopinto, E., Vassallo, A.: Bottomonium self-energies due to the coupling to the meson-meson continuum. Phys. Rev. C **86**, 015204 (2012)
75. Kalashnikova, Y.S.: Coupled-channel model for charmonium levels and an option for X(3872). Phys. Rev. D **72**, 034010 (2005)
76. García-Tecocoatzi, H., Bijker, R., Ferretti, J., Santopinto, E.: Self-energies of ground-state octet and decuplet baryon states due to the coupling to the baryon-meson continuum. arXiv:1603.07526

Few-Body Syst (2016) 57:1009–1017
DOI 10.1007/s00601-016-1142-9

M. M. Giannini

High Q^2 Helicity Amplitudes in the Hypercentral Constituent Quark Model

Received: 14 February 2016 / Accepted: 26 July 2016 / Published online: 18 August 2016
© Springer-Verlag Wien 2016

Abstract The predictions of the hypercentral constituent quark model for the helicity amplitudes are briefly discussed, with particular emphasis for their high Q^2 behaviour.

1 Introduction

In order to describe the internal nucleon structure various constituent quark models (CQM) have been introduced in the past. Among them, the hypercentral CQM (hCQM) [1] turned out to be able to describe with fair accuracy many baryon properties (for a review see [2]). In the hCQM, there are few free parameters, which are fixed by fitting the observed spectrum of the baryon resonances. In this way the model is completely defined and allows to calculate any quantity of interest. In particular, the model has been used to predict the photocouplings [3], the helicity amplitudes for the transverse excitation of negative parity resonances [4], the nucleon elastic form factors [5–8], the systematic behaviour of the transverse and longitudinal helicity amplitudes up to $Q^2 = 5 (GeV/c)^2$ for both proton and neutron [9] and the strong decays [10]. In this paper, after a brief description of the model and its results, the attention will be concentrated on the Q^2 behaviour of the helicity amplitudes for the excitation of some of the most important resonances.

2 The Model and the Spectrum

The internal motion of quarks within the baryons is described by the Jacobi coordinates ρ and λ

$$\rho = \frac{1}{\sqrt{2}}(\mathbf{r}_1 - \mathbf{r}_2), \quad \lambda = \frac{1}{\sqrt{6}}(\mathbf{r}_1 + \mathbf{r}_2 - 2\mathbf{r}_3). \tag{1}$$

It si convenient to introduce the so called hyperspherical coordinates which are obtained by substituting $\rho = |\rho|$ and $\lambda = |\lambda|$ with the hyperradius, x, and the hyperangle, ξ, defined respectively by

$$x = \sqrt{\rho^2 + \lambda^2}, \quad \xi = arctg\left(\frac{\rho}{\lambda}\right). \tag{2}$$

keeping the angular coordinates unchanged. In all the CQMs considered in the past decades, the three quark interaction is split into two parts, as prescribed already by the first LQCD calculations [11,12], one spin-independent SU(6) invariant part containing the confinement potential and a short range SU(6) violating

---

This article belongs to the special issue "Nucleon Resonances".

M. M. Giannini (✉)
Dipartimento di Fisica, Università di Genova, Genova, Italy
E-mail: giannini@ge.infn.it

part. In the hCQM, the SU(6) invariant interaction is assumed to be hypercentral, that is to depend on the hyperradius x only, in agreement with the observation [13] that two-body potentials, when applied to baryons, behave approximately as a hypercentral one. In this respect, the three quark potential can be considered as the hypercentral approximation of a two-body potential. On the other hand, a hypercentral potential may contain contributions from three-body forces, which have been taken into account also in the calculations by Ref. [14] and in the relativized version of the Isgur–Karl model [15]. In any case, the dominance of the SU(6) invariant potential allows to group the baryons states into SU(6) multiplets, while the presence of a spin dependent interaction produces a splitting of the states within the SU(6) multiplets: such feature is quite evident in the structure of the experimental spectrum [2].

The three quark hamiltonian in the hCQM is [1]

$$H = \frac{p_\lambda^2}{2m} + \frac{p_\rho^2}{2m} - \frac{\tau}{x} + \alpha x + H_{hyp} \tag{3}$$

where $\mathbf{p}_\rho$ and $\mathbf{p}_\lambda$ are the momenta conjugated to the coordinates ρ, λ and m is the quark mass taken as $1/3$ of the nucleon mass. In Eq. (3) the SU(6) violation is given by the hyperfine interaction H_{hyp} [16], which has been widely used, in particular in the pioneering work by Isgur and Karl [17,18]. It should be noted that the x-dependent terms in Eq. (3) can be considered as the hypercentral approximation to a two-body potential of the form suggested by LQCD calculations [19,20].

The three free parameters are fitted to the experimental spectrum of the non strange resonances. The strength of the hyperfine interaction is determined by the $\Delta - N$ mass difference and the remaining ones are given by $\tau = 4.59$ and $\alpha = 1.61 \, fm^{-2}$ [1].

The advantage of the hypercentral hypothesis is that the Schrödinger equation for the SU(6) invariant hamiltonian is reduced to one single differential equation for the hyperradial wave function $\psi_{\gamma\nu}(x)$, labeled by the grand angular quantum number [21] $\gamma = 2n + l_\rho + l_\lambda$ (with n a non negative integer) and the number of nodes ν.

In Fig. 1, the qualitative structure of the spectrum up to the first three shells is reported. For each level, the possible SU(6) configurations are also given, using the notation L_t^P, where P is the parity and L the total angular momentum of the three quark system, while t is the permutation symmetry type of the space wave function. The combination of L with the total quark spin S determines which three quark states belong to the various levels.

With respect to the spectrum given by a harmonic oscillator (h.o.), there are some features which are beneficial for the description of the experimental spectrum [1,2,22]. In fact, in the h.o. case, as in any two-

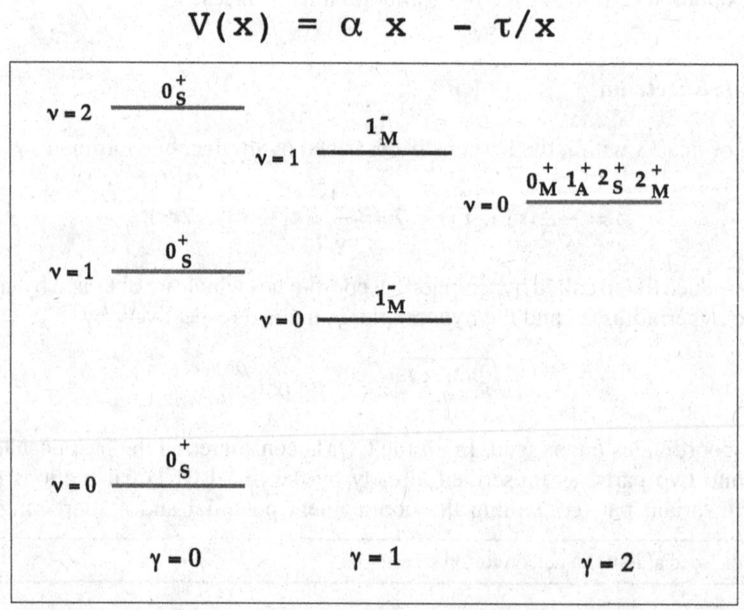

Fig. 1 The structure of the baryon spectrum provided by the SU(6) invariant part of the Hamiltonian of Eq. 3, showing the respective contents of SU(6) configurations

body potential, the positive and negative parity states belong to different shells and the negative states are then predicted to be well below the first nucleon radial excitation (the Roper), at variance with the data, while in the spectrum of Fig. 1, in each level, apart from the fundamental one, there are both positive and negative parity states, in better agreement with phenómenology.

Furthermore, the hypercentral spectrum contains two levels more than in the h.o. case, namely one 0_S^+ and one 1_M^-. The first one implies the existence of a second Roper resonance, for which there are certainly candidates in the observed spectrum. The presence of a second negative parity level, is particularly interesting [22]. Up to a few years ago, the PDG [23] reported seven negative parity resonances with the four- or three-star status and seven is just the number of states which are predicted to belong to the 1_M^- configuration. However, in the recent PDG editions [24,25] a new three star state $N(1875)3/2^-$ is reported, which has an energy comparable to the resonances belonging to the third shell and can then belong, in the case of the hypercentral potential, to the second 1_M^- configuration. If one considers also the other new negative parity states [24,25] with two stars, the level 1_M^- becomes almost full. In the h.o. case, all these states should be allocated in the fourth shell, which has a too high energy.

The solution of Eq. 3 allows to build all the SU(6) configurations of interest and the physical baryon states are the result of the mixing produced by the hyperfine interaction, treated as a perturbation. In this way, the model baryon states are completely defined and can be used for the calculation of the physical properties of baryons, in particular the elastic form factors [5–8] and the helicity amplitudes for the electromagnetic excitation of the baryon resonances [3,4,9,26].

3 The Helicity Amplitudes

The electromagnetic excitation of the baryons resonances is described by the helicity amplitudes $A_{1/2}$, $A_{3/2}$ and $S_{1/2}$, defined as the matrix elements of the quark electromagnetic interaction, $A_\mu J^\mu$, between the nucleon, N, and the resonance, B, states:

$$A_M = \sqrt{\frac{2\pi\alpha}{k}} \left\langle B, J', J_z' = M \,|J_+|\, N, J = \frac{1}{2}, J_z = M - 1 \right\rangle, \quad M = \frac{1}{2}, \frac{3}{2}$$

$$S_{1/2} = \sqrt{\frac{2\pi\alpha}{k}} \left\langle B, J', J_z' = \frac{1}{2} \,|J_z|\, N, J = \frac{1}{2}, J_z = -\frac{1}{2} \right\rangle \tag{4}$$

where J_+ and J_z are the $+$ and z components of the electromagnetic current J_μ carried by quarks and will be used in their non relativistic form [27,28]; k is the photon momentum in the Breit frame.

The hCQM predictions for the photocouplings [3,9] are shown in Fig. 2 in comparison with the experimental data [25]. In some cases the agreement between theoretical and experimental values is good, for the rest the overall trend is fairly reproduced, although a lack of strength is often present, but this is a problem common to all CQM calculations. Similar results are valid for the neutron excitation. The reason of the agreement rely also on the fact that the predicted proton radius is about $0.5\,fm$, a value too low with respect to real proton dimensions but it is just the value which has been seen to be necessary in order to fit the helicity amplitudes [27,28].

The hCQM allows a systematic calculation of the helicity amplitudes for the transverse and longitudinal excitation and the predictions for all the resonances having a relevant excitation strength have been reported in Ref. [9] in comparison with the available experimental data.

For many resonances there are still few data, however, thanks to the recent Jlab experiments at higher Q^2, one can make a significant comparison between the extracted amplitudes and the hCQM predictions (see [9] and references quoted therein). It should be mentioned that the predictions for the transverse excitation to the negative parity states had been published [4] before the appearance of the new Jlab data.

In the low Q^2 region there is the already mentioned lack of strength. This discrepancy is attributed to the missing quark-antiquark pair or meson production mechanisms [4] and actually there is a general consensus on such statement. A support is provided by the calculation of the meson cloud effects in the DMT model [29], which shows that their contribution is relevant for low Q^2, just in the region where the hCQM fails to reproduce the strength [30].

For medium-high Q^2, the overall behaviour is in general fairly reproduced, specially for the 1/2 amplitudes. In some cases the agreement is fairly good, unexpectedly, if one considers the non relativistic character of the model.

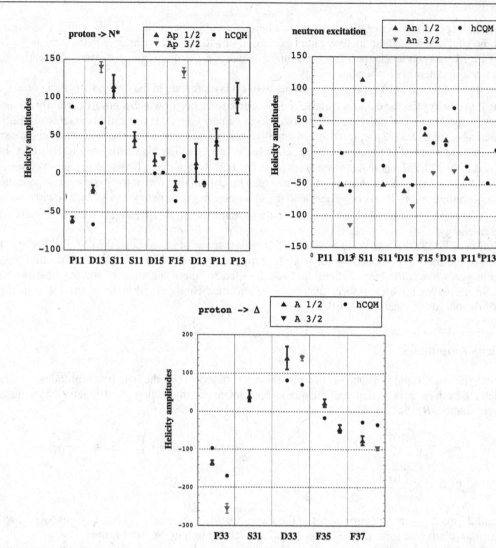

Fig. 2 The photocouplings predicted by the hCQM [3,9], in comparison with the experimental data [25]. *Upper left* proton excitation to N^* resonances; *upper right* the same for neutron excitation; *lower* proton excitation to Δ resonances

A significant tool for testing the high Q^2 behaviour of the transverse helicity amplitudes is provided by the asymmetry ratio [31,32]

$$Z = \frac{|A_{1/2}|^2 - |A_{3/2}|^2}{A_{1/2}|^2 + |A_{3/2}|^2} \tag{5}$$

Because of helicity conservation in the virtual photon-quark interaction, QCD predicts that Z should reach the value 1 as long as $Q^2 \to \infty$ [33]. In Fig. 3 the Q^2 dependence of Z for the resonances $N(1520)3/2^-$, $N(1680)5/2^+$ and $\Delta(1232)3/2^+$ is shown. For the $N(1520)3/2^-$, the data seem to reach an asymptotic value in agreement with the QCD prescription as well as the hCQM predictions; one should not forget that in the case of the negative parity resonances the hCQM description is particularly good [4], specially in the medium-high Q^2 range. In the case of the $N(1680)5/2^+$ resonance, there is some inconsistency of the data, nevertheless there seems to be an asymptotic value but at 0.5 instead of 1. The situation for the $\Delta(1232)3/2^+$ is peculiar: the hCQM predictions and the data are in agreement, but the behaviour is far from being in agreement with QCD. The helicity amplitudes can be expressed in terms of the electric $E2$ and magnetic $M1$ multipoles [32,38]

$$A_{1/2} = -C_W \frac{1}{\sqrt{3}}(G_{M1} - 3G_{E2}) \quad A_{3/2} = C_W(G_{M1} + G_{E2}) \tag{6}$$

 Springer

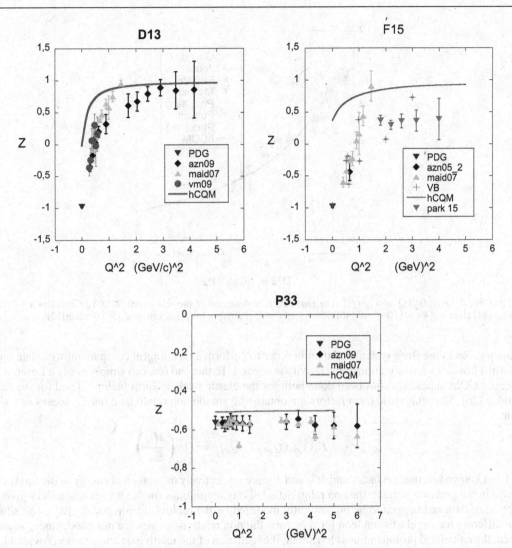

Fig. 3 The ratio Z of Eq. 5 for the resonances $N(1520)3/2^-$ (*upper left*), $N(1680)5/2^+$ (*upper right*) and $\Delta(1232)3/2^+$ (*lower*). The data are from [25] (PDG), [34] (azn09), [35] (maid07), [36] (vm09), [37] (park15). The theoretical curves are the predictions [4,9] of the hCQM [1,2]

where C_W is a kinematical factor. The asymmetry ratio can then be written as [38]

$$Z = -\frac{1}{2} + 3\frac{G_{E2}(G_{E2} - G_{M1})}{G_{M1}^2 + 3G_{E2}^2} \tag{7}$$

the ratio turns out to be near to $1/2$ in agreement with the fact that the quadrupole transition is very small. In CQMs, the quadrupole transition is made possible by the mixing with the 2_S^+ configuration (see Fig. 1), which gives rise to a small D component in the Δ. The value $Z = 1$ would be obtained for $G_{E2} = -G_{M1}$, that is the electric and magnetic transitions should have the same strength, a feature not respected neither by the CQMs nor by the data. Here we are faced with the problem of the presence of higher orbital states in the nucleon wave function, as discussed for instance in [39], but in order to get a reasonable asymptotic behaviour, such states should be of the same order of magnitude as the standard S one.

4 Introducing Relativity

The hCQM allows a fair description of many baryon properties [2] notwithstanding its non relativistic character. The CQM leads to the determination of the three quark wave function in the c.m.s., however, the form factors

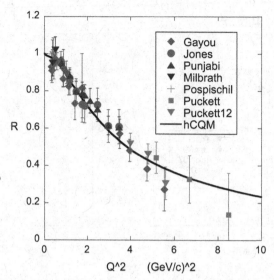

Fig. 4 The ratio $R = \mu_p G_E^p(Q^2)/G_M^p(Q^2)$. The *curve* is the calculation of the relativized hCQM [7,8]. Data are from [43] (Milbrath), [44] (Jones), [45,47] (Gayou), [46] (Pospischil), [48] (Punjabi), [49] (Puckett) and [50] (Puckett12)

are usually given in the Breit system and then in order to perform a meaningful comparison with data one has to perform a Lorentz transformation and relativity is needed. To this end one can simply apply a Lorentz boost to the c.m. hCQM states. This has been done both for the elastic nucleon form factor [5] and for the helicity amplitudes [26]. The relativistic form factors are obtained from the non relativistic one by means of a simple relation:

$$A_{rel}(Q^2) = F A_{n.rel}(Q_{eff}^2) \quad Q_{eff}^2 = Q^2 \left(\frac{M_N}{E_N} \right) \tag{8}$$

where F is a known kinematical factor and M_N and E_N are respectively the mass and energy of the final hadron. As shown in the previous section, the non relativistic helicity amplitudes for the higher resonances give a fair description of data and happen to be only slightly modified by the relativistic corrections [26]. The situation is quite different for the elastic nucleon form factors: the non relativistic ones are not good, since, as already observed, the calculated proton radius is too small, the inclusion of the relativistic corrections provided by Eq. (8) bring the theoretical curves nearer to the experimental data [5], but the agreement is not yet satisfactory. Therefore, in the case of the elastic form factors a consistent relativistic formulation is necessary and can be achieved using the relativistic dynamics proposed by Dirac [40].

The relativistic version of hCQM is based on the point form, which allows to combine the angular momentum states as in the non relativistic case and leads to a good description of the four elastic form factors of the nucleon [7,8]. Some problem still remains for the ratio $R = \mu_p G_E^p(Q^2)/G_M^p(Q^2)$, recently measured at JLab. In fact, thanks to the relativistic effects [6,41], the calculation predicts a decrease but not as strong as in the data [7]. However, from the deep inelastic scattering of electrons off protons, there is a phenomenological evidence that quarks behave as non point-like structures [42], a feature which can be taken into account by introducing intrinsic quark form factors. In this way, one obtains a splendid agreement with data [7,8], in particular for the ratio R, as it is shown in Fig. 4.

An important issue in connection with the decrease of R is the possibility that the proton electric form factor $G_E^p(Q^2)$ passes through zero at some point. The hCQM curve seems to decrease continuously but remaining definitely positive, while the data taken up to 2005 [43–48] showed an apparent tendency to arrive at a dip. The recent high Q^2 results [49] apparently reinforces such behaviour, even if the crucial point at $Q^2 = 8.49 GeV^2$ has a very large error. However, a reanalysis [50] of the higher points [47] suggests that the behaviour of the data should be more compatible with a fitted curve having no dip [50].

It should be noted that using the relativistic version [51] of the quark-diquark model [52,53], one obtains a good description of the nucleon elastic form factors [54], again using the point form dynamics and introducing intrinsic quark form factors.

Another important aspect of relativity is the possibility of the creation of quark-antiquark pairs. As mentioned above, the lack of this mechanism is supposed [4] to be responsible for the missing strength at low

Q^2 observed in the helicity amplitudes. The inclusion of the quark pair mechanism in the CQMs, that is the unquenching of the CQM, is a hard task, which, for the meson sector, has been started some time ago [55–57], showing that it is compatible with a linear confinement and the OZI rule, and completed only recently [58–61] and recently performed also for baryons [62–66]. In these works, after having verified that the old results for the CQM on the baryon magnetic moments are preserved [63], the Unquenched CQM (UCQM) has been applied to the calculation of the flavour asymmetry [64], the strangeness radius [65,66] and the strange magnetic moment [66]. Thanks to the UCQM it has been possible to calculate the relative strength of electroproduction channels of various strange resonances, allowing to explain a strangeness suppression [67] recently observed in measurements performed at JLab [68]. A particular attention has been devoted to the problem of the proton spin [62,63,65], showing that the contributions coming from orbital angular momentum account for about 1/3 of the total spin and are therefore crucial for its explanation.

5 Conclusions

This brief overview shows that the hCQM provides a *simple* and *systematic* approach to baryon properties, such as spectrum, helicity amplitudes and nucleon elastic form factors. In particular, the structure of the hCQM levels allows to take into account all the new parity resonances without invoking higher shells.

Relativity is certainly an important issue for the elastic form factors and the helicity amplitudes for the $N \to \Delta$ transitions, but not so for the electroexcitation to the higher resonances.

The medium-high Q^2 behaviour of the helicity amplitudes is fairly well described by the hCQM, showing the relevance of the quark core contribution in this range. However, as in all CQMS, there is missing strength at low Q^2, a fact that is considered due to the lack of quark-antiquark pair mechanisms.

Such mechanisms may be important also for the high Q^2 behaviour of the elastic form factors, but also for the spectrum and the strong decays. In the latter case, the use of an UCQM will allow to calculate in a consistent way the elctroproduction of mesons on a quark basis only.

References

1. Ferraris, M., Giannini, M.M., Pizzo, M., Santopinto, E., Tiator, L.: A three body force model for the baryon structure. Phys. Lett. B **364**, 231–238 (1995)
2. Giannini, M.M., Santopinto, E.: The hypercentral constituent quark model and its application to baryon properties. Chin. J. Phys. **53**, 020301-1–020301-74 (2015)
3. Aiello, M., Ferraris, M., Giannini, M.M., Pizzo, M., Santopinto, E.: A three-body force model for the electromagnetic excitation of the nucleon. Phys. Lett. B **387**, 215–221 (1996)
4. Aiello, M., Giannini, M.M., Santopinto, E.: Electromagnetic transition form factors of negative parity resonances. J. Phys. G: Nucl. Part. Phys. **24**, 753–762 (1998)
5. De Sanctis, M., Santopinto, E., Giannini, M.M.: A relativistic study of the nucleon form factors. Eur. Phys. J. A **1**, 187–192 (1998)
6. De Sanctis, M., Giannini, M.M., Repetto, L., Santopinto, E.: The proton form factor in the hypercentral constituent quark model. Phys. Rev. C **C62**, 025208-1–025208-7 (2000)
7. De Sanctis, M., Giannini, M.M., Santopinto, E., Vassallo, A.: Electromagnetic form factors and the hypercentral constituent quark model. Phys. Rev. C **76**, 062201-1–062201-5 (2007)
8. Santopinto, E., Vassallo, A., Giannini, M.M., De Sanctis, M.: High Q^2 behaviour of the electromagnetic form factors in the relativistic hypercentral constituent quark model. Phys. Rev. C **82**, 065204-1–065204-3 (2010)
9. Santopinto, E., Giannini, M.: Systematic study of longitudinal and transverse helicity amplitudes in the hypercentral Constituent Quark Model. Phys. Rev. C **86**, 065202-1–065202-14 (2012)
10. Ferretti, J., Bijker, R., Galatà, G., García-Tecocoatzi, H., Santopinto. E.: Strong decays of baryons and missing resonances. (2015). arXiv:1506.07469 [hep-ph]
11. Wilson, K.: Confinement of quarks. Phys. Rev. D **10**, 2445–2459 (1974)
12. Kogut, J., Susskind, L.: Vacuum polarization and the absence of free quarks in four dimensions. Phys. Rev. D **9**, 3501–3512 (1974)
13. Hasenfratz, P., Horgan, R.R., Kuti, J., Richard, J.M.: Heavy baryon spectroscopy in the QCD bag model. Phys. Lett. **94B**, 401–404 (1980)
14. Carlson, J., Kogut, J., Pandharipande, V.R.: Quark model for baryons based on quantum chromodynamics. Phys. Rev. D **27**, 233–243 (1983)
15. Capstick, S., Isgur, N.: Baryons in a relativized quark model with chromodynamics. Phys. Rev. D **34**, 2809–2835 (1986)
16. De Rújula, A., Georgi, H., Glashow, S.L.: Hadron masses in a gauge theory. Phys. Rev. D **12**, 147–162 (1975)
17. Isgur, N., Karl, G.: P-wave baryons in the quark model. Phys. Rev. D **18**, 4187–4205 (1978)
18. Isgur, N., Karl, G.: Positive-parity excited baryons in a quark model with hyperfine interactions. Phys. Rev. D **19**, 2653–2677 (1979)

19. Bali, G.S., Schlichter, C., Schilling, K.: Observing long color flux tubes in SU(2) lattice gauge theory. Phys. Rev. D **51**, 5165–5198 (1995)
20. Bali, G.S.: QCD forces and heavy quark bound states. Phys. Rep. **343**, 1–136 (2001)
21. Ballot, J., Fabre de la Ripelle, M.: Application of the hyperspherical formalism to the trinucleon bound state problems. Ann. Phys. **127**, 62–125 (1980)
22. Giannini, M.M., Santopinto, E.: The baryon spectrum and the hypercentral constituent quark model. (2015). arXiv:1510.00582 [nucl-th]
23. Nakamura, K., et al.: [Particle Data Group] Review of particle physics. J. Phys. G: Nucl. Part. Phys. **37**, 075021-1–075021-1422 (2010)
24. Beringer, J., et al.: [Particle Data Group] Review of particle physics. Phys. Rev. D **86**, 010001-1–010001-1526 (2012)
25. Olive, K.A., et al.: [Particle Data Group] Review of particle physics. Chin. Phys. C **38**, 090001-1–090001-1676 (2014)
26. De Sanctis, M., Santopinto, E., Giannini, M.M.: A relativistic study of the nucleon helicity amplitudes Eur. Phys. J. A **2**, 403–409 (1998)
27. Copley, L.A., Karl, G., Obryk, E.: Single pion photoproduction in the quark model. Nucl. Phys. B **13**, 303–319 (1969)
28. Koniuk, R., Isgur, N.: Baryon decays in a quark model with chromodynamics. Phys. Rev. D **21**, 1868–1886 (1980)
29. Kamalov, S.S., Yang, S.N.: Pion cloud and the Q^2 dependence of $\gamma^* N \to \Delta$ transition form factors. Phys. Rev. Lett. **83**, 4494–4497 (1999)
30. Tiator, L., Drechsel, D., Kamalov, S., Giannini, M.M., Santopinto, E., Vassallo, A.: Electroproduction of nucleon resonances. Eur. Phys. J. A **19**, 55–60 (2004)
31. Warns, M., Pfeil, W., Rollnik, H.: Helicity and isospin asymmetries in the electroproduction of nucleon resonances. Phys. Rev. D **42**, 2215–2225 (1991)
32. Warns, M., Schröder, H., Pfeil, W., Rollnik, H.: Calculations of electromagnetic nucleon form factors and electroexcitation amplitudes of isobars. Z. Phys. C **45**, 627–644 (1991)
33. Carlson, C.E.: Electromagnetic $N - \Delta$ transition at high Q^2. Phys. Rev. D **34**, 2704–2709 (1986)
34. Aznauryan, I.G., et al.: (CLAS collaboration) Electroexcitation of nucleon resonances from CLAS data on single pion electroproduction. Phys. Rev. C **80**, 055203-1–055203-22 (2009)
35. Drechsel, D., Kamalov, S.S., Tiator, L.: Unitary isobar model —MAID2007. Eur. Phys. J. A **34**, 69–97 (2007)
36. Mokeev, V.I., et al.: (CLAS collaboration) Model analysis of the $ep \to e'p\pi^+\pi^-$ electroproduction reaction on the proton. Phys. Rev. C **80**, 045212-1–045212-23 (2009)
37. Park, K. et al. (CLAS Collaboration) (2015) Measurements of ep → e'+n at 1.6 < W < 2.0 GeV and extraction of nucleon resonance electrocouplings at CLAS. Phys. Rev. C **91**, 045203-1–045203-21
38. Giannini, M.M.: Electromagnetic excitations in the constituent quark model. Rep. Prog. Phys. **54**, 453–529 (1991)
39. Altarelli, G., Cabibbo, N., Maiani, L., Petronzio, R.: The nucleon as a bound state of three quarks and deep inelastic phenomena. Nucl. Phys. B **69**, 531–556 (1974)
40. Dirac, P.A.M.: Forms of relativistic dynamics. Phys. Rev. **21**, 291–399 (1949)
41. Giannini M.M: Models of the nucleon. Nucl. Phys. A **666-667**:321c–329c (2000)
42. Petronzio, R., Simula, S., Ricco, G.: Possible evidence of extended objects inside the proton. Phys. Rev. D **67**, 094004-1–094004-11 (2003). Erratum: Phys. Rev. D **67**, 094004-1 (2003)
43. Milbrath, B.D., et al.: [Bates FPP collaboration]: A comparison of polarization observables in electron scattering from the proton and deuteron. Phys. Rev. Lett. **80**, 452–455 (1998). Erratum. Phys. Rev. lett. **82**, 2221 (1999)
44. Jones, M.K., et al.: [Jefferson Lab Hall A Collaboration] G(E(p)) / G(M(p)) ratio by polarization transfer in polarized e p → e polarized p. Phys. Rev. Lett. **84**, 1398–1402 (2000)
45. Gayou, O., et al.: Measurements of the elastic electromagnetic form-factor ratio mu(p) G(Ep)/G(Mp) via polarization transfer. Phys. Rev. C **64**, 038202-1–038202-4 (2001)
46. Pospischil, T. et al.: [A1 Collaboration] Measurement of G(E(p))/G(M(p)) via polarization transfer at $Q^2 = 0.4 - GeV/c^2$. Eur. Phys. J. A **12**, 125–127 (2001)
47. Gayou O et al.: [Jefferson Lab Hall A Collaboration] Measurement of G(Ep)/G(Mp) in polarized-e p → e polarized-p to $Q^2 = 5.6 - GeV^2$. Phys. Rev. Lett. **88**, 092301-1–092301-5
48. Punjabi, V. et al.: Proton elastic form-factor ratios to $Q^2 = 3.5 - GeV^2$ by polarization transfer. Phys. Rev. C **71**, 055202-1-055202-27 (2005). Erratum 069902-1 (2005)
49. Puckett, A.J.R., et al.: Recoil polarization measurements of the proton electromagnetic form factor ratio to $Q^2 = 8.5 GeV^2$. Phys. Rev. Lett. **104**, 242301-1–242301-6 (2010)
50. Puckett, A.J.R., et al.: Final analysis of proton form factor ratio data at $Q^2 = 4.0, 4.8$ and $5.6 GeV^2$. Phys. Rev. C **85**, 045203-1–045203-26 (2012)
51. Ferretti, J., Vassallo, A., Santopinto, E.: Relativistic quark-diquark model of baryons. Phys. Rev. C **83**, 065204-1–065204-5 (2011)
52. Santopinto, E.: Interacting quark-diquark model of baryons. Phys. Rev. C **72**, 022201(R)-1–022201(R)-5 (2005)
53. Santopinto, E., Ferretti, J.: Strange and nonstrange baryon spectra in the relativistic interacting quark-diquark model with a Gürsey and Radicati-inspired exchange interaction. Phys. Rev. C **92**, 025202-1–025202-9 (2015)
54. De Sanctis, M., Ferretti, J., Santopinto, E., Vassallo, A.: Electromagnetic form factors in the relativistic interacting quark-diquark model of baryons. Phys. Rev. C **84**, 055201-1–055201-5 (2011)
55. Geiger, P., Isgur, N.: The Quenched Approximation in the Quark Model. Phys. Rev. D **41**, 1595–1605 (1990)
56. Geiger, P., Isgur, N.: How the Okubo–Zweig–Iizuka rule evades large loop corrections. Phys. Rev. Lett. **67**, 1066–1069 (1991)
57. Geiger, P., Isgur, N.: Reconciling the OZI rule with strong pair creation. Phys. Rev. D **44**, 799–808 (1991)
58. Ferretti, J., Galatà, G., Santopinto, E.: Bottomonium self-energies due to the coupling to the meson-meson continuum. Phys. Rev. C **86**, 015204-1–015204-6 (2012)

59. Ferretti, J., Galatà, G., Santopinto, E.: Interpretation of the X(3872) as a charmonium state plus an extra component due to the coupling to the meson-meson continuum. Phys. Rev. C **88**, 015207-1–015207-9 (2013)
60. Ferretti, J., Galatà, G., Santopinto, E.: Quark structure of the X(3872) and χ_b (3P) resonances. Phys. Rev. D **90**, 054010-1–054010-9 (2014)
61. Ferretti, J., Santopinto, E.: Higher mass bottomonia. Phys. Rev. D **90**, 094022-1–094022-13 (2014)
62. Santopinto, E., Bijker, R.: Quark-antiquark effects in baryons. Few Body Syst. **44**, 95–97 (2008)
63. Bijker, R., Santopinto, E.: Unquenched quark model for baryons: magnetic moments, spins and orbital angular momenta. Phys. Rev. C **80**, 065210-1–065210-9 (2009)
64. Santopinto, E., Bijker, R.: Flavor asymmetry of sea quarks in the unquenched quark model. Phys. Rev. C **82**, 062202-1–062202-5 (2010)
65. Santopinto, E., Bijker, R., Ferretti, J.: Unquenching the quark model. Few Body Syst. **50**, 199–201 (2011)
66. Bijker, R., Ferretti, J., Santopinto, E.: s$\bar{s}$ sea pair contribution to electromagnetic observables of the proton in the unquenched quark model. Phys. Rev. C **85**, 035204-1–035204-5 (2012)
67. Santopinto, E., Bijker, R., García-Tecocoatzi, H.: Electroproduction of Baryon resonances and strangeness suppression. (2016). arXiv:1601.06987 [hep-ph]
68. Mestayer, M.D.: (CLAS Collaboration) Strangeness suppression of q$\bar{s}$ creation observed in exclusive reactions. Phys. Rev. Lett. **113**, 152004-1–152004-6 (2014)

Few-Body Syst (2016) 57:925–932
DOI 10.1007/s00601-016-1129-6

Guy F. de Téramond

The Spectroscopy and Form Factors of Nucleon Resonances from Superconformal Quantum Mechanics and Holographic QCD

Received: 15 February 2016 / Accepted: 2 June 2016 / Published online: 16 June 2016
© Springer-Verlag Wien 2016

Abstract The superconformal algebraic approach to hadronic physics is used to construct a semiclassical effective theory for nucleons which incorporates essential nonperturbative dynamical features, such as the emergence of a confining scale and the Regge resonance spectrum. Relativistic bound-state equations for nucleons follow from the extension of superconformal quantum mechanics to the light front and its holographic embedding in a higher dimensional gravity theory. Superconformal algebra has been used elsewhere to describe the connections between the light mesons and baryons, but in the present context it relates the fermion positive and negative chirality states and uniquely determines the confinement potential of nucleons. The holographic mapping of multi-quark bound states also leads to a light-front cluster decomposition of form factors for an arbitrary number of constituents. The remarkable analytical structure which follows incorporates the correct scaling behavior at high photon virtualities and also vector dominance at low energies.

1 Introduction

The study of the dynamics and internal structure of nucleons is an intricate problem in hadronic physics. In fact, lattice QCD calculations of the excitation spectrum of the light hadrons, and particularly nucleons, represent a formidable task due to the enormous computational complexity beyond the leading ground state configuration [1]. On the other hand, holographic methods provide new analytical tools for the study of strongly correlated quantum systems, which are complementary to other nonperturbative approaches to strongly coupled gauge theories [2]. The best known example is the AdS/CFT correspondence between gravity in anti-de Sitter (AdS) five-dimensional space and conformal field theories (CFT) in physical space-time [3], which leads to new insights into the nonperturbative dynamics of QCD.

There is a remarkable connection of light-front quantized theories [4,5] to gravity theory in AdS: The Hamiltonian equations in AdS space can be precisely mapped to the relativistic semiclassical bound-state equations in the light front [6,7]. This connection gives an exact relation between the holographic variable z of AdS space and the invariant impact light-front variable ζ in physical space-time [6,8]. This connection also implies that the light-front (LF) effective potential U in the LF Hamiltonian equations, corresponds to the modification of AdS space—described in terms of a dilaton profile φ in the string frame. The light-front effective potential U acts on the valence state and incorporates and infinite number of multiple-component higher Fock states [9]. To compute U one must systematically express higher Fock components as functionals of the lower ones. Its actual derivation remains an unsolved problem.

Faced with the enormous complexity of nonperturbative QCD, other methods, which encompass the essential features of the strong interaction dynamics, are needed to gain further understanding into the nature of

This article belongs to the special issue "Nucleon Resonances".

G. F. de Téramond (✉)
Universidad de Costa Rica, San José, Costa Rica
E-mail: gdt@asterix.crnet.cr

confinement physics. This complexity can be understood from the increase of the QCD coupling in the infrared domain, which implies that an infinite number of quark and gluons are dynamically coupled.[1] Recent progress along these lines has followed from the study of conformal quantum mechanics (QM) [11] and its mapping to the light front, which determines the form of the LF confinement potential U and thus the dilaton profile φ [12]. This procedure allows the introduction of a scale $\sqrt{\lambda}$ in the Hamiltonian while the action remains conformal invariant [11].

The supersymmetric extension of conformal QM, namely superconformal quantum mechanics [13,14], can also be mapped to the semiclassical LF effective theory [15–17]—a one dimensional QFT, and consequently to gravity theory in AdS. This new approach to hadronic physics incorporates confinement, the appearance of a massless pion in the limit of zero-mass quarks, and the Regge excitation spectrum consistent with experimental data. Furthermore, this framework gives remarkable connections between the light meson and nucleon spectra [16]. It also gives predictions for the heavy-light hadron spectra, where heavy charm and bottom quark masses break the conformal invariance, but the underlying supersymmetry holds [17].[2]

Following Ref. [15] we discuss in this article how the superconformal framework leads to relativistic bound-state equations for nucleons from the mapping to light-front physics and its embedding in a higher dimensional AdS space. In this case, the superconformal algebra relates the nucleon positive and negative chirality states and determines the effective confinement potential of nucleons. In turn, this allows us to study the holographic embedding of the semiclassical effective theory for nucleons. In fact, in contrast to mesons, a dilaton term in the AdS fermionic action has no dynamical effects since it can be rotated away by a redefinition of the fermion fields [18] and a specific Yukawa-like interaction term has to be introduced in the AdS action to break conformal invariance [19], but its form is left unspecified. The superconformal approach has thus the advantage that mesons and nucleons are treated on the same footing, and the confinement potential is uniquely determined by the formalism—including additional spin-dependent constant terms which are critical to describe the hadronic spectrum.

For a multi-quark bound state the light front invariant impact variable ζ corresponds to a system of an active quark plus an spectator cluster. The holographic embedding is also characterized by a single variable ζ which is mapped to the AdS variable z [8]. For example, for a three quark system, the three-body problem is reduced to an effective two-body problem where two of the constituents form a diquark cluster. However, in the present framework the diquark is not a tightly bound state. It is also important to notice that the reduction to a single variable is also crucial in the superconformal formulation of hadron physics where mesons and nucleons are in the same multiplet.

The light-front cluster decomposition of hadronic bound states is also important to solve the standing problem of the twist assignment of the proton in holographic QCD [20]. Since the lowest bound-state solution to the holographic Dirac equation corresponds to twist 2, the nucleon is described by the wave function of a quark-diquark cluster. At high energies, however, all the constituents in the proton are resolved and therefore the fall-off of the form factor is governed by the number of all constituents, *i.e.*, it is twist 3. A related problem was found in the study of sequential decay chains in baryons [21], which are sensitive to the short distance behavior of the wave function. The solution to this problem follows from the LF cluster decomposition for bound states [22–24]. It will be discussed below.

2 Light-Front Holographic Embedding

As a brief review, we first examine the embedding of the semiclassical light-front wave equations for mesons in AdS space. We study then the general structure of the LF equations for nucleons starting from the Dirac AdS action. The actual confinement potential is determined in the next section from the superconformal algebraic structure.

Our starting point is an effective action in AdS$_5$ space for the spin-J tensor field $\Phi_{N_1...N_J}$

$$S_{eff} = \int d^4x\, dz\, \sqrt{g}\, e^{\varphi(z)}\, g^{N_1 N_1'} \ldots g^{N_J N_J'} \left(g^{MM'} D_M \Phi^*_{N_1...N_J} D_{M'} \Phi_{N_1'...N_J'} - \mu^2_{eff}(z)\, \Phi^*_{N_1...N_J}\, \Phi_{N_1'...N_J'} \right),$$

(1)

---

[1] The QCD confinement problem is of such complexity that it could be undecidable: This means that it is not possible to know, starting from the fundamental degrees of freedom of the QCD Lagrangian, whether the system is gapped or not [10].

[2] In hadronic physics supersymmetry is an emergent dynamical property from color $SU(3)_C$ since a diquark is in the same color representation as an antiquark, namely a $\bar{\mathbf{3}} \sim \mathbf{3} \times \mathbf{3}$.

with coordinates $x^M = (x^\mu, z)$. The holographic variable is z and x^μ are Minkowski flat space-time coordinates. The metric determinant is $\sqrt{g} = (R/z)^5$ and D_M is the covariant derivative which includes the affine connection (R is the AdS radius). The dilaton field φ breaks the maximal symmetry of AdS, and the effective AdS mass μ_{eff} is determined by the mapping to light-front physics [7].

A hadron with momentum P and physical polarization $\epsilon_{\nu_1...\nu_J}(P)$ is represented by

$$\Phi_{\nu_1...\nu_J}(x, z) = e^{i P \cdot x} \epsilon_{\nu_1...\nu_J}(P) \Phi_J(z), \tag{2}$$

with invariant hadron mass $P_\mu P^\mu \equiv \eta^{\mu\nu} P_\mu P_\nu = M^2$. Variation of the action (1) leads to the wave equation

$$\left[-\frac{z^{3-2J}}{e^{\varphi(z)}} \partial_z \left(\frac{e^{\varphi(z)}}{z^{3-2J}} \partial_z \right) + \frac{(\mu R)^2}{z^2} \right] \Phi_J = M^2 \Phi_J, \tag{3}$$

where $(\mu R)^2 = (\mu_{eff}(z) R)^2 - J z \varphi'(z) + J(5 - J)$ is a constant determined by kinematical conditions in the light front [7].

We now compare the wave equations in the dilaton-modified AdS space with LF bound-state equations in the semiclassical approximation described in [6]. In the light front the hadron four-momentum generator is $P = (P^+, P^-, \mathbf{P}_\perp)$, $P^\pm = P^0 \pm P^3$, and the hadronic spectrum is computed from the invariant Hamiltonian $P^2 = P_\mu P^\mu = P^- P^+ - \mathbf{P}_\perp^2$:

$$P^2 |\psi(P)\rangle = M^2 |\psi(P)\rangle, \tag{4}$$

where $|\psi(P)\rangle$ is expanded in multi-particle Fock states $|n\rangle$: $|\psi\rangle = \sum_n \psi_n |n\rangle$.

In the limit of zero-quark masses, the bound-state dynamics of the constituents can be separated from the longitudinal kinematics and the orbital dependence in the transverse LF plane leading to the wave equation [6, 7]:

$$\left(-\frac{d^2}{d\zeta^2} - \frac{1 - 4L^2}{4\zeta^2} + U(\zeta) \right) \phi(\zeta) = M^2 \phi(\zeta), \tag{5}$$

where the invariant transverse variable in impact space $\zeta^2 = x(1 - x)\mathbf{b}_\perp^2$ is conjugate to a two-body LF invariant mass $M_{q\bar{q}}^2 = \mathbf{k}_\perp^2 / x(1 - x)$. The critical value of the orbital angular momentum $L = 0$ corresponds to the lowest possible stable solution [25]. Equation (5) is a relativistic and frame-independent LF Schrödinger equation: The confinement potential U is instantaneous in LF time and comprises all interactions, including those with higher Fock states. Upon the substitution $\Phi_J(z) \sim z^{(d-1)/2-J} e^{-\varphi(z)/2} \phi_J(z)$ and $z \to \zeta$ in Eq. (3), we find Eq. (5) with

$$U(\zeta) = \tfrac{1}{2} \varphi''(\zeta) + \tfrac{1}{4} \varphi'(\zeta)^2 + \frac{2J-3}{2\zeta} \varphi'(\zeta),$$

and $(\mu R)^2 = -(2 - J)^2 + L^2$ [7,26]. The effective LF confining potential $U(\zeta)$ thus corresponds to the IR modification of AdS space.

Nucleons with arbitrary half-integer spin $J = T + \tfrac{1}{2}$ are described in AdS by an effective action for Rarita–Schwinger spinors $\Psi_{N_1...N_T}$

$$S_{eff} = \tfrac{1}{2} \int d^d x \, dz \, \sqrt{g} \, g^{N_1 N_1'} \ldots g^{N_T N_T'} \tag{6}$$

$$\left[\bar{\Psi}_{N_1...N_T} \left(i \, \Gamma^A e_A^M D_M - \mu - \rho(z) \right) \Psi_{N_1'...N_T'} + h.c. \right], \tag{7}$$

where curved space indices are $M, N = 0, \ldots, 4$ and tangent indices are $A, B = 0, \ldots, 4$, with e_A^M the inverse vielbein, $e_A^M = \left(\frac{z}{R} \right) \delta_A^M$. The covariant derivative D_M includes the affine connection and the spin connection. The tangent-space Dirac matrices obey the usual anticommutation relation $\{ \Gamma^A, \Gamma^B \} = 2\eta^{AB}$. The effective interaction $\rho(z)$ breaks conformal symmetry and generates a baryon spectrum [19].

A nucleon with four-momentum P and chiral spinors $u_{\nu_1...\nu_T}^\pm(P)$ is represented by

$$\Psi_{\nu_1...\nu_T}^\pm(x, z) = e^{i P \cdot x} u_{\nu_1...\nu_T}^\pm(P) \left(\frac{R}{z} \right)^{T-2} \Psi_T^\pm(z). \tag{8}$$

Variation of the action (6) leads to a system of linear coupled equations

$$-\frac{d}{dz}\psi_- - \frac{\nu + \frac{1}{2}}{z}\psi_- - V(z)\psi_- = M\psi_+,$$

$$\frac{d}{dz}\psi_+ - \frac{\nu + \frac{1}{2}}{z}\psi_+ - V(z)\psi_+ = M\psi_-, \tag{9}$$

where $\mu R = \nu + \frac{1}{2}$, $\psi_\pm \equiv \Psi_T^\pm$, and

$$V(z) = \frac{R}{z}\rho(z), \tag{10}$$

a J-independent potential. Thus, independently of the specific form of the potential, the value of the nucleon masses along a given Regge trajectory depends only on the LF orbital angular momentum L,[3] in agreement with the observed near-degeneracy in the baryon spectrum [28].

Mapping to the light front, $z \to \zeta$, Eq. (9) is equivalent to the system of second order equations

$$\left(-\frac{d^2}{d\zeta^2} - \frac{1 - 4L^2}{4\zeta^2} + U^+(\zeta)\right)\psi_+ = M^2\psi_+, \tag{11}$$

$$\left(-\frac{d^2}{d\zeta^2} - \frac{1 - 4(L+1)^2}{4\zeta^2} + U^-(\zeta)\right)\psi_- = M^2\psi_-, \tag{12}$$

where

$$U^\pm(\zeta) = V^2(\zeta) \pm V'(\zeta) + \frac{1 + 2\nu}{\zeta}V(\zeta),$$

and $L = \nu = \mu R - \frac{1}{2}$. The plus and minus component wave Eqs. (11) and (12) correspond, respectively, to LF orbital angular momentum L and $L + 1$.

3 Superconformal Quantum Mechanics and Nucleon Bound-State Equations

We follow Ref. [15] to construct nucleon bound-state equations by extending the superconformal algebraic structure of Fubini and Rabinovici [14] to the light front. Superconformal quantum mechanics is a one-dimensional quantum field theory invariant under conformal and supersymmetric transformations. Imposing conformal symmetry leads to a unique choice of the superpotential and thus to a unique confinement potential in the light front. In addition to the Hamiltonian H and the usual fermionic operators Q and $Q^\dagger$ of supersymmetric quantum mechanics [29], an additional generator S, which is related to the generator of conformal transformations K, is introduced.

We use the representation of the operators

$$Q = \chi\left(\frac{d}{dx} + \frac{f}{x}\right), \quad Q^\dagger = \chi^\dagger\left(-\frac{d}{dx} + \frac{f}{x}\right),$$

where f is a dimensionless constant, and $S = \chi x$, $S^\dagger = \chi^\dagger x$. It is now simple to verify the closure of the enlarged algebraic structure. In a Pauli matrix representation:

$$\frac{1}{2}\{Q, Q^\dagger\} = H, \qquad\qquad \frac{1}{2}\{S, S^\dagger\} = K,$$

$$\frac{1}{2}\{Q, S^\dagger\} = \frac{f}{2} + \frac{\sigma_3}{4} - iD, \qquad \frac{1}{2}\{Q^\dagger, S\} = \frac{f}{2} + \frac{\sigma_3}{4} + iD, \tag{13}$$

where the operators

$$H = \frac{1}{2}\left(-\frac{d^2}{dx^2} + \frac{f^2 - \sigma_3 f}{x^2}\right), \quad D = \frac{i}{4}\left(\frac{d}{dx}x + x\frac{d}{dx}\right), \quad K = \frac{1}{2}x^2, \tag{14}$$

---

[3] This result was also found in Ref. [27].

satisfy the conformal algebra:

$$[H, D] = iH, \quad [H, K] = 2iD, \quad [K, D] = -iK. \tag{15}$$

Following Ref. [14] we define a new fermionic operator R, a linear combination of the generators Q and S,

$$R = \sqrt{u}\, Q + \sqrt{w}\, S, \qquad R^\dagger = \sqrt{u}\, Q^\dagger + \sqrt{w}\, S^\dagger, \tag{16}$$

which generates a new Hamiltonian G

$$\tfrac{1}{2}\{R, R^\dagger\} = G, \tag{17}$$

where by construction $\{R_\lambda, R_\lambda\} = \{R_\lambda^\dagger, R_\lambda^\dagger\} = 0$ and $[R_\lambda, G] = [R_\lambda^\dagger, G] = 0$. We find

$$G = uH + wK + \tfrac{1}{2}\sqrt{uw}\,(2f + \sigma_3), \tag{18}$$

which is a compact operator for $uw > 0$. Since the new Hamiltonian G commutes with R and $R^\dagger$, it follows that $|\phi\rangle$ and $R|\phi\rangle$ have identical eigenvalues.

We now extend the Hamiltonian G (18) to a relativistic LF Hamiltonian by performing the substitutions

$$x \to \zeta, \quad f \to L + \tfrac{1}{2}, \quad \sigma_3 \to \gamma_5, \quad 2G \to H_{LF}.$$

We obtain:

$$H_{LF} = \{R, R^\dagger\} = -\frac{d^2}{d\zeta^2} + \frac{\left(L + \tfrac{1}{2}\right)^2}{\zeta^2} - \frac{L + \tfrac{1}{2}}{\zeta^2}\gamma_5 + \lambda^2\zeta^2 + \lambda(2L + 1) + \lambda\gamma_5, \tag{19}$$

where L is the relative LF angular momentum between the active quark and the spectator cluster and the arbitrary coefficients u and w in (18) are fixed to $u = 1$ and $w = \lambda^2$. In a 2×2 block-matrix form the light-front Hamiltonian (19) can be expressed as

$$H_{LF} = \begin{pmatrix} -\frac{d^2}{d\zeta^2} - \frac{1-4L^2}{4\zeta^2} + \lambda^2\zeta^2 + 2\lambda(L+1) & 0 \\ 0 & -\frac{d^2}{d\zeta^2} - \frac{1-4(L+1)^2}{4\zeta^2} + \lambda^2\zeta^2 + 2\lambda L \end{pmatrix}. \tag{20}$$

Since H_{LF} commutes with R, the eigenvalues for the chirality plus and minus eigenfunctions are identical. Comparing (20) with (11) and (12) we obtain the effective confining potential (10) in AdS space: It is the linear potential $V = \lambda z$.

The light-front eigenvalue equation $H_{LF}|\psi\rangle = M^2|\psi\rangle$ has eigenfunctions

$$\psi_+(\zeta) \sim \zeta^{\frac{1}{2}+L} e^{-\lambda\zeta^2/2} L_n^L(\lambda\zeta^2), \tag{21}$$

$$\psi_-(\zeta) \sim \zeta^{\frac{3}{2}+L} e^{-\lambda\zeta^2/2} L_n^{L+1}(\lambda\zeta^2),$$

and eigenvalues,

$$M^2 = 4\lambda(n + L + 1). \tag{22}$$

Both components have identical normalization [15]:

$$\int d\zeta\, \psi_+^2(\zeta) = \int d\zeta\, \psi_-^2(\zeta).$$

The nucleon spin is thus carried by the orbital angular momentum [20].

The predictions for the resonance spectrum of the positive-parity internal spin-$\frac{1}{2}$ light nucleons is shown in Fig. 1a for the parent Regge trajectory, $n = 0$, and the daughter trajectories for $n = 1$, $n = 2$, ... The lowest stable state, the nucleon $N\frac{1}{2}^+(940)$, corresponds to $n = 0$ and $L = 0$. The Roper state $N\frac{1}{2}^+(1440)$ and the $N\frac{1}{2}^+(1710)$ are described as the first and second radial excited states of the nucleon. The model is also successful in explaining the J-degeneracy for states with the same orbital angular momentum, such as the $L = 2$ positive-parity doublet $N\frac{3}{2}^+(1720) - N\frac{5}{2}^+(1680)$. Only confirmed PDG [30] states are shown.

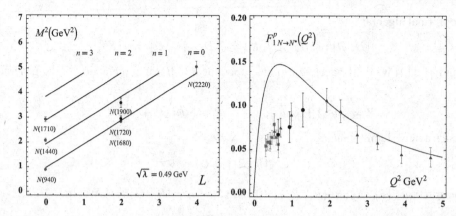

Fig. 1 *Left* Nucleon resonance spectrum of positive parity nucleons. *Right* Proton transition form factor $F_{1\ N\to N^*}^p(Q^2)$ for the radial transition $N\frac{1}{2}^+(940) \to N\frac{1}{2}^+(1440)$. Nucleon spectrum data from PDG [30] (*left*) and JLAB [34–36] (*right*)

4 Light-Front Cluster Decomposition and Form Factors

The problem of the twist assignment of the proton in holographic QCD has been addressed in Ref. [24]. The solution to this outstanding problem is based on the LF cluster decomposition for bound states given in Refs. [22,23], where the factorization properties of the deuteron were extensively analyzed. As a result of the cluster decomposition, the deuteron wave function factorizes into two distinct nucleon wave functions convoluted with a two-body factor f_d, $F_d(Q^2) = f_d(Q^2) F_N^2(\frac{1}{4}Q^2)$, where $Q^2 f_d(Q^2) \simeq const$ at large Q^2. The nucleon form factors F_N are evaluated at $Q/2$, since both nucleons share the momentum transferred to the bound state by the incoming probe.[4]

For simplicity we examine here spin-non-flip transition amplitudes. On the higher dimensional gravity theory it corresponds to the coupling of an external electromagnetic (EM) field $A^M(x, z)$ propagating in AdS with a fermionic mode $\Psi_P(x, z)$, given by the left-hand side of the equation

$$\int d^4x\, dz\, \sqrt{g}\, \bar{\Psi}_{P'}(x, z)\, e_M^A\, \Gamma_A\, A^M(x, z)\Psi_P(x, z) \tag{23}$$

$$\sim (2\pi)^4 \delta^4\left(P' - P - q\right) \epsilon_\mu \bar{u}(P')\gamma^\mu F_1(q^2)u(P), \tag{24}$$

The expression on the right-hand side represents the Dirac EM form factor in physical space-time. It is the EM spin-conserving matrix element of the quark current $J^\mu = e_q \bar{q}\gamma^\mu q$ with local coupling to the constituents. A precise mapping of the matrix elements can be carried out at fixed LF time, providing an exact correspondence between the holographic variable z and the LF impact variable ζ in ordinary space-time.[5]

For an N constituent bound state the mapping expressed by Eq. (23) leads to the analytic form [20,32]

$$F_{\tau=N}(Q^2) = \frac{1}{\left(1 + \frac{Q^2}{M_{n=0}^2}\right)\left(1 + \frac{Q^2}{M_{n=1}^2}\right)\cdots\left(1 + \frac{Q^2}{M_{n=\tau-2}^2}\right)}, \tag{25}$$

where the twist τ is equal to the number of constituents, *i. e.*, $\tau = 3$ for the proton valence state. To compare with the data one has to shift the poles to their physical location at $-Q^2 = 4\lambda(n + \frac{1}{2})$, with $\sqrt{\lambda} = M_\rho/2$: The predicted bound-state poles of the ρ ground state and its radial excitations [20]. Following this procedure we obtain the cluster form for the twist-τ elastic form factor

$$F_{\tau=N}(Q^2) = F_{\tau=2}(Q^2)\, F_{\tau=2}(\tfrac{1}{3}Q^2) \cdots F_{\tau=2}(\tfrac{1}{2N-3}Q^2), \tag{26}$$

which is the product of twist-two (pion) form factors evaluated at different scales. For the proton elastic Dirac form factor the corresponding expression is

$$F_1^p(Q^2) = F_\pi(Q^2)\, F_\pi(\tfrac{1}{3}Q^2). \tag{27}$$

---

[4] The form factors of the deuteron where studied in the framework of AdS/QCD in Ref. [31].

[5] For n partons the invariant LF variable ζ is the x-weighted definition of the transverse impact variable of the $n - 1$ spectator system [8]: $\zeta = \sqrt{\frac{x}{1-x}}\left|\sum_{j=1}^{n-1} x_j \mathbf{b}_{\perp j}\right|$ where $x = x_n$ is the longitudinal momentum fraction of the active quark.

It corresponds to the factorization of the proton form factor as a product of a point-like quark and composite diquark form factors with the correct scaling behavior at large Q^2.

A similar LF cluster decomposition for arbitrary twist can be obtained, for example, for the radial transition $n = 0 \to n = 1$ [24]. The Dirac transition form factor for the radial transition $N\frac{1}{2}^{+}(940) \to N\frac{1}{2}^{+}(1440)$ is given by the twist-3 expression [24,33]

$$F^p_{1\,N\to N^*}(Q^2) = \frac{2\sqrt{2}}{3}\, F_{\pi\to\pi'}(Q^2)\, F_\pi(\tfrac{1}{5}Q^2), \tag{28}$$

which is the product of the radial transition form factor of the pion $F_{\pi\to\pi'}(Q^2)$,

$$F_{\pi\to\pi'}(Q^2) = \frac{1}{2}\,\frac{Q^2/M_\rho^2}{\left(1+\frac{Q^2}{M_\rho^2}\right)\left(1+\frac{Q^2}{M_{\rho'}^2}\right)}, \tag{29}$$

and the pion elastic form factor. The predictions are shown in Fig. 1.[6] The twist-3 transition form factor behaves as $F_{\tau=3} \sim \frac{1}{Q^4}$ at large momentum transfer and reproduces quite well the data above $Q^2 \geq 2\,\mathrm{GeV^2}$. The low energy data, however, is not well reproduced and possibly indicates the necessity to include higher Fock components which give important contributions at low Q^2, but vanish rapidly at large Q^2. For example a $\tau = 5$ higher Fock component $|qqq\bar{q}q\rangle$, in addition to the $\tau = 3$ valence $|qqq\rangle$ state, contributes to the low energy region and vanishes as $F_{\tau=5} \sim 1/Q^8$ at higher energies. Indeed, it was found for the pion form-factor that the inclusion of higher Fock components are required to have a meaningful comparison with experiment [20].

5 Concluding Remarks

The connection of light-front dynamics, classical gravity in AdS space and superconformal quantum mechanics leads to a semiclassical approximation which describes the dynamics and internal structure of nucleons. The superconformal algebraic structure determines uniquely the effective confinement potential and connects mesons to baryons. The emerging confinement scale $\sqrt{\lambda}$ is directly related to physical observables, such as a hadron mass, and can be related to scheme dependent perturbative scales, such as the QCD renormalization scale Λ_s [38,39]. The light-front holographic mapping also leads to a cluster decomposition of hadron bound states, which is particularly useful for understanding the structure of transition amplitudes. The analytical structure of the form factors incorporates the short distance scaling behavior dictated by the number of constituents N and the transition to the nonperturbative region determined by vector dominance. For a hadron with angular momentum L, the expression (25) is still valid but the twist is $\tau = N + L$. This result could be used to test supersymmetric QM connections at the amplitude level: It implies that the higher value of the orbital angular momentum L of a partner meson, $L_M = L_B + 1$, is compensated by the additional constituent in the baryon [16].

Extension of the results described here to the full light-hadron spectrum has been carried out recently by enforcing superconformal symmetry in the holographic embedding of the light-front bound-state equations, including internal spin and quark masses [24].[7] Other relevant questions left open in Ref. [16], such as the nature of the spin interaction and the correct twist assignment of the baryon wave functions, have also been addressed in Ref. [24]. The new framework based on the superconformal algebraic structure and its mappings to light front holographic bound-state equations, provides a set of new analytic tools which can be particularly useful for the theoretical interpretation of the upcoming results at the new energy scales and kinematic regions which are about to be explored with the JLab 12 GeV Upgrade Project [41].

Acknowledgments The results presented here are based on collaborations with Stanley J. Brodsky, Alexandre Deur, Hans Guenter Dosch and Cédric Lorce. I want to thank the organizers of the ECT* 2015 Workshop on Nucleon Resonances for their hospitality at Trento.

---

[6] An AdS/QCD computation including the $A^p_{1/2}$ and $S^p_{1/2}$ amplitudes is given in Ref. [37].

[7] A recent extension of light front-holographic QCD to describe octet and decouplet baryons is given in Ref. [40].

References

1. Edwards, R.G., Dudek, J.J., Richards, D.G., Wallace, S.J.: Excited state baryon spectroscopy from lattice QCD. Phys. Rev. D **84**, 074508 (2011)
2. Brambilla, N., et al.: QCD and strongly coupled gauge theories: challenges and perspectives. Eur. Phys. J. C **74**, 2981 (2014)
3. Maldacena, J.M.: The large-N limit of superconformal field theories and supergravity. Int. J. Theor. Phys. **38**, 1113 (1999)
4. Dirac, P.A.M.: Forms of relativistic dynamics. Rev. Mod. Phys. **21**, 392 (1949)
5. Brodsky, S.J., Pauli, H.C., Pinsky, S.S.: Quantum chromodynamics and other field theories on the light cone. Phys. Rep. **301**, 299 (1998)
6. de Teramond, G.F., Brodsky, S.J.: Light-front holography: a first approximation to QCD. Phys. Rev. Lett. **102**, 081601 (2009)
7. de Teramond, G.F., Dosch, H.G., Brodsky, S.J.: Kinematical and dynamical aspects of higher-spin bound-state equations in holographic QCD. Phys. Rev. D **87**, 075005 (2013)
8. Brodsky, S.J., de Teramond, G.F.: Hadronic spectra and light-front wave functions in holographic QCD. Phys. Rev. Lett. **96**, 201601 (2006)
9. Pauli, H.C.: On confinement in a light cone Hamiltonian for QCD. Eur. Phys. J. C **7**, 289 (1999)
10. Cubitt, T.S., Perez-Garcia, D., Wolf, M.M.: Undecidability of the spectral gap. Nature **528**, 207 (2015)
11. de Alfaro, V., Fubini, S., Furlan, G.: Conformal invariance in quantum mechanics. Nuovo Cim. A **34**, 569 (1976)
12. Brodsky, S.J., de Teramond, G.F., Dosch, H.G.: Threefold complementary approach to holographic QCD. Phys. Lett. B **729**, 3 (2014)
13. Akulov, V.P., Pashnev, A.I.: Quantum superconformal model in (1,2) space. Theor. Math. Phys. **56**, 862 (1983)
14. Fubini, S., Rabinovici, E.: Superconformal quantum mechanics. Nucl. Phys. B **245**, 17 (1984)
15. de Teramond, G.F., Dosch, H.G., Brodsky, S.J.: Baryon spectrum from superconformal quantum mechanics and its light-front holographic embedding. Phys. Rev. D **91**, 045040 (2015)
16. Dosch, H.G., de Teramond, G.F., Brodsky, S.J.: Superconformal baryon-meson symmetry and light-front holographic QCD. Rev. D **91**, 085016 (2015)
17. Dosch, H.G., de Teramond, G.F., Brodsky, S.J.: Supersymmetry across the light and heavy-light hadronic spectrum. Phys. Rev. D **92**, 074010 (2015)
18. Kirsch, I.: Spectroscopy of fermionic operators in AdS/CFT. JHEP **0609**, 052 (2006)
19. Abidin, Z., Carlson, C.E.: Nucleon electromagnetic and gravitational form factors from holography. Phys. Rev. D **79**, 115003 (2009)
20. Brodsky, S.J., de Teramond, G.F., Dosch, H.G., Erlich, J.: Light-front holographic QCD and emerging confinement. Phys. Rep. **584**, 1 (2015)
21. Thiel, A., et al. (CBELSA/TAPS Collaboration): Three-body nature of N^* and Δ^* resonances from sequential decay chains. Phys. Rev. Lett. **114**, 091803 (2015)
22. Brodsky, S.J., Ji, C.R., Lepage, G.P.: Quantum chromodynamic predictions for the deuteron form factor. Phys. Rev. Lett. **51**, 83 (1983)
23. Brodsky, S.J., Ji, C.R.: Factorization property of the deuteron. Phys. Rev. D **33**, 2653 (1986)
24. Brodsky, S.J., de Teramond, G.F., Dosch, H.G., Lorce, C.: Universal effective hadron dynamics from superconformal algebra. Phys. Lett. B. **759**, 171 (2016)
25. Breitenlohner, P., Freedman, D.Z.: Stability in gauged extended supergravity. Ann. Phys. **144**, 249 (1982)
26. de Teramond, G.F., Brodsky, S.J.: Gauge/gravity duality and hadron physics at the light front. AIP Conf. Proc. **1296**, 128 (2010)
27. Gutsche, T., Lyubovitskij, V.E., Schmidt, I., Vega, A.: Dilaton in a soft-wall holographic approach to mesons and baryons. Phys. Rev. D **85**, 076003 (2012)
28. Klempt, E., Richard, J.M.: Baryon spectroscopy. Rev. Mod. Phys. **82**, 1095 (2010)
29. Witten, E.: Dynamical breaking of supersymmetry. Nucl. Phys. B **188**, 513 (1981)
30. Olive, K.A., et al. (Particle Data Group Collaboration): Review of particle physics. Chin. Phys. C **38**, 090001 (2014)
31. Gutsche, T., Lyubovitskij, V.E., Schmidt, I., Vega, A.: Nuclear physics in soft-wall AdS/QCD: deuteron electromagnetic form factors. Phys. Rev. D **91**, 114001 (2015)
32. Brodsky, S.J., de Teramond, G.F.: Light-front dynamics and AdS/QCD correspondence: the pion form factor in the space- and time-like regions. Phys. Rev. D **77**, 056007 (2008)
33. de Teramond, G.F., Brodsky, S.J.: Excited baryons in holographic QCD. AIP Conf. Proc. **1432**, 168 (2012)
34. Aznauryan, I.G., et al. (CLAS Collaboration): Electroexcitation of nucleon resonances from CLAS data on single pion electroproduction. Phys. Rev. C **80**, 055203 (2009)
35. Mokeev, V.I., et al. (CLAS Collaboration): Experimental study of the $P_{11}(1440)$ and $D_{13}(1520)$ resonances from CLAS data on $ep \to e'\pi^+\pi^- p'$. Phys. Rev. C **86**, 035203 (2012)
36. Mokeev, V.I. et al. (CLAS Collaboration): New results from the studies of the $N(1440)1/2^+$, $N(1520)3/2^-$, and $\Delta(1620)1/2^-$ resonances in exclusive $ep \to e'p'\pi^+\pi^-$ electroproduction with the CLAS detector. Phys. Rev. C **93**(2), 025206 (2016) arXiv:1509.05460 [nucl-ex]
37. Gutsche, T., Lyubovitskij, V.E., Schmidt, I., Vega, A.: Nucleon resonances in AdS/QCD. Phys. Rev. D **87**, 016017 (2013)
38. Deur, A., Brodsky, S.J., de Teramond, G.F.: Connecting the hadron mass scale to the fundamental mass scale of quantum chromodynamics. Phys. Lett. B **750**, 528 (2015)
39. Deur, A., Brodsky, S.J., de Teramond, G.F.: On the interface between perturbative and nonperturbative QCD. Phys. Lett. B **757**, 275 (2016)
40. Liu, T., Ma, B.Q.: Baryon properties from light-front holographic QCD. Phys. Rev. D **92**(9), 096003 (2015)
41. Aznauryan, I.G., et al.: Studies of nucleon resonance structure in exclusive meson electroproduction. Int. J. Mod. Phys. E **22**, 1330015 (2013)

Springer

Few-Body Syst (2016) 57:1001–1008
DOI 10.1007/s00601-016-1141-x

Igor T. Obukhovsky · Amand Faessler · Thomas Gutsche ·
Valery E. Lyubovitskij

Nucleon Resonance Electrocouplings from Light-Front Quark Models at Q^2 up to 12 GeV^2

Received: 13 February 2016 / Accepted: 26 July 2016 / Published online: 17 August 2016
© Springer-Verlag Wien 2016

Abstract A relativistic light-front quark model is used to describe both the elastic nucleon and nucleon-Roper transition form factors in a large Q^2 range, up to 35 GeV^2 for the elastic and up to 12 GeV^2 for the resonance case. Relativistic three-quark configurations satisfying the Pauli exclusion principle on the light-front are used for the derivation of the current matrix elements. The Roper resonance is considered as a mixed state of a three-quark core configuration and a molecular $N + \sigma$ hadron component. Based on this ansatz we obtain a realistic description of both processes, elastic and inelastic, in the sector of positive parity and show that existing experimental data are indicative of a composite structure of the Roper resonance. A useful generalization of this technique is suggested for description of negative parity nucleon resonances $1/2^-$, $3/2^-$, $5/2^-$.

1 Introduction

Last decade has been marked by significant progress in the experimental study of low-lying baryonic resonances (the radial/orbital nucleon excitations with $J^P = \frac{1}{2}^\pm, \frac{3}{2}^\pm$). Specifically new insights have been obtained in π [1] and 2π [2] electroproduction on the proton with the polarized electron beam at JLab (CLAS Collaboration) followed by a combined analysis of pion- and photo-induced reactions made by CB-ELSA and the A2-TAPS Collaborations [3]. Electro- and photoproduction of these resonances is recognized as an important tool which

This work was supported by the RFBR-DFG Grant No. 16-52-12019, by the DFG Grants Nos. FA-67-42-1 and GU-267/3-1, by the German Bundesministerium für Bildung und Forschung (BMBF) under Project 05P2015—ALICE at High Rate (BMBF-FSP 202): "Jet- and fragmentation processes at ALICE and the parton structure of nuclei and structure of heavy hadrons", by Tomsk State University Competitiveness Improvement Program and the Russian Federation program "Nauka" (Contract No. 0.1526.2015, 3854).

This article belongs to the special issue "Nucleon Resonances".

I. T. Obukhovsky (✉)
Institute of Nuclear Physics, Moscow State University, Moscow, Russia 119991
E-mail: obukh@nucl-th.sinp.msu.ru

A. Faessler · T. Gutsche · V. E. Lyubovitskij
Institut für Theoretische Physik, Kepler Center for Astro and Particle Physics, Universität Tübingen,
Auf der Morgenstelle 14, 72076 Tübingen, Germany

V. E. Lyubovitskij
Department of Physics, Tomsk State University, Tomsk, Russia 634050

V. E. Lyubovitskij
Laboratory of Particle Physics, Mathematical Physics Department, Tomsk Polytechnic University,
Lenin Avenue 30, Tomsk, Russia 634050

allows to study the relevant degrees of freedom, wave functions and interactions between constituents and the transition to perturbative quantum chromodynamics (pQCD).

From the theoretical side many approaches have been applied to the study of electromagnetic nucleon elastic and transition form factors including their light-quark flavor decomposition — quark and quark-diquark models, QCD sum rules, lattice QCD, Bethe-Salpeter approaches, the hypercentral constituent quark model, the AdS/QCD, etc. (see, e.g., the recent works [4–10] and references therein).

In the context of projected extensive studies of baryons with $J^P = 1/2^\pm, 3/2^\pm, 5/2^\pm$, etc, there is an interest in calculation of electrocouplings of baryons at large Q^2. Rough estimates can be made on the basis of light-front quark models which are founded on a consistent relativistic theory for composite systems with a fixed number of constituents. Such approach implies the construction of a good basis of quark light-front configurations possessing the definite value of the total angular momentum $\mathbf{J}$ and satisfying the Pauli exclusion principle.

We discuss a successful method that allow to construct such good basis for resonances of negative parity. Starting from this we suggest some evaluations of high Q^2 transition amplitudes. Our approach is to fit parameters of light-front quark configurations to the elastic nucleon form factors extracted from recent data on polarized electron scattering and use these to calculate the transition form factors at large Q^2 up to 12 GeV^2.

2 Light Front Approach

In Ref. [11] we have generalized our late results [12] on the Roper resonance electroproduction at $Q^2 < 4\,\mathrm{GeV^2}$ by going to more high Q^2. It has taken to rewrite our old non-relativistic model in terms of quark configurations at the light front.

The decisive advantage of the light-front approach is the separation of the total momentum of the system $\mathscr{P}^+ = p_1^+ + p_2^+ + p_3^+$, $\mathscr{P}_\perp = \boldsymbol{p_{1\perp}} + \boldsymbol{p_{2\perp}} + \boldsymbol{p_{3\perp}}$ from the inner (relative) moments of quarks.

$$\lambda^\mu = \frac{x_2 p_1^\mu - x_1 p_2^\mu}{x_1 + x_2}, \qquad \Lambda^\mu = (x_1 + x_2)p_3^\mu - x_3(p_1^\mu + p_2^\mu), \tag{1}$$

where $x_i = p_i^+/\mathscr{P}^+$. It is important that the values $\lambda_\perp$, $\boldsymbol{\Lambda}_\perp$ and x_i are invariants of Lorentz transformations $\ell(\mathscr{P}' \leftarrow \mathscr{P})$ defined at the light front as two-step Lorentz boosts

$$\ell(p' \leftarrow p) = \lambda(p' \leftarrow p_\infty)\lambda(p_\infty \leftarrow p), \tag{2}$$

where $\lambda(p' \leftarrow p)$ is a conventional rotationless Lorentz boost and $\lambda(p_\infty \leftarrow p) = \lim_{p_z'' \to \infty} \lambda(p'' \leftarrow p)$.

Starting from the relative moments λ^μ, Λ^μ one can define the inner (orbital) angular momentum L in the rest frame of the $3q$ system (or in the rest frame of a two-quark subsystem), and add up the orbital momentum L and the quark spin S using the relativistic technique developed long ago [13–15] to define the total angular momentum $\mathbf{J} = \mathbf{L} + \mathbf{S}$. However, it should be noted (see, e.g., Ref. [16,17]) that the Melosh [18] transformation from the canonical (c) spin to the light-front (f) spin generates relativistic spin wave functions with only approximately correct $\mathbf{J}^2$ because of the interaction dependence of the light-front rotation generators $\mathbf{J}_\perp$. It seems reasonably to say that the baryons of negative parity $N_{j_-}^*$ can only be approximated by a free-quark basis with $L = 1$.

In the case of $L = 0$ the "true" (i.e. dynamically justified) form of quark configurations is also the basic challenge in the description of the large Q^2 form factors. It is well known that at low and moderate values of $Q^2 \lesssim 2$–$3\,\mathrm{GeV^2}/c^2$ both relativistic and nonrelativistic description of elastic or transition nucleon form factors with quark wave functions of the Gaussian type

$$\Phi_0(\lambda, \Lambda, \xi, \eta) = \mathscr{N}_0\, e^{-b^2 \mathscr{M}_{123}^2}, \quad x_1 = \xi\eta, \ x_2 = (1-\xi)\eta, \ x_3 = 1-\eta \tag{3}$$

could be correlated well with the data by fitting free parameters of models. Here

$$\mathscr{M}_{123}^2 = \frac{\mathscr{M}_{12}^2}{\eta} + \frac{\Lambda_\perp^2 + \eta M_q^2}{\eta(1-\eta)}, \quad \mathscr{M}_{12}^2 = \frac{\lambda_\perp^2 + M_q^2}{\xi(1-\xi)} \tag{4}$$

are the invariant masses of $3q$ and $2q$ clusters of free quarks and M_q is the constituent quark mass. It is apparent that the Gaussian anzatz artificially reduces the role of the high-momentum component of trial functions and underestimates the role of the quark core of the nucleon in the high-Q^2 transitions. The form factors defined by the Gaussian (3) with the size parameter $b \simeq 0.5$–$0.6 \ fm$ is quickly dying out at $Q^2 \gtrsim 3$–$4\text{GeV}^2/c^2$.

Possible alternatives to the Gaussian wave function are:

(a) a superposition of many Gaussians modified with polynomial factors (i.e an extended shell-model basis as it was considered in Ref. [19]);

(b) a non-Gaussian wave function, e.g. a pole-like function

$$\Phi_0(\lambda, \Lambda, \xi, \eta) = \frac{1}{(1 + \mathcal{M}_{123}^2/\beta^2)^\gamma} \tag{5}$$

that was firstly fitted to the elastic nucleon form factors by Schlumpf [20] with $\gamma = 3.5$ and $\beta \simeq 2M_q$;

(c) a model with the running quark mass suggested in Ref. [4] following the QCD predictions.

We have chosen a pole-like form of the wave function and in the present brief paper we report on the elastic and inelastic nucleon electromagnetic form factors using a combined, light-front quark and hadron-molecular, model recently developed by us and discussed in [11].

3 Quark Configurations at Light Front

First we briefly review the elements of the light-front quark model (see details in [11]), which is based on ideas given in [14,17,19–21]. The canonical and front spin states of i-th particle are defined by Lorentz boosts of two types:

$$|p_i; s_i\mu_i\rangle_c = U(\lambda(p_i \leftarrow \overset{\circ}{p}_i))| \ \overset{\circ}{p}_i; s_i\mu_i\rangle, \quad |p_i; s_i\mu_i\rangle_f = U(\ell(p_i \leftarrow \overset{\circ}{p}_i))| \ \overset{\circ}{p}_i; s_i\mu_i\rangle, \tag{6}$$

where $\overset{\circ}{p}_i = \{M_q, 0, 0, 0\}$ and μ_i are the momentum and the spin projection in the rest frame, $U(\lambda)$ and $U(\ell)$ are the unitary representations of Lorentz boosts. The states (6) are related by a specific rotation R_{p_i}

$$|p_i; s_i\mu_i\rangle_c = \sum_{\bar{\mu}_i} |p_i; s_i\bar{\mu}_i\rangle_f \, D_{\bar{\mu}_i\mu_i}^{s_i}(R_{p_i}^\dagger), \quad R_{p_i} = \lambda(\overset{\circ}{p}_i \leftarrow p_\infty)\lambda(p_\infty \leftarrow p_i)\lambda(p_i \leftarrow \overset{\circ}{p}_i) \tag{7}$$

which is known as the Melosh transformation. For the quark ($s_i = 1/2$) one can write:

$$D_{\bar{\mu}_i\mu_i}^{\frac{1}{2}}(R_{p_i}) = \frac{M + p_{iz} + i\hat{n}_z \cdot [\boldsymbol{\sigma}_i \times \mathbf{p}_{i\perp}]}{\sqrt{(M + p_{iz})^2 + p_{i\perp}^2}} \tag{8}$$

Boosts $\ell(p_i' \leftarrow p_i)$ does not rotate the front states $|p_i; s_i\mu_i\rangle_f \to U(\ell(p_i' \leftarrow p_i))|p_i; s_i\mu_i\rangle_f = |p_i'; s_i\mu_i\rangle_f$ that is very convenient to use in moving reference frames. But the front states are not convenient to use in construction of rotationally covariant states [17]. For the quark pair "12" such state can be constructed in the proper rest frame (cm), where $\mathcal{P}_{12cm} = \overset{\circ}{\mathcal{P}}_{12} = \{\mathcal{M}_{12}, 0, 0, 0\}$, starting from canonical spins and using the instant form of rotation generators

$$|\overset{\circ}{\mathcal{P}}_{12}, j_{12}j_{12}^z(LS_{12})\rangle = \mathcal{N} \sum_{\mu_1\mu_2\mu_{12}M} (s_1\mu_1s_2\mu_2|S_{12}\mu_{12})(LMS_{12}|j_{12}j_{12}^z) \int d\hat{\lambda}_{cm} Y_{LM}(\hat{\lambda}_{cm})|p_{1cm}\mu_1\rangle_c|p_{2cm}\mu_2\rangle_c \tag{9}$$

This equation transformed by Lorentz boosts (the known "Shirokov formula" [13]) is much used in light-front models (see, e.g. [14,15]). Here we follow the formalism developed in Refs. [14,15].

To construct the rotationally covariant basis for 3q configurations $|\overset{\circ}{\mathcal{P}}, J(LS)M\rangle$ with definite values of total (JM), orbital (L) and spin (S) angular moments in the proper rest frame (CM), where $\overset{\circ}{\mathcal{P}} = \{\mathcal{M}_{123}, 0, 0, 0, 0\}$, one should change the canonical spin states in the r.h.s. to the front ones by Eqs. (7)–(8) and act with the boost $U[\ell(\mathcal{P}_{12CM} \leftarrow \mathcal{P}_{12cm})]$ on each front spin state. After that one can consider the obtained front state $|\mathcal{P}_{12CM}, j_{12}j_{12}^z(L_{12}S_{12})\rangle_f$ as a state of an elementary particle with the total spin j_{12} and add up the "spin" j_{12}, the spin of the third quark and the orbital momentum L_3 (described by spherical functions $Y_{L_3 M_3}(\hat{\Lambda}_{CM})$) using the same scheme of angular momentum couplings as in the case of two-particle system in Eq. (9).

Here we consider two simple cases: (Λ) $L_{12} = 0$, $L_3 = L = 1$ and (λ) $L_{12} = L = 1$, $L_3 = 0$, which are actual for negative parity resonances with $J^P = 1/2^-, 3/2^-, 5/2^-$. Starting from (9) and (3)–(5) the spin-orbital (SO) wave functions for these resonances may be written in two variants as

$$\psi_\Lambda \doteq \mid \mathring{\mathscr{P}}, J(LS(S_{12}))J_z \rangle_c = \mathcal{N}_\Lambda \sum_{\bar{\mu}_i} \Bigg\{ \Phi_0 \sum_{M\mu} \Lambda Y_{1M}(\hat{\Lambda})(1MS\mu|JJ_z) \Bigg[\sum_{\mu_i\mu_{12}} (1/2\mu_1 1/2\mu_2 | S_{12}\mu_{12})$$

$$\times (S_{12}\mu_{12} 1/2\mu_3 | S\mu) D^{1/2}_{\bar{\mu}_1\mu_1}(R^\dagger_{P1}) D^{1/2}_{\bar{\mu}_2\mu_2}(R^\dagger_{P2}) D^{1/2}_{\bar{\mu}_3\mu_3}(R^\dagger_{P3}) \Bigg] \Bigg\} |p_1, \bar{\mu}_1\rangle_f |p_1, \bar{\mu}_1\rangle_f |p_1, \bar{\mu}_1\rangle_f, \qquad (10)$$

$$\psi_\lambda \doteq \mid \mathring{\mathscr{P}}, J(j_{12}(LS_{12}))J_z \rangle_c = \mathcal{N}_\lambda \sum_{\bar{\mu}_i} \Bigg\{ \Phi_0 \sum_{M\mu_{12}j_{12}^z} \lambda Y_{1M}(\hat{\lambda})(1MS_{12}\mu_{12}|j_{12}j_{12}^z) \sum_{\mu_i} (1/2\mu_1 1/2\mu_2 | S_{12}\mu_{12})$$

$$\times (j_{12}j_{12}^z 1/2\mu_3 | JJ_z) D^{1/2}_{\bar{\mu}_1\mu_1}(R^\dagger_{P1cm} R^\dagger_{\mathscr{P}12}) D^{1/2}_{\bar{\mu}_2\mu_2}(R^\dagger_{P2cm} R^\dagger_{\mathscr{P}12}) D^{1/2}_{\bar{\mu}_3\mu_3}(R^\dagger_{P3}) \Bigg\} |p_1, \bar{\mu}_1\rangle_f |p_1, \bar{\mu}_1\rangle_f |p_1, \bar{\mu}_1\rangle_f,$$

$$(11)$$

where $\Lambda = |\boldsymbol{\Lambda}_{CM}|$, $\lambda = |\boldsymbol{\lambda}_{CM}|$, $p_i = p_{iCM}$, $\mathscr{P}_{12} = p_{1CM} + p_{2CM}$, $\Phi_0 = \Phi_0(\mathscr{M}_{123})$. Here the expressions in curl brackets are the inner wave functions that remain invariant at Lorentz boosts $\ell(\mathscr{P} \leftarrow \mathring{\mathscr{P}})$. Specifically, at the transitions to the special Breit frame $\mathring{\mathscr{P}} \to \mathscr{P}_B = \mathring{\mathscr{P}} \pm q_B/2$, where the photon momentum q_B is pure transverse, $q_B^\nu = \{0, \boldsymbol{q}_\perp, 0\}$, $q^2 = -\boldsymbol{q}_\perp^2$. In the Breit frame any electromagnetic transition amplitude can be calculated by using the expression for matrix element of the "plus" component of the quark current $I_q^{(i)+}$

$$_f\langle p_i', \mu_i' | I_q^{(i)+} | p_i, \mu_i \rangle_f = \delta_{\mu_i'\mu_i} 2\varepsilon_i (2\pi)^3 \delta^{(3)}(\mathbf{p}_i' - \mathbf{p}_i - \mathbf{q})$$

$$= \delta_{\mu_i'\mu_i} 2p_i^+ (2\pi)^3 \delta^{(2)}(\boldsymbol{p}_{i\perp}' - \boldsymbol{p}_{i\perp} - \boldsymbol{q}_\perp) \delta(p_i'^+ - p_i^+) \qquad (12)$$

between front states normalized by equation $_f\langle p_i', \mu_i' | p_i, \mu_i \rangle_f = \delta_{\mu_i'\mu_i} 2p_i^+ (2\pi)^3 \delta^{(2)}(\boldsymbol{p}_{i\perp}' - \boldsymbol{p}_{i\perp}) \delta(p_i'^+ - p_i^+)$.

3.1 Pauli Exclusion Principle

The next problem is to construct the quark configurations satisfying the Pauli exclusion principle. The functions (10)–(11) are symmetric ($P_{12}^o \psi_\Lambda = \psi_\Lambda$) or antisymmetric ($P_{12}^o \psi_\lambda = -\psi_\lambda$) with respect of permutation of the first two quarks in the orbital (O) space P_{12}^o. It is evident that starting from ψ_Λ and ψ_λ and using the all pair permutations P_{ij}^o one can construct the irreducible representations (IR) of symmetric group $S(3)$ with the Young schemes $[3]_o$, $[21]_o$ and $[1^3]_o$ in the orbital space. The symmetric ($[3]_o$) or antisymmetric ($[1^3]_o$) states can be obtained by acting the symmetric ($S_3 = \frac{1}{6}(I + P_{12} + P_{13} + P_{23} + P_{13}P_{12})$) or antisymmetric ($A_3 = \frac{1}{6}(I - P_{12} - P_{13} - P_{23} + P_{13}P_{12})$) projectors onto $\psi_{\Lambda/\lambda}$: $S_3\psi_\Lambda \sim |[3]_o\rangle$, $A_3\psi_\lambda \sim |[1^3]_o\rangle$.

However, the basis vectors of mixed symmetry $[21]_o$ with definite values of Yamanouchi symbols $y^{(1)} = [12, 3]$ and $y^{(2)} = [13, 2]$ require more complicated calculations including the orthogonalization to the states of trivial symmetry, $[3]_o$ and $[1^3]_o$. Starting from Eq. (10) and using the relations $P_{13}\Lambda^\nu = -\xi\Lambda^\nu + \lambda^\nu$, $P_{13}\lambda^\nu = \frac{1-\xi}{1-\xi\eta}\Lambda^\nu + \frac{1-\eta}{1-\xi\eta}\lambda^\nu$, $P_{23}\Lambda^\nu = -(1-\xi)\Lambda^\nu - \lambda^\nu$, etc we found that the most convenient expressions are:[1]

$$|[21]_o y_o^{(1)}, M\rangle = -\mathcal{N}_1 \Lambda Y_{1M}(\hat{\Lambda})\Phi_0, \quad |[21]_o y_o^{(2)}, M\rangle = \mathcal{N}_2 \underset{\sim}{\lambda} Y_{1M}(\underset{\sim}{\hat{\lambda}})\Phi_0 \quad \text{with} \quad \underset{\sim}{\lambda}^\nu = \lambda^\nu + \frac{1-2\xi}{2}\Lambda^\nu. \quad (13)$$

---

[1] If Eq. (11) with the fixed value of j_{12} were used as the initial state, the final expressions (13) would be more complicated. Note that the values of $\mathcal{N}_1$ and $\mathcal{N}_2$ are restricted by equation $\mathcal{N}_2 = \sqrt{3/4}\mathcal{N}_1$.

One can directly check that the functions (13) realize the IR $[21]_o$ of the symmetric group $S(3)$, e.g.,

$$P_{12}\begin{bmatrix} |[21]_o y_o^{(1)}\rangle \\ |[21]_o y_o^{(2)}\rangle \end{bmatrix} = \begin{pmatrix} 1 & 0 \\ 0 & -1 \end{pmatrix}\begin{bmatrix} |[21]_o y_o^{(1)}\rangle \\ |[21]_o y_o^{(2)}\rangle \end{bmatrix}, \quad P_{13}\begin{bmatrix} |[21]_o y_o^{(1)}\rangle \\ |[21]_o y_o^{(2)}\rangle \end{bmatrix} = \begin{pmatrix} -\frac{1}{2} & -\sqrt{\frac{3}{4}} \\ -\sqrt{\frac{3}{4}} & \frac{1}{2} \end{pmatrix}\begin{bmatrix} |[21]_o y_o^{(1)}\rangle \\ |[21]_o y_o^{(2)}\rangle \end{bmatrix}$$

The obtained orbital states (13) are convenient to use for description (construction) of quark configurations in terms of the standard LS coupling, i.e. with definite values of both the orbital momentum $L = 1$ and the total canonical spin $S = 1/2, 3/2$ at $J^P = 1/2^-, 3/2^-, 5/2^-$.

The basis of spin states is complicated by the effect of Melosh transformations. Note that the expression in square brackets in the r.h.s. of Eq. (10) is the spin state $|S(S_{12}), \mu\rangle$ modified by the Melosh transformation $\mathscr{R}_{\mathscr{M}} = \Pi_{i=1}^3 D_{\bar{\mu}_i \mu_i}^{1/2}(R_{p_i}^\dagger)$. The modification does not conserve the total spin S, i.e. $\langle S'(S_{12}), \mu | \mathscr{R}_{\mathscr{M}} | S(S_{12}), \mu\rangle \neq 0$ at $S' \neq S$, but the Young scheme in the modified spin $(S_{\mathscr{M}})$ space $[21]_{S_{\mathscr{M}}}$ is the same one as the initial Young scheme $[21]_S$ in the canonical spin space (it is shown in Ref. [11]). Hence we can consider the modified state $\mathscr{R}_{\mathscr{M}}|S(S_{12})\rangle$ as a proper basis state of IR of the $S(3)$ group, $|[21]_{S_{\mathscr{M}}} y_{S_{\mathscr{M}}}^{(1)}\rangle$ at $S_{12} = 1$ and $|[21]_{S_{\mathscr{M}}} y_{S_{\mathscr{M}}}^{(2)}\rangle$ at $S_{12} = 0$. The isospin basis for nucleon resonances N^* $(T = 1/2)$ with the charge $e_{N^*} = 1/2 + T_z$ we also denote with the Young schemes, e.g. $|T = 1/2(T_{12} = 1), T_z\rangle = |[21]_T y_T^{(1)}, T_z\rangle$ and $|T = 1/2(T_{12} = 0), T_z\rangle = |[21]_T y_T^{(2)}, T_z\rangle$.

4 Wave Functions of the Lowest Orbital and Radial Excitations of the Nucleon

The full basis of quark configurations for the first orbital excitation $(L = 1)$ includes five states:

$$|J(LS)J_z, TT_z\rangle = \begin{cases} |N_{J^-}^{[21]}\rangle : [21]_{S_{\mathscr{M}}}, \; S = 1/2, \; L = 1, \quad J = 1/2, 3/2, \quad T = 1/2 \\ |N_{J^-}^{[3]}\rangle : \; [3]_{S_{\mathscr{M}}}, \; S = 3/2, \; L = 1, \; J = 1/2, 3/2, 5/2, \; T = 1/2 \end{cases}$$

The wave functions $|N_{J^-}^{[f]}\rangle$ constructed by the standard quark shell-model technique (e.g., see Ref. [22]) are:

$$|N_{J^-}^{[21]}\rangle = \mathscr{N}^{[21]}\Phi_0\sqrt{\frac{1}{2}}\sum_{M\mu}(1M\tfrac{1}{2}\mu|JJ_z)\left\{ -\sqrt{\frac{1}{2}}\Lambda Y_{1M}(\hat{\Lambda})\left[|[21]_{S_{\mathscr{M}}} y_{S_{\mathscr{M}}}^{(1)}, \mu\rangle|[21]_T y_T^{(1)}\rangle\right.\right.$$
$$-|[21]_{S_{\mathscr{M}}} y_{S_{\mathscr{M}}}^{(2)}, \mu\rangle|[21]_T y_T^{(2)}\rangle\Big] - \sqrt{\frac{3}{4}}\lambda Y_{1M}(\hat{\lambda})\Big[|[21]_{S_{\mathscr{M}}} y_{S_{\mathscr{M}}}^{(1)}, \mu\rangle|[21]_T y_T^{(2)}\rangle$$
$$\left. + |[21]_{S_{\mathscr{M}}} y_{S_{\mathscr{M}}}^{(2)}, \mu\rangle|[21]_T y_T^{(1)}\rangle\Big]\right\}, \tag{14}$$

$$|N_{J^-}^{[3]}\rangle = \mathscr{N}^{[3]}\Phi_0\sqrt{\frac{1}{2}}\sum_{M\mu}(1M\tfrac{3}{2}\mu|JJ_z)\left\{ -\Lambda Y_{1M}(\hat{\Lambda})|[3]_{S_{\mathscr{M}}}, \mu\rangle|[21]_T y_T^{(1)}\rangle\right.$$
$$\left. +\sqrt{\frac{3}{4}}\lambda Y_{1M}(\hat{\lambda})|[3]_{S_{\mathscr{M}}}, \mu\rangle|[21]_T y_T^{(2)}\rangle\right\} \tag{15}$$

where λ is defined in Eq. (13). The quark configurations with $L = 0$ used in Ref. [11] for calculation of electrocoupling of the nucleon $(N_{1/2^+})$ and the Roper resonance $(N_{1/2^+}^*)$ are the specific case of Eqs. (14)-(15)

$$|N_{1/2^+}\rangle = \mathscr{N}_N\Phi_0\sqrt{\frac{1}{2}}\left[|[21]_{S_{\mathscr{M}}} y_{S_{\mathscr{M}}}^{(1)}\rangle|[21]_T y_T^{(1)}\rangle + |[21]_{S_{\mathscr{M}}} y_{S_{\mathscr{M}}}^{(2)}\rangle|[21]_T y_T^{(2)}\rangle\right] \tag{16}$$

$$|N_{1/2^+}^*\rangle = \mathscr{N}_R\Phi_2\sqrt{\frac{1}{2}}\left[|[21]_{S_{\mathscr{M}}} y_{S_{\mathscr{M}}}^{(1)}\rangle|[21]_T y_T^{(1)}\rangle + |[21]_{S_{\mathscr{M}}} y_{S_{\mathscr{M}}}^{(2)}\rangle|[21]_T y_T^{(2)}\rangle\right], \tag{17}$$

where $\Phi_2 = \left(1 - C_R\frac{\mathscr{M}_{123}^2}{\beta^2}\right)\Phi_0$ is the "radial" wave function of the first radially excited state in the $3q$ system.

The configurations (14)–(15) written in terms of LS coupling cannot be proper states of a relativistic Hamiltonian for which the jj coupling is the most natural. In realistic quark models the energy splitting for states (14) and (15) is about 100–150 MeV as the result of strong spin-spin interaction, while the pairs of basis vectors that has got the same $J = 1/2$ or $3/2$ should be mixed in physical baryons by virtue of moderate spin-orbit coupling. Hence in calculations of electrocouplings of the physical baryons S_{11}, D_{13} and D_{15} their wave functions should be approximated with mixed states, e.g. dependent on two mixing angles, θ_1 and θ_3:

$$S_{11}(1535) = \cos\theta_1 |N^{[21]}_{\frac{1}{2}-}\rangle + \sin\theta_1 |N^{[3]}_{\frac{1}{2}-}\rangle, \quad D_{13}(1520) = \cos\theta_3 |N^{[21]}_{\frac{3}{2}-}\rangle + \sin\theta_3 |N^{[3]}_{\frac{3}{2}-}\rangle,$$

$$S_{11}(1650) = -\sin\theta_1 |N^{[21]}_{\frac{1}{2}-}\rangle + \cos\theta_1 |N^{[3]}_{\frac{1}{2}-}\rangle, \quad D_{13}(1700) = -\sin\theta_3 |N^{[21]}_{\frac{3}{2}-}\rangle + \cos\theta_3 |N^{[3]}_{\frac{3}{2}-}\rangle,$$

$$D_{15}(1675) = |N^{[3]}_{\frac{5}{2}-}\rangle$$

In principle, high-quality data on electrocouplings of these resonances at large Q^2 (i.e. in the region of quark dominance) could give some restrictions on the values of θ_1 and θ_3.

It should be noted that the inner structure of these resonances was considered in detail in Ref. [23] on the basis of the hypercentral constituent quark model and the theoretical predictions for helicity amplitudes and transition form factors in the region of $0 \leq Q^2 \lesssim 3$ GeV^2 has been obtained [23–25]. However in the region of more large $Q^2 \gtrsim 3$–4 GeV^2 the relativistic models were preferable. Then one runs into the problem of covariant description of inner quark moments in the three-quark system moving as a whole with a relativistic velocity. For example, the covariant description of the inner quark states with a fixed orbital momentum and satisfying the Pauli exclusion principle becomes a nontrivial problem.

At the light front the electrocouplings sought are given by matrix elements of the "plus" component of quark current (12) for transitions $N_{1/2^+} \to N_{J^-}, R_{1/2^+}$ in the Breit frame (i.e. for $q^\nu = P'^\nu - P^\nu = \{0, \boldsymbol{q}_\perp, 0\}$)

$$\langle N^{[f]}_{J^-}, P'| \sum_{i=1}^3 e^{(i)}_q I^{(i)+}_q |N_{1/2^+}, P\rangle = 3\langle N^{[f]}_{J^-}, P'|e^{(3)}_q I^{(3)+}_q |N_{1/2^+}, P\rangle, \quad e^{(i)}_q = \frac{1}{6} + \frac{1}{2}\tau^{(i)}_z \quad (18)$$

Substitution into the l.h.s. of Eq. (18) the quark configurations (14)–(17) satisfying the Pauli exclusion principle simplifies the calculation. First, the states (14)–(17) are fully symmetric (the Young scheme is $[3]_{oS_{\mathscr{M}}T}$), hence only the current of one quark, e.g. the 3rd one as in Eq. (18), should be taken into account. Second, in calculation of the current matrix element one can use the elegant technique of fractional parentage coefficients (f.p.c.) (see, e.g. [22]) with the result that the final expression of matrix element (18) for each baryon N_{J^-} or $R_{1/2^+}$ is a superposition of a few basic integrals of the type

$$\int \Phi_0(\mathscr{M}'_{123})\Phi_0(\mathscr{M}_{123})\Omega^{[f]\dagger}_{J^-}(\mathscr{R}'_{\mathscr{M}})\Omega_{1/2^+}(\mathscr{R}_{\mathscr{M}})d\Gamma, \quad d\Gamma = \frac{d\xi d^2\lambda_\perp}{\xi(1-\xi)}\frac{d\eta d^2\Lambda_\perp}{\eta(1-\eta)}, \quad (19)$$

where $\Omega^{[f]}_{J^-}(\mathscr{R}'_{\mathscr{M}})$ and $\Omega_{1/2^+}(\mathscr{R}_{\mathscr{M}})$ are symbolic notations for full Melosh rotations of spin parts of wave functions $N^{[f]}_{J^-}$ and $N_{1/2^+}$ respectively (for detail see Ref. [11], where matrices $\Omega_{1/2^+}$ for the transition $N_{1/2^+} \to R_{1/2^+}$ are given in terms of specific coefficients $C_{S'_{12}S_{12}}(\mu', \mu)$). Here $\boldsymbol{\Lambda}'_\perp = \boldsymbol{\Lambda}_\perp + \eta\boldsymbol{q}_\perp$ and $\mathscr{M}'_{123}$ are defined by Eq. (4) with the substitution $\boldsymbol{\Lambda}_\perp \to \boldsymbol{\Lambda}'_\perp$. Integrals of such type for the case of transition $N_{1/2^+} \to R_{1/2^+}$ were studied in our recent work [11]. The basic integrals (19) are common for all baryons, but they enter into the final expression for the current matrix element in the product with f.p.c.'s dependent on the quantum numbers of the given baryon $N^{[f]}_{J^-}$ and with the cos / sin of mixing angle θ_1/θ_3 of the given physical baryon.

5 Results on the Nucleon form Factors and on the Roper Electroproduction Helicity Amplitudes

In Ref. [11] we have shown that the given LF quark model allows for a good description of all the new data on nucleon form factors in a large interval of Q^2 from 0 up to 35 GeV^2. The model has only 5 free parameters (see Eq. (5)) γ, β, M_q, and $\varkappa_q$ (the quark anomalous magnetic moment), which are fitted to the data. For the values $\gamma = 3.51$, $\beta = 0.579$ GeV, $M_q = 0.251$ GeV, $\varkappa_u = -0.0028$, and $\varkappa_d = 0.0224$ an optimal description of the elastic nucleon data is obtained (see, e.g., Figs. 1 and 4 in our recent work [11], where we

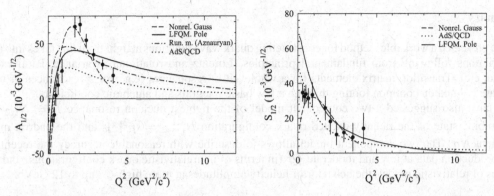

Fig. 1 Transverse helicity amplitude $A_{1/2}$ (*left panel*) and longitudinal helicity amplitude $S_{1/2}$ (*right panel*) of the Roper resonance electroproduction in different approaches. *Solid line* the two-component model (20) with the LF quark configuration (17), $\cos\theta = 0.57$ (modified value as compared with Ref. [11]). *Dashed line* the same model (20), but with the nonrelativistic Gaussian wave function [12], $\cos\theta = 0.7$. *Pointed line* results from the AdS/QCD approach [9,10]. *Dash-pointed line* the comparison with results of the model with running quark mass [4]. The data are from [1]

presented a comparison with known data and the soft-wall AdS/QCD approach [9,10]). The model generates a Q^2 behavior for $Q^4 F_{1p}(Q^2)$, $Q^4 F_{1n}(Q^2)$ and $Q^2 F_{2p}(Q^2)/F_{1p}(Q^2)$ which should tend to a constant at high Q^2. At low and moderate values of Q^2 the model is compatible not only with the magnetic moments of the nucleons, $\mu_p = 2.79$ and $\mu_n = -1.91$ but also with the known negative slope for the ratio $G_E^p(Q^2)/G_M^p(Q^2)$. The absolute theory values for $G_M^p(Q^2)$, $G_M^n(Q^2)$ and $G_E^p(Q^2)$ do also correlate well with the data. Only in the case of the neutron charge form factor G_E^n this model is not entirely adequate to describe data at low and moderate values of Q^2. But in this Q^2 region the pion cloud contribution to G_E^n, neglected in the present work, can be considerable. This contribution can also be important for inelastic nucleon form factors at $Q^2 \lesssim 1$ GeV^2, as was firstly shown in Refs. [24,25] in a nonrelativistic quark model and recently noted in Ref. [4] in terms of the light front quark model.

In the case of the Roper resonance the main problem is that its inner structure cannot be adequately described in terms of only constituent quark degrees of freedom, and thus other (more soft) degrees of freedom should be taken into consideration along with the quark core. It is evident that at high Q^2 the contribution of such soft components to the transition form factors is quickly dying out with rising of the momentum transfer, and only the contribution of the quark core survives.

We considered the Roper resonance $R = N_{1/2^+}(1440)$ as a mixed state of the radially excited quark configuration (17) $|N_{1/2^+}^*\rangle$ and the "hadron molecule" (a loosely bound state of nucleon and σ meson) $(N\sigma)_{mol} = |N + \sigma\rangle$

$$R = \cos\theta\, |N^*\rangle + \sin\theta\, |N + \sigma\rangle, \tag{20}$$

where θ is a mixing angle, the value of which is adjusted to optimize the description of the data on the Roper resonance electroproduction (the helicity amplitude $A_{1/2}$ only). The dynamic of $N\sigma$ component is considered in the framework of the hadronic molecular approach (e.g., see Ref. [26] and references therein) which is manifestly Lorentz invariant. The hadron loop gives a negative contribution $\Sigma_{N\sigma}$ to the mass of the Roper resonance, and the $RM\sigma$ coupling constant $g_{RN\sigma}$ is defined by the compositeness condition $Z_R = 1 - \Sigma'_{N\sigma}(m_R) = 0$, i.e. the elementary particle R has a zero weight in the hadron molecule.

Results on transverse ($A_{1/2}$) and longitudinal ($S_{1/2}$) helicity amplitudes of the Roper electroproduction obtained in our model (20) are shown in Fig. 1 in comparison with predictions of other models, the light-front model with running quark mass [4], the AdS/QCD model [9,10] and our old nonrelativistic model with the Gaussian wave function [12]. We use for the Roper resonance the same values of free parameters as for the elastic nucleon form factors and do only fit the mixing angle θ (in both relativistic and nonrelativistic variants we have obtained similar values of this parameter, $\cos\theta \approx 0.6$–0.7). One can see that the predictions of three different relativistic models for the high Q^2 behavior of $A_{1/2}$ are in qualitative agreement, while they become in quantitative disagreement with rising Q^2. It is evident that new data at large values of $Q^2 \gtrsim 5$–6 GeV^2 could be an important factor in choosing the most realistic model.

6 Summary

We have presented a workable method for constructing quark wave functions at light front that takes into account all restrictions following from fundamental principles (Lorentz and rotational invariance, Pauli exclusion principle, etc). Transition matrix elements for a whole set of resonances (e.g., for 5 resonances of negative parity) are defined on common footing through a few basic integrals and algebraic coefficients.

We have also suggested a two-component model of the lightest nucleon resonance $R = N_{1/2+}(1440)$ as a combine state of the radially excited quark configuration $N_{1/2+}^* = sp^2[3]_o$ and the hadron molecule component $N\sigma$. The two-component model allows to describe with reasonable accuracy the recent CLAS electroproduction data at low and moderate Q^2 (in terms of nonrelativistic quark configurations) and predict (in terms of relativistic model) the behavior of helicity amplitudes at more high Q^2 up to 12 GeV^2.

Acknowledgments I.T.O. (speaker) would like to thank V. Mokeev and V. Burkert for very useful discussions and to express his gratitude to the organisers of the ECT* Workshop in Trento *Nucleon Resonances: from Fotoproduction to high Photon Virtualities*, whose support helped his participation.

References

1. Aznauryan, I.G., et al. [CLAS Collaboration]: Electroexcitation of nucleon resonances from CLAS data on single pion electroproduction. Phys. Rev. C **80**, 055203 (2009)
2. Mokeev, V.I., et al. [CLAS Collaboration]: Experimental study of the $P_{11}(1440)$ and $D_{13}(1520)$ resonances from CLAS data on $ep \rightarrow e'\pi^+\pi^-p'$. Phys. Rev. C **86**, 035203 (2012)
3. Sarantsev, A.V., et al.: New results on the Roper resonance and the P(11) partial wave. Phys. Lett. B **659**, 94 (2008)
4. Aznauryan, I.G., Burkert, V.D.: Nucleon electromagnetic form factors and electroexcitation of low lying nucleon resonances in a light-front relativistic quark model. Phys. Rev. C **85**, 055202 (2012)
5. Roberts C.D.: Three Lectures on Hadron Physics. arXiv:1509.02925 [nucl-th]
6. Ramalho, G., Pena, M.T.: A covariant model for the gamma $N \rightarrow N(1535)$ transition at high momentum transfer. Phys. Rev. D **84**, 033007 (2011)
7. Giannini, M., Santopinto, E.: The hypercentral constituent quark model and its application to baryon properties. Chin. J. Phys. **53**, 020301 (2015)
8. De Teramond, G.F., Dosch, H.G., Brodsky, S.J.: Baryon spectrum from superconformal quantum mechanics and its light-front holographic embedding. Phys. Rev. D **91**, 045040 (2015)
9. Gutsche, T., Lyubovitskij, V.E., Schmidt, I., Vega, A.: Nucleon structure including high Fock states in AdS/QCD. Phys. Rev. D **86**, 036007 (2012)
10. Gutsche, T., Lyubovitskij, V.E., Schmidt, I., Vega, A.: Nucleon resonances in AdS/QCD. Phys. Rev. D **87**, 016017 (2013)
11. Obukhovsky, I.T., Faessler, A., Gutsche, T., Lyubovitskij, V.E.: Electromagnetic structure of the nucleon and the Roper resonance in a light-front quark approach. Phys. Rev. D **89**(1), 014032 (2014)
12. Obukhovsky, I.T., Faessler, A., Fedorov, D.K., Gutsche, T., Lyubovitskij, V.E.: Electroproduction of the Roper resonance on the proton: the role of the three-quark core and the molecular $N\sigma$ component. Phys. Rev. D **84**, 014004 (2011)
13. Shirokov Iu.M.: Relativistic theory of polarisation effects. Soviet Physics JETP **8**, 703 (1959)
14. Kondratyuk L.A., Terent'ev M.V.: Preprint of the Institute of Theoretical and Experimental Physics, ITEP-48 (1979) **(in Russian)**
15. Bakker, B.L.G., Kondratyuk, L.A., Terent'ev, M.V.: On the formulation of two- and three-body relativistic equations employing light-front dynamics. Nucl. Phys. B **158**, 497 (1979)
16. Konen, W., Weber, H.J.: Electromagnetic $N \rightarrow N^*(1535)$ transition in the relativistic constituent quark model. Phys. Rev. D **41**, 2201 (1990)
17. Keister, B.D.: Rotational covariance and light front current matrix elements. Phys. Rev. D **49**, 1500 (1994)
18. Melosh, H.J.: Quarks: currents and constituents. Phys. Rev. D **9**, 1095 (1974)
19. Capstick, S., Keister, B.D., Morel, D.: Nucleon to resonance form factor calculations. J. Phys. Conf. Ser. **69**, 012016 (2007)
20. Schlumpf, F.: Nucleon form-factors in a relativistic quark model. J. Phys. G **20**, 237 (1994)
21. Cardarelli, F., Pace, E., Salme, G., Simula, S.: Electroproduction of the Roper resonance and the constituent quark model. Phys. Lett. B **397**, 13 (1997)
22. Obukhovsky, I.T.: Algebraic technique in the quark-cluster approach to N N interaction. Prog. Part. Nucl. Phys. **36**, 359 (1996)
23. Santopinto, E., Iachello, F., Giannini, M.M.: Nucleon form factors in a simple three-body quark model. Eur. Phys. J **A1**, 307 (1998)
24. Aiello, M., et al.: Electromagnetic transition form factors of negative parity nucleon resonances. J. Phys. G **24**, 753 (1998)
25. Tiator, L., et al.: Electroproduction of nucleon resonances. Eur. Phys. J **A19**, 55 (2004)
26. Faessler, A., Gutsche, T., Lyubovitskij, V.E., Ma, Y.L.: Strong and radiative decays of the D(s0)*(2317) meson in the DK-molecule picture. Phys. Rev. D **76**, 014005 (2007)

Few-Body Syst (2016) 57:1095–1101
DOI 10.1007/s00601-016-1160-7

E. Santopinto · J. Ferretti

Effective Degrees of Freedom in Baryon Spectroscopy

Received: 21 September 2016 / Accepted: 22 September 2016 / Published online: 8 October 2016
© Springer-Verlag Wien 2016

Abstract Three quark and quark–diquark models are characterized by several missing resonances, even if in the latter case the state space is a reduced one. Moreover, even quark–diquark models show some differences in their predictions for missing states. After several years of discussion, we still do not know whether baryons can be completely described in terms of three quark models or if diquark correlations have to be taken into account; another possibility, suggested in Santopinto (Phys Rev C 72:022201, 2005), Ferretti et al. (Phys Rev C 83:065204, 2011) and Galatà and Santopinto (Phys Rev C 86:045202, 2012), is that the previous pictures (three-quark and quark–diquark) represent the dominant descriptions of baryons at different energy scales. New experiments may be planned at Jlab (JLab12), Bes, Belle and LHCb in order to answer this fundamental open question.

1 Missing States

As discussed Ref. [4], distinct three-quark models are characterized by different missing states. Also quark–diquark models still exhibit missing states [2,5,6], even if they have a reduced state space; moreover, each version of quark–diquark model is characterized by a different set or spectrum of missing states (for example, one can compare old results [7,8] with more recent ones [2,5,6,9]). After many years of discussion, we are still unable to answer the question if nature is completely described by three-quark models, if diquark correlations in quark–diquark models have to be dismissed, or if each of the two previous pictures dominates at a certain energy scale, as suggested in Ref. [1–3]. New experiments at JLab (JLab12), Bes, Belle and LHCb might be able to answer this open question.

In parallel, theoretical approaches based on QCD have been strongly developed. Lattice QCD performs ab initio calculation for hadron spectroscopy, even if it is not easy to approach hadron states at the physical pion mass or with heavy flavor. See Refs. [10,11].

The most recent LQCD studies have predicted a three-quark–like clustering at least at lower energy scales, even if we know that LQCD results are not at the pion mass physical point, thus they are still unable to encode the complexity and richness of the chiral symmetry breaking. It is also worthwhile noting that any kind of interaction which binds a quark–antiquark pair to a colorless state is also able to produce attraction between a quark pair in baryons. For example, this was shown in the Dyson-Swinger approach to QCD within the rainbow-

---

This article belongs to the special issue "Nucleon Resonances".

E. Santopinto (✉)
INFN, Sezione di Genova, via Dodecaneso 33, 16146 Genoa, Italy
E-mail: santopinto@ge.infn.it

J. Ferretti
Dipartimento di Fisica and INFN, 'Sapienza' Università di Roma, P.le Aldo Moro 5, 00185 Rome, Italy

ladder approximation of the DSE in Ref. [12]. Nevertheless, because of the approximations introduced in the calculations, we are still dealing with a model, even if rooted in QCD.

On the contrary, quark–diquark models are by definition phenomenological models from the beginning and correspond, in first approximation, to the leading Regge trajectories. Many of the resonances belonging to those trajectories are still waiting to be discovered, thus it is reasonable to expect that at least the resonances corresponding to the leading Regge trajectories should exist.

Up to 2 GeV, the interacting quark–diquark model has 8 missing Λ's in the octet and 6 in the singlet, so that many more can be expected up to 10 GeV. If we believe in a string-like Regge behavior at higher excitations, where the quark–diquark picture should be the dominant one, then it seems reasonable to expect that at least the quark–diquark subset of states will be found by the experiments, but also these resonances have still to be discovered. In this respect, the study of the higher energy part of the spectrum will shed light on the confinement mechanism [13,14].

From a three-quark follower point of view, we can argue in another way, but still the conclusions will be the same: the number of Λ states (but the same can be said for Σ's or Ω's) is expected to be equal to that of N^* or Δ resonances (around 26), if we believe in three-quark SU(3) flavor-symmetry. As, up to now, only a few strange states have been observed, for sure a 10 GeV experiment at JLab should provide relevant results. For example, a better knowledge of Λ excited states may help to carry out a more precise analysis of charmonium-like Pentaquark states [15]. Comparing the number of Λ states predicted by the relativistic Interacting Quark–Diquark model (8 for the octet and 6 for the singlet under 2 GeV), which is only a subset of that predicted by three-quark models, we can also suggest a next generation Pentaquark analysis that evaluates the systematic error on the background due to the missing Λ's (see Ref. [5]). Thus, the future discovering of missing Λ resonances by new experiments at JLab (JLab12), Bes, Belle and LHCb, might not change the structures seen in the Dalitz Plot by the LHCb analysis, but eventually modify some parameters.

Various aspects of the hadron structures have been investigated experimentally and theoretically in the last years. The observation of the hadron states with an exotic structure have attracted a lot of interest. In particular, regarding the light flavor region, we can remind exotic states such as the scalar mesons $a_0(980)$ and $f_0(980)$, or the $\Lambda(1405)$, which are expected to have an exotic structure and behave as multiquarks, hadronic molecules or hybrid states, and so on [16,17]. On the other hand, in the heavy counterpart, there are now accumulating evidences of exotic heavy hadrons, like Z_c [18,19] and $Z_b^{(\prime)}$ [20], which cannot be explained by the simple quark model picture.

The chiral effective field theory respecting the chiral symmetry provides the hadron-hadron scattering amplitudes at low energy with the Nambu–Goldstone bosons exchange. This is a powerful tool to investigate hadronic molecules as meson–meson [21–23], meson–baryon [24,25], and baryon–baryon [26,27] states appearing near thresholds; also in this case, a fine tuning of the model parameters may only be obtained with high precision experiments.

2 Phenomenological Motivation for a Quark–Diquark Model

The notion of diquark is as old as the quark model itself. Gell-Mann [28] mentioned the possibility of diquarks in his original paper on quarks, along with the possibility of tetra- and pentaquarks. Soon afterwards, Ida and Kobayashi [7] and Lichtenberg and Tassie [8] introduced effective degrees of freedom of diquarks in order to describe baryons as composed of a constituent diquark and quark. Moreover different phenomenological indications for diquark correlations have been collected during the years, including some regularities in hadron spectroscopy, the $\Delta I = \frac{1}{2}$ rule in weak nonleptonic decays [29], some regularities in parton distribution functions and in spin-dependent structure functions [30] and in the $\Lambda(1116)$ and $\Lambda(1520)$ fragmentation functions. Although the phenomenon of color superconductivity [31,32] in dense quark matter cannot be considered an argument in support of diquarks in the vacuum, it is nevertheless of interest since it stresses the important role of Cooper pairs of color superconductivity, which are color antitriplet, flavor antisymmetric, scalar diquarks. The concept of diquark in hadronic physics has some similarities to that of correlated pairs in condensed matter physics (superconductivity [33]) and in nuclear physics (interacting boson model [34]), where effective bosons emerge from pairs of electrons [35] and nucleons [36], respectively.

The microscopic origin of the diquark as an effective degree of freedom is not completely clear, nevertheless one may attempt to correlate the data in terms of a phenomenological model, the interacting quark–diquark model. In this short contribution, we will review the model in its original formulation [1] and discuss also its Point Form relativistic reformulation [2,6]. We will show results for strange and nonstrange baryon spectra

[2,6] and discuss some important consequences of a quark–diquark model calculation of the ratio between electric and magnetic form factors of the proton [37], namely the presence of a zero at $Q^2 = 8$ GeV^2, which is impossible within three-quark models. New experiments at JLab (JLab12), Bes, Belle and LHCb will eventually shed light on the three-quark versus quark–diquark structure of the nucleon.

3 The Interacting Quark–Diquark Model

Up to an energy of about 2 GeV, the diquark can be described as two correlated quarks with no internal spatial excitations [1,2]. Thus, its color-spin-flavor wave function must be antisymmetric. Moreover, as we consider only light baryons, made up of u, d, s quarks, the internal group is restricted to $SU_{sf}(6)$. If we denote spin by its value, flavor and color by the dimension of the representation, the quark has spin $s_2 = \frac{1}{2}$, flavor $F_2 = \mathbf{3}$, and color $C_2 = \mathbf{3}$. The diquark must transform as $\overline{\mathbf{3}}$ under $SU_c(3)$, hadrons being color singlets. Then, one only has the symmetric $SU_{sf}(6)$ representation $\mathbf{21}_{sf}(S)$, containing $s_1 = 0$, $F_1 = \overline{\mathbf{3}}$, and $s_1 = 1$, $F_1 = \mathbf{6}$, i.e. the scalar and axial-vector diquarks, respectively. We assume that baryons are composed of a valence quark and a valence diquark.

The relative two-body configurations can be described in terms of a relative coordinate $\mathbf{r}$ with conjugate momentum $\mathbf{p}$. The Hamiltonian contains a direct and an exchange interaction. The direct interaction is a Coulomb plus linear interaction, while the exchange one is of the spin–spin, isospin–isospin, and spin–isospin type. A contact term is included in order to describe the splitting between the nucleon and the Δ:

$$
\begin{aligned}
H = E_0 + \frac{p^2}{2m} - \frac{\tau}{r} + \beta r + (B + C\delta_0)\delta_{S_{12},1} \\
+(-1)^{l+1}2Ae^{-\alpha r}[\mathbf{s_{12}} \cdot \mathbf{s_3} + \mathbf{t_{12}} \cdot \mathbf{t_3} + 2\mathbf{s_{12}} \cdot \mathbf{s_3}\,\mathbf{t_{12}} \cdot \mathbf{t_3}],
\end{aligned}
\tag{1}
$$

For a purely Coulomb-like interaction the problem is analytically solvable. The solution is trivial, with eigenvalues

$$
E_{n,l} = -\frac{\tau^2 m}{2\,n^2}, \quad n = 1, 2 \ldots.
\tag{2}
$$

Here m is the reduced mass of the diquark–quark configuration and n the principal quantum number. The eigenfunctions are the usual Coulomb functions

$$
R_{n,l}(r) = \sqrt{\frac{(n-l-1)!(2g)^3}{2n[(n+l)!]^3}}(2gr)^l\,e^{-gr}L_{n-l-1}^{2l+1}(2gr),
\tag{3}
$$

where $g = \frac{\tau m}{n}$. We treat all the other interactions as perturbations, so the model is completely analytical. The matrix elements of βr can be evaluated in closed form as

$$
\Delta E_{n,l} = \int_0^\infty \beta r[R_{n,l}(r)]^2 r^2 \mathrm{d}r = \frac{\beta}{2m\tau}[3n^2 - l(l+1)].
\tag{4}
$$

To complete the evaluation, we need the matrix elements of the exponential. These can be obtained in analytic form

$$
I_{n,l}(\alpha) = \int_0^\infty e^{-\alpha r}[R_{n,l}(r)]^2 r^2 \mathrm{d}r .
\tag{5}
$$

Our results are in present in Tables 1 and 2.

Table 1 Mass spectrum of N-type resonances (up to 2.1 GeV) in the interacting quark diquark model [1]

Baryon $L_{2I,2J}$	Status	Mass (MeV)	J^p	M_{cal} (MeV)
$N(939)P_{11}$	****	939	$1/2^+$	940
$N(1440)P_{11}$	****	1410–1450	$1/2^+$	1538
$N(1520)D_{13}$	****	1510–1520	$3/2^-$	1543
$N(1535)S_{11}$	****	1525–1545	$1/2^-$	1538
$N(1650)S_{11}$	****	1645–1670	$1/2^-$	1673
$N(1675)D_{15}$	****	1670–1680	$5/2^-$	1673
$N(1680)F_{15}$	****	1680–1690	$5/2^+$	1675
$N(1700)D_{13}$	***	1650–1750	$3/2^-$	1673
$N(1710)P_{11}$	***	1680–1740	$1/2^+$	1640
$N(1720)P_{13}$	****	1700–1750	$3/2^+$	1675
$N(1860)F_{15}$	**	1820–1960	$5/2^+$	1975
$N(1875)D_{13}$	***	1820–1920	$3/2^-$	1838
$N(1880)P_{11}$	**	1835–1905	$1/2^+$	1838
$N(1895)S_{11}$	**	1880–1910	$1/2^-$	1838
$N(1900)P_{13}$	***	1875–1935	$3/2^+$	1967
$N(1990)F_{17}$	**	1995–2125	$7/2^+$	2015
$N(2000)F_{15}$	**	1950–2150	$5/2^+$	2015
$N(2040)P_{13}$	*	2031–2065	$3/2^+$	2015
$N(2060)D_{15}$	**	2045–2075	$5/2^-$	2078
$N(2100)P_{11}$	**	2050–2200	$1/2^+$	2015
$N(2120)D_{13}$	**	2090–2210	$3/2^-$	2069

The value of the parameters are reported in Ref. [1]. The table reports also the prediction for the remaining resonances, including the recent upgraded 3* $P13(1900)$. The experimental values are taken from Ref. [38]

Table 2 As Table 1, but for Δ-type resonances [1]

Baryon $L_{2I,2J}$	Status	Mass (MeV)	State	M_{cal} (MeV)
$\Delta(1232)P_{33}$	****	1230–1234	$3/2^+$	1235
$\Delta(1600)P_{33}$	***	1500–1700	$3/2^+$	1709
$\Delta(1620)S_{31}$	****	1600–1660	$1/2^-$	1673
$\Delta(1700)D_{33}$	****	1670–1750	$3/2^-$	1673
$\Delta(1900)S_{31}$	**	1840–1920	$1/2^-$	2003
$\Delta(1905)F_{35}$	****	1855–1910	$5/2^+$	1930
$\Delta(1910)P_{31}$	****	1860–1910	$1/2^+$	1967
$\Delta(1920)P_{33}$	***	1900–1970	$3/2^+$	1930
$\Delta(1930)D_{35}$	***	1900–2000	$5/2^-$	2003
$\Delta(1940)D_{33}$	**	1940–2060	$3/2^-$	2003
$\Delta(1950)F_{37}$	****	1915–1950	$7/2^+$	1930
$\Delta(2000)F_{35}$	**	≈ 2000	$5/2^+$	2015

4 The Relativistic Interacting Quark–Diquark Model

The extension of the Interacting quark–diquark model [1] in Point Form dynamics [39] was carried out in Refs. [2,6]. As in Ref. [1], the Hamiltonian contains two basic ingredients: a Coulomb-like plus linear confining interaction and an exchange one, depending on the spin and isospin of the quark and the diquark. The mass operator is given by

$$M = E_0 + \sqrt{\mathbf{q}^2 + m_1^2} + \sqrt{\mathbf{q}^2 + m_2^2} + M_{\text{dir}}(r) + M_{\text{ex}}(r) \ , \tag{6}$$

where E_0 is a constant, $M_{\text{dir}}(r)$ and $M_{\text{ex}}(r)$ the direct and the exchange diquark–quark interaction, respectively, m_1 and m_2 stand for diquark and quark masses. For more details, see also the contribution by Ferretti *et al*. at this workshop.

The results for the nonstrange baryon spectrum from Ref. [2] (see Fig. 1) were obtained by diagonalizing the mass operator of Eq. (6) by means of a numerical variational procedure, based on harmonic oscillator trial wave functions. With a basis of 150 harmonic oscillator shells, the results converge very well. In a second stage, the mass operator of Eq. (6) was generalized to the strange sector by substituting the model exchange

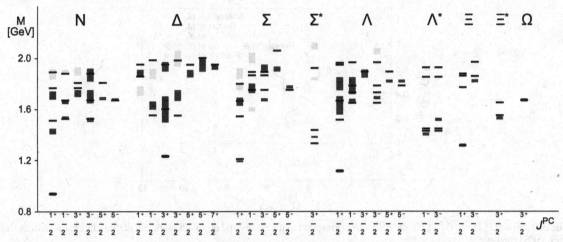

Fig. 1 The calculated masses (*black lines*) from Refs. [2,6] are compared to three-/four-star resonances (*dark boxes*) and one-/two-star resonances (*pale boxes*) from PDG [38]. We include theoretical predictions up to an energy of 2 GeV

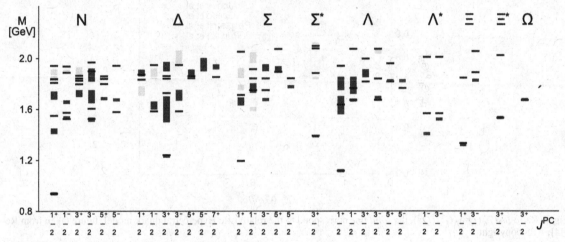

Fig. 2 The hypercentral QM results (*black lines*) of Refs. [40,41] are compared to three-/four-star resonances (*dark boxes*) and one-/two-star resonances (*pale boxes*) from PDG [38]

operator, $M_{ex}(r)$, with a more general Gürsey-Radicati inspired interaction [6]. The results for the strange sector are reported in Fig. 1. See also Fig. 2, where we compare the hypercentral quark model results of Refs. [40,41] with data.

It is clear that a larger number of experiments and analyses, looking for missing resonances, are necessary because many aspects of hadron spectroscopy are still unclear. In particular, the number of Λ states reported by the PDG is small with respect to the predictions of Lattice QCD and models. In this respect, it is worthwhile to note that the relativistic version of the interacting quark–diquark model predicts seven Λ missing states belonging to the octet and other six missing states belonging to the singlet (considering only states under 2 GeV).

It is also worthwhile noting that in our model [6] $\Lambda(1116)$ and $\Lambda^*(1520)$ are described as bound states of a scalar diquark $[n, n]$ and a quark s, where the quark–diquark system is in S or P-wave, respectively [6]. This is in accordance with the observations of Refs. [42,43] on Λ's fragmentation functions, that the two resonances can be described as $[n, n] - s$ systems.

We should also underline that the interacting quark–diquark model provides wave functions that can describe in a reasonable way the elastic electromagnetic form factors of the nucleon. In particular, they provide a reproduction of the existing data for the ratio of the electric and magnetic form factor of the proton characterized by a zero at $Q^2 = 8$ GeV^2 (see Fig. 3), like in vector meson parametrizations. On the contrary, it was impossible to get this zero with a three-quark model calculation [44] (see Fig. 4). New experiments at Jlab (JLab12), Bes,

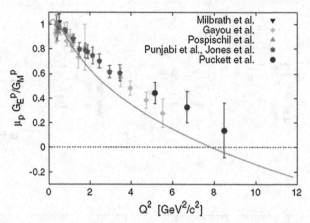

Fig. 3 Ratio $\mu_p G_E^p(Q^2)/G_M^p(Q^2)$, the *solid line* correspond to the relativistic quark–diquark calculation, figure taken from Ref. [37]; APS copyright

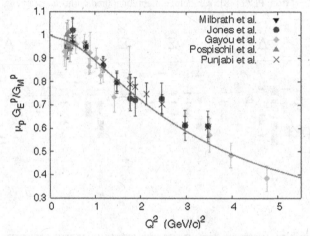

Fig. 4 Ratio $\mu_p G_E^p(Q^2)/G_M^p(Q^2)$, the *solid line* correspond to the relativistic Hypercetral quark model, figure taken from Ref. [44]; APS copyright

Belle and LHCb will be able to distinguish between these two scenarios and, possibly, rule out one of the two models.

References

1. Santopinto, E.: An interacting quark-diquark model of baryons. Phys. Rev. C **72**, 022201 (2005)
2. Ferretti, J., Vassallo, A., Santopinto, E.: Relativistic quark-diquark model of baryons. Phys. Rev. C **83**, 065204 (2011)
3. Galatà, G., Santopinto, E.: Hybrid quark-diquark baryon model. Phys. Rev. C **86**, 045202 (2012)
4. Bijker, R., Ferretti, J., Galatà, G., García-Tecocoatzi, H., Santopinto, E.: Strong decays of baryons and missing resonances. Accepted on Phys. Rev. D arXiv:1506.07469
5. Santopinto, E.: Diquark correlations in baryons: the Interacting Quark Diquark Model. JPS Conf. Proc. **10**, 010010 (2016)
6. Santopinto, E., Ferretti, J.: Strange and nonstrange baryon spectra in the relativistic interacting quark-diquark model with a Gürsey and Radicati-inspired exchange interaction. Phys. Rev. C **92**, 025202 (2015)
7. Ida, M., Kobayashi, R.: Baryon resonances in a quark model. Prog. Theor. Phys. **36**, 846 (1966)
8. Lichtenberg, D.B., Tassie, L.J.: Baryon mass splitting in a boson-fermion model. Phys. Rev. **155**, 1601 (1967)
9. De Sanctis, M., Ferretti, J., Santopinto, E., Vassallo, A.: Relativistic quark-diquark model of baryons with a spin-isospin transition interaction: non-strange baryon spectrum and nucleon magnetic moments. Eur. Phys. J. A **52**, 121 (2016)
10. Dudek, J.J., Edwards, R.G., Peardon, M.J., Richards, D.G., Thomas, C.E.: Highly excited and exotic meson spectrum from dynamical lattice QCD. Phys. Rev. Lett. **103**, 262001 (2009)

11. Aoki S. et al.: [PACS-CS Collaboration]. Lattice quantum chromodynamics at the physical point and beyond. PTEP 2012:01A102 (2012)
12. Cahill, R.T., Roberts, C.D., Praschifka, J.: Calculation of diquark masses in QCD. Phys. Rev. D **36**, 2804 (1987)
13. Ostrander, A., Santopinto, E., Szczepaniak, A.P., Vassallo, A.: Gluon chain formation in presence of static charges. Phys. Rev. D **86**, 114015 (2012)
14. Roberts, C.D., Segovia, J.: Baryons and the Borromeo. arXiv:1603.02722 [nucl-th]
15. Aaij, R., et al.: [LHCb Collaboration]. Observation of $J/\Psi p$ resonances consistent with Pentaquark States in $\Lambda_b^0 \to J/\Psi K p$ decays. Phys. Rev. Lett. **115**, 072001 (2015)
16. Klempt, E., Zaitsev, A.: Glueballs, hybrids, multiquarks. experimental facts versus QCD inspired concepts. Phys. Rept. 454:1 (2007)
17. Brambilla, N., et al.: Heavy quarkonium: progress, puzzles, and opportunities. Eur. Phys. J. C **71**, 1534 (2011)
18. Ablikim, M., et al.: [BESIII Collaboration]. Observation of a charged charmonium-like structure in e^+e^- to $\pi^+\pi^- J/\Psi$ at $\sqrt{s} = 4.26$ GeV. Phys. Rev. Lett. **110**, 252001 (2013)
19. Liu, Z.Q., et al.: [Belle Collaboration]. Study of $e^+e^- \to \pi^+\pi^- J/\psi$ and observation of a charged charmonium-like State at Belle. Phys. Rev. Lett. **110**, 252002 (2013)
20. Bondar, A., et al.: [Belle Collaboration]. Observation of two charged bottomonium-like resonances in $\Upsilon(5S)$ decays. Phys. Rev. Lett. **108**, 122001 (2012)
21. Oller, J.A., Oset, E.: Chiral symmetry amplitudes in the S wave isoscalar and isovector channels and the sigma, $f_0(980)$, $a_0(980)$ scalar mesons. Nucl. Phys. A 620:438 (1997) [Nucl. Phys. A 652:407 (1999)]
22. Wang, P., Wang, X.G.: Study on $X(3872)$ from effective field theory with pion exchange interaction. Phys. Rev. Lett. **111**, 042002 (2013)
23. Baru, V., Epelbaum, E., Filin, A.A., Guo, F.-K., Hammer, H.-W., Hanhart, C., Meissner, U.-G., Nefediev, A.V.: Remarks on study of $X(3872)$ from effective field theory with pion-exchange interaction. Phys. Rev. D **91**, 034002 (2015)
24. Hyodo, T., Jido, D.: The nature of the $\Lambda(1405)$ resonance in chiral dynamics. Prog. Part. Nucl. Phys. **67**, 55 (2012)
25. Yamaguchi, Y., Ohkoda, S., Yasui, S., Hosaka, A.: Exotic baryons from a heavy meson and a nucleon—negative parity states. Phys. Rev. D **84**, 014032 (2011)
26. Machleidt, R., Entem, D.R.: Chiral effective field theory and nuclear forces. Phys. Rept. **503**, 1 (2011)
27. Haidenbauer, J., Meissner, U.G.: Exotic bound states of two baryons in light of chiral effective field theory. Nucl. Phys. A **881**, 44 (2012)
28. Gell-Mann, M.: A schematic model of baryons and mesons. Phys. Lett. **8**, 214 (1964)
29. Neubert, M., Stech, B.: Quark quark correlations and the $\Delta I = 1/2$ Rule. Phys. Lett. B **231**, 477 (1989)
30. Close, F.E., Thomas, A.W.: The spin and flavor dependence of parton distribution functions. Phys. Lett. B **212**, 227 (1988)
31. Bailing, D., Love, A.: Superfluidity and superconductivity in relativistic fermion systems. Phys. Rept. **107**, 325 (1984)
32. Alford, M.G., Rajagopal, K., Wilczek, F.: Color flavor locking and chiral symmetry breaking in high density QCD. Nucl. Phys. B **537**, 443 (1999)
33. Bardeen, J., Cooper, L.N., Schrieffer, J.R.: Theory of superconductivity. Phys. Rev. **108**, 1175 (1957)
34. Iachello, F., Arima, A.: The Interacting Boson Model. Cambridge University Press, Cambridge (1987)
35. Cooper, L.N.: Bound electron pairs in a degenerate fermi gas. Phys. Rev. **104**, 1189 (1956)
36. Otsuka, T., Arima, A., Iachello, F., Talmi, I.: Electron scattering in the interacting boson model. Phys. Lett. B **76**, 135 (1978)
37. De Sanctis, M., Ferretti, J., Santopinto, E., Vassallo, A.: Electromagnetic form factors in the relativistic interacting quark-diquark model of baryons. Phys. Rev. C **84**, 055201 (2011)
38. Olive, K.A., et al.: [Particle Data Group]. Review of particle physics. Chin. Phys. C **38**, 090001 (2014)
39. Klink, W.H.: Relativistic simultaneously coupled multiparticle states. Phys. Rev. C **58**, 3617 (1998)
40. Giannini, M.M., Santopinto, E., Vassallo, A.: A new application of the Gürsey and radicati mass formula. Eur. Phys. J. A **25**, 241 (2005)
41. Giannini, M.M., Santopinto, E.: The hypercentral Constituent Quark Model and its application to baryon properties. Chin. J. Phys. **53**, 020301 (2015)
42. Jaffe, R.L.: Exotica. Phys. Rept. **409**, 1 (2005). doi:10.1016/j.physrep.2004.11.005
43. Selem, A., Wilczek, F.: Hadron systematics and emergent diquarks. arXiv:hep-ph/0602128
44. Santopinto, E., Vassallo, A., Giannini, M.M., De Sanctis, M.: High Q^2 behavior of the electromagnetic form factors in the relativistic hypercentral constituent quark model. Phys. Rev. C **82**, 065204 (2010)

Few-Body Syst (2016) 57:1077–1085
DOI 10.1007/s00601-016-1154-5

G. Ramalho

A Relativistic Model for the Electromagnetic Structure of Baryons from the 3rd Resonance Region

Received: 15 February 2016 / Accepted: 29 August 2016 / Published online: 24 September 2016
© Springer-Verlag Wien 2016

Abstract We present some predictions for the $\gamma^* N \to N^*$ transition amplitudes, where N is the nucleon, and N^* is a nucleon excitation from the third resonance region. First we estimate the transition amplitudes associated with the second radial excitation of the nucleon, interpreted as the $N(1710)$ state, using the covariant spectator quark model. After that, we combine some results from the covariant spectator quark model with the framework of the single quark transition model, to make predictions for the $\gamma^* N \to N^*$ transition amplitudes, where N^* is a member of the $SU(6)$-multiplet $[70, 1^-]$. The results for the $\gamma^* N \to N(1520)$ and $\gamma^* N \to N(1535)$ transition amplitudes are used as input to the calculation of the amplitudes $A_{1/2}$, $A_{3/2}$, associated with the $\gamma^* N \to N(1650)$, $\gamma^* N \to N(1700)$, $\gamma^* N \to \Delta(1620)$, and $\gamma^* N \to \Delta(1700)$ transitions. Our estimates are compared with the available data. In order to facilitate the comparison with future experimental data at high Q^2, we derived also simple parametrizations for the amplitudes, compatible with the expected falloff at high Q^2.

1 Introduction

One of the challenges in the modern physics is the description of the internal structure of the baryons and mesons. The electromagnetic structure of the nucleon N and the nucleon resonances N^* can be accessed through the $\gamma^* N \to N^*$ reactions, which depend of the (photon) momentum transfer squared Q^2 [1–5]. The data associated with those transitions are represented in terms of helicity amplitudes and have been collected in the recent years at Jefferson Lab, with increasing Q^2 [1]. The new data demands the development of theoretical models based on the underlying structure of quarks and quark-antiquark states (mesons) [1,2]. Those models may be used to guide future experiments as the ones planned for the Jlab–12 GeV upgrade, particularly for resonances in the second and third resonance region [energy $W = 1.4$–1.8 GeV] (see Fig. 1) [1].

An example of a model appropriated for the study of the electromagnetic structure of resonances at large Q^2 is the covariant spectator quark model [6–10]. In the covariant spectator quark model the baryons are described in terms of covariant wave functions based on quarks with internal structure (constituent quarks). Following previous studies for the nucleon and the first radial excitation of the nucleon [6,7,11], we use the covariant spectator quark model to calculate the transition amplitudes associated with the second radial excitation of the nucleon [12].

In a different work, we use the results of the covariant spectator quark model for the $\gamma^* N \to N(1520)$ and $\gamma^* N \to N(1535)$ transition amplitudes [13–15] to estimate the transition amplitudes associated with four

This work was supported by Brazilian Ministry of Science, Technology and Innovation (MCTI-Brazil).

This article belongs to the special issue "Nucleon Resonances".

G. Ramalho (✉)
International Institute of Physics, Federal University of Rio Grande do Norte, Campus Lagoa Nova - Anel Viário da UFRN, Lagoa Nova, Natal, RN 59070-405, Brazil
E-mail: gilberto.ramalho@iip.ufrn.br

negative parity nucleon resonances from the $SU(6)$-multiplet $[70, 1^-]$, in the third resonance region. This study is possible due to the combination with the single quark transition model, which allows the parametrization of the amplitudes $A_{1/2}$, $A_{3/2}$ for six resonances from the $SU(6)$-multiplet $[70, 1^-]$, based on only three coefficients dependent of Q^2 [16].

2 Covariant Spectator Quark Model

In the covariant spectator quark model, baryons are treated as three-quark systems. The baryon wave functions are derived from the quark states according with the $SU(6) \otimes O(3)$ symmetry group. A quark is off-mass-shell, and free to interact with the photon fields, and other two quarks are on-mass-shell [6–10]. Integrating over the quark-pair degrees of freedom we reduce the baryon to a quark-diquark system, where the diquark can be represented as an on-mass-shell spectator particle with an effective mass m_D [6,7,10,13–15].

The electromagnetic interaction with the baryons is described by the photon coupling with the constituent quarks in the relativistic impulse approximation. The quark electromagnetic structure is represented in terms of the quark form factors parametrized by a vector meson dominance mechanism [6,7,10,17]. The parametrization of the quark current was calibrated in the studies of the nucleon form factors data [6,7] and by the lattice QCD data for the decuplet baryon [10]. The quark electromagnetic form factors encodes effectively the gluon and quark-antiquark substructure of the constituent quarks. The quark current is decomposed as $j_q^\mu = j_1 \gamma^\mu + j_2 \frac{i\sigma^{\mu\nu} q_\nu}{2M}$, where j_i ($i = 1, 2$) are the Dirac and Pauli quark form factors, and M is the nucleon mass. In the $SU(2)$-flavor sector the functions j_i can also be decomposed into the isoscalar (f_{i+}) and the isovector (f_{i-}) components: $j_i = \frac{1}{6} f_{i+} + \frac{1}{2} f_{i-} \tau_3$, where τ_3 acts on the isospin states of baryons (nucleon or resonance). The details can be found in Ref. [6–10].

When the nucleon wave function (Ψ_N) and the resonance wave function (Ψ_R) are both expressed in terms of the single quark and quark-pair states, the transition current is calculated in the relativistic impulse approximation, integrating over the diquark on-mass-shell momentum, and summing over the the intermediate diquark polarizations [6,7,10]. In the study of inelastic transitions we use the Landau prescription to ensure the conservation of the transition current [13–16].

Using the relativistic impulse approximation, we can express the transition current in terms of the quark electromagnetic form factor $f_{i\pm}$ ($i = 1, 2$) and the radial wave functions ψ_N and ψ_R [6,7,13–15]. The radial wave functions are scalar functions that depend on the baryon (P) and diquark (k) momenta and parametrize the momentum distributions of the quark-diquark systems. From the transition current we can extract the form factors and the helicity transition amplitudes, defined in the rest frame of the resonance (final state), for the reaction under study [1,2,13–15].

The covariant spectator quark model was used already in the study of several nucleon excitations including isospin 1/2 systems $N(1410)$, $N(1520)$, $N(1535)$ [11,13–15] and the isospin 3/2 systems [17–23]. In Fig. 1,

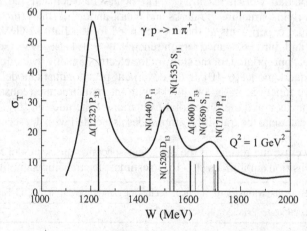

Fig. 1 Representation of the $\gamma p \to n\pi^+$ cross section. The graph define the three resonance regions. The *vertical lines* represent resonant states described by the covariant spectator quark model, including the resonance $N(1710)$. At *red* we indicate the states studded in this work in the context of the single quark transition model. At *blue* are the states used as input: $N(1520)$ and $N(1535)$ (color figure online)

the position of the nucleon excitations are represented and compared with the bumps of the cross sections. The model generalized to the $SU(3)$-flavor sector was also used to study the octet and decuplet baryons as well as transitions between baryons with strange quarks [24–27]. Based on the parametrization of the quark current j_q^μ in terms of the vector meson dominance mechanism, the model was extended to the lattice QCD regime (heavy pions and no meson cloud) [17,18], to the nuclear medium [8,9] and to the timelike regime [28,29]. The model was also used to study the nucleon deep inelastic scattering [6,7,30] and the axial structure of the octet baryon [31].

Most of the works refereed below, are based on the valence quarks degrees of freedom, as consequence of the relativistic impulse approximation. There are however some processes such as the meson exchanged between the different quarks inside the baryon, which cannot be reduced to simple diagrams with quark dressing. Those processes are regarded as arising from a meson exchanged between the different quarks inside the baryon and can be classified as meson cloud corrections to the hadronic reactions [8,9,15,24,25]. In some cases one can use the covariant spectator quark model to infer the effect of the meson cloud effects based on empirical information or lattice QCD simulations [13,14,17,18,24,25,31,32].

3 Resonance $N(1710)$

We discuss now the $\gamma^* N \to N^*$ transitions, where N is the nucleon state and N^* is a radial excitation of the nucleon based on a S-state wave function. We are excluding where the $J^P = \frac{1}{2}^+$ N^* states based on spin-isospin wave functions with mixed symmetry. The N^* state share then with the nucleon, the structure of spin and isospin, and differ only in the radial structure (radial wave function ψ_R). Therefore, if we exclude the meson cloud effects, in principle relevant only at low Q^2, we can estimate the transition form factors using the formalism already developed for the nucleon [6,7], replacing the radial wave function of the nucleon, labeled here as ψ_{N0}, in the final state, by the radial wave function of the resonances. Since the spin and isospin structure is the same for N and N^* the orthogonality between the radial excitations is a consequence of the orthogonality of the radial wave functions. Labeling the radial wave functions of the first and second excitations, as ψ_{N1} and ψ_{N2} respectively, we can express the orthogonality condition as [11,12]

$$\int_k \psi_{N1}\psi_{N0}\bigg|_{Q^2=0} = 0, \qquad \int_k \psi_{N2}\psi_{N0}\bigg|_{Q^2=0} = 0, \qquad \int_k \psi_{N2}\psi_{N1}\bigg|_{Q^2=0} = 0, \qquad (1)$$

where the subindex $Q^2 = 0$ indicates that the integral is calculated in the limit $Q^2 = 0$.

The functions ψ_{N1}, ψ_{N2} can be defined in terms of the momentum range parameters β_1, β_2 of the nucleon radial wave function, $(\beta_2 > \beta_1)$, where β_1 regulates the long-range structure and β_2 regulates the short-range structure. If we choose a radial wave function that preserves the short-range structure of the nucleon wave function, we can determine all parameters of the functions ψ_{N1}, ψ_{N2} using Eq. (1) [12]. In this case no parameters have to be adjusted, and the model provide true predictions for the transition form factors or the helicity amplitudes. This method was already used for the $N(1440)$ state (the Roper), where we concluded, that, the model gives a very good description of the $Q^2 > 1.5$ GeV^2 data [11,33–36], the region where the valence quark degrees of freedom dominate.

We tested then if the model could also be extended for the second radial excitation of the nucleon. Based on the quantum numbers (P_{11}), we assumed that the second radial excitation of the nucleon could be the $N(1710)$ state. The model predictions for the $N(1710)$ transition amplitudes are presented in Fig. 2 up to 12 GeV^2, pointing for the upper limit of the Jlab-12 GeV upgrade. The results are compared with the equivalent results for the $N(1440)$ state. In order to obtain a comparison between elastic (nucleon) and inelastic transition amplitudes we also present the results of the nucleon *equivalent amplitudes*. The nucleon *equivalent amplitudes* are defined by $A_{1/2} = \sqrt{2}\mathcal{R}G_M$ and $S_{1/2} = \sqrt{2}\frac{\mathcal{R}}{\sqrt{2}}\sqrt{\frac{1+\tau}{\tau}}G_E$, where G_M, G_E are the nucleon form factors, $\tau = \frac{Q^2}{4M^2}$, and $\mathcal{R}$ is a function dependent of the $N(1440)$ variables (mass M_R), $\mathcal{R} = \frac{e}{2}\sqrt{\frac{(M_R-M)^2+Q^2}{M_R M K}}$, where $K = \frac{M_R^2-M^2}{2M_R}$ (e is the elementary electric charge). In the amplitudes the factor $\sqrt{2}$ was included for convenience.

In Fig. 2, one can see, that, the results for the amplitudes are similar for all the system for $Q^2 > 4$ GeV^2 (large Q^2). The exception is the result for amplitude $S_{1/2}$ for the nucleon, which vanishes for $Q^2 \approx 7$ GeV^2, because $G_E \simeq 0$ [6,7]. One can interpret the approximated convergence of results for large Q^2 as a consequence of the correlations between the excited radial wave functions and the nucleon radial wave function [12].

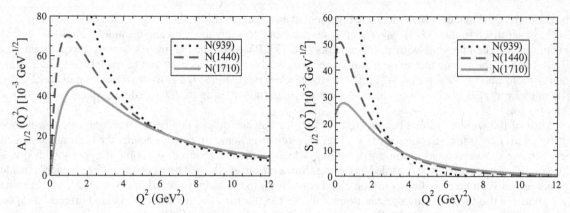

Fig. 2 Helicity amplitudes for the nucleon, $N(1440)$ and $N(1710)$. See definition of the nucleon *equivalent amplitudes* in the text

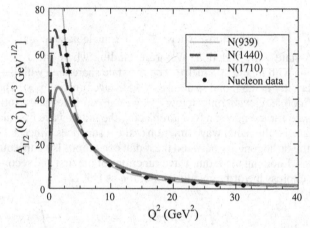

Fig. 3 Amplitude $A_{1/2}$ for the nucleon, $N(1440)$ and $N(1710)$, compared with the data extracted from the nucleon magnetic form factor [37]

Since the *equivalent* amplitude $A_{1/2}$ for the nucleon is known already for very large Q^2 (using the G_M data), this result can be compared with the estimates of the amplitudes for the $N(1440)$ and $N(1710)$ systems, for very large Q^2, as presented in Fig. 3. As shown in the figure, one predicts that the amplitude $A_{1/2}$ follows closely the *equivalent amplitude* of the nucleon, for large Q^2. Future experiments in the range $Q^2 = 4$–10 GeV^2 can confirm or deny this prediction.

The $\gamma^* N \to N(1710)$ transition amplitudes were determined for the first time in Ref. [38]. The data is compared with our estimates in Fig. 4. From the figure, one can conclude that our estimate differs from the data in magnitude for $A_{1/2}$ and in sign for $S_{1/2}$, at least for $Q^2 \leq 4$ GeV^2. The new data suggests that, or, our estimate is valid only for larger values of Q^2, to be confirmed by new data for $Q^2 > 4$ GeV^2, or, our interpretation of $N(1710)$ as the second radial resonance of the nucleon, is not valid. It is worth to mention that our calculations are consistent with others estimates based on the assumption of the second nucleon radial excitation, where $A_{1/2} > 0$ and $S_{1/2} > 0$ [39,40]. The signs, and magnitudes of the data, are however consistent with the estimates of the hypercentral quark model [38,41].

From the theoretical point of view there are other possible interpretations for the $N(1710)$ state. Although the interpretation of the states from the third resonance region based on the quark structure are partially tentative, the state $N(1710)$ can be interpreted as an excitation associated with a state with mixed symmetry [42–45]. Several other suggestions have been made for the composition of the $N(1710)$ state, such as the mixture of states $\pi N - \pi\pi N$, the mixture $\pi N - \sigma N$, a $\sigma_v N$ state, where σ_v represent a vibrational state of the σ, and even a gN mixture, where g is a gluon [12]. Another possibility is that $N(1710)$ is a dynamically generated resonance as suggested by Ref. [46], although the estimated mass is larger (1820 MeV). Finally in the interacting quark diquark model the $N(1710)$ is described as a quark-axial diquark system in a relative S-wave configuration.

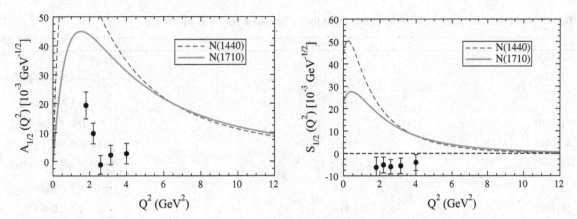

Fig. 4 $\gamma^* N \to N(1710)$ transition amplitudes compared with the data from Park et al. [38]. See discussion in the text

The prediction of the mass is respectively, 1640 MeV in the non relativistic version of the model [47] and 1776 MeV in the relativistic version of the same model [48].

Contrarily to the nucleon and the $N(1440)$ that are part of the $SU(6)$-multiplet $[56, 0^+]$, the $N(1710)$ state is more frequently associated with the multiplet $[70, 0^+]$ [41–45,49]. In those conditions the second radial excitation of the nucleon should correspond to an higher mass resonance. The state $N(1880)$ listed by the particle data group (PDG) [42] is at the moment the best candidate. Within the quark models it is interesting to note that hypercentral quark model is the only model which admits two radial excitations of the nucleon belonging to $[56, 0^+]$ multiplets [41,49,50].

We expect, that, future experiments provide accurate data for the transverse ($A_{1/2}$) and longitudinal ($S_{1/2}$) amplitudes, which may be used to test the expected falloff associated with the valence quark effects: $A_{1/2} \propto 1/Q^3$ and $S_{1/2} \propto 1/Q^3$, for large Q^2 [51]. If this is not case, and $N(1710)$ is in fact a mixture of meson-baryon states, the $1/Q^3$ falloff cannot be observed, and the interpretation of $N(1710)$ as a state dominated by valence quark has to be questioned.

4 Resonances from the $[70, 1^-]$ $SU(6)$-Multiplet

The combination of the wave functions of a baryon (three-quark system) given by $SU(6) \otimes O(3)$ group and the description of electromagnetic interaction in impulse approximation leads to the so-called single quark transition model (SQTM) [52–54]. In this context *single* means that only one quark couples with the photon. In these conditions the SQTM can be used to parametrize the transition current between two multiplets, in an operational form that includes only four independent terms, with coefficients exclusively dependent of Q^2.

In particular, the SQTM can be used to parametrize the $\gamma^* N \to N^*$ transitions, where N^* is a N (isospin 1/2) or a Δ (isospin 3/2) state from the $[70, 1^-]$ multiplet, in terms on three independent functions of Q^2: A, B, and C [52–54]. The relations between the functions A, B, and C and the amplitudes are presented in the Table 1. In this analysis we do not take into account the constraints at the pseudo-threshold when $Q^2 = -(M_R - M)^2$ as in Ref. [55–57]. Using the results for the $\gamma^* N \to N(1535)$ and $\gamma^* N \to N(1520)$ amplitudes, respectively $A_{1/2}^{S11}$, $A_{1/2}^{D13}$, and $A_{3/2}^{D13}$ in the spectroscopic notation, we can write, using $\cos \theta_D = 0.995 \simeq 1$

$$A = 2\frac{A_{1/2}^{S11}}{\cos \theta_S} + \sqrt{2}A_{1/2}^{D13} + \sqrt{6}A_{3/2}^{D13}, \qquad B = 2\frac{A_{1/2}^{S11}}{\cos \theta_S} - 2\sqrt{2}A_{1/2}^{D13},$$

$$C = -2\frac{A_{1/2}^{S11}}{\cos \theta_S} - \sqrt{2}A_{1/2}^{D13} + \sqrt{6}A_{3/2}^{D13}. \tag{2}$$

We use then the amplitudes $A_{1/2}^{S11}$, $A_{1/2}^{D13}$ and $A_{3/2}^{D13}$, determined by the covariant spectator quark model for the $\gamma^* N \to N(1520)$ and $\gamma^* N \to N(1535)$ transitions [13–16], to calculate the coefficients A, B and C. After that we can predict the amplitudes associated with the remaining transition for $[70, 1^-]$ states, namely for the the transitions $\gamma^* N \to N(1650)$, $\gamma^* N \to N(1700)$, $\gamma^* N \to \Delta(1620)$ and $\gamma^* N \to \Delta(1700)$. Since the covariant spectator quark model breaks the $SU(2)$-flavor symmetry, we restrict our study to reactions with

Table 1 Amplitudes $A_{1/2}$ and $A_{3/2}$ estimated by SQTM for the proton targets ($N = p$) [16,52]

State	Amplitude	
$N(1535)$	$A_{1/2}$	$\frac{1}{6}(A + B - C) \cos\theta_S$
$N(1520)$	$A_{1/2}$	$\frac{1}{6\sqrt{2}}(A - 2B - C) \cos\theta_D$
	$A_{3/2}$	$\frac{1}{2\sqrt{6}}(A + C) \cos\theta_D$
$N(1650)$	$A_{1/2}$	$\frac{1}{6}(A + B - C) \sin\theta_S$
$\Delta(1620)$	$A_{1/2}$	$\frac{1}{18}(3A - B + C)$
$N(1700)$	$A_{1/2}$	$\frac{1}{6\sqrt{2}}(A - 2B - C) \sin\theta_D$
	$A_{3/2}$	$\frac{1}{2\sqrt{6}}(A + C) \sin\theta_D$
$\Delta(1700)$	$A_{1/2}$	$\frac{1}{18\sqrt{2}}(3A + 2B + C)$
	$A_{3/2}$	$\frac{1}{6\sqrt{6}}(3A - C)$

The angle θ_S is the mixing angle associated with the $N\frac{1}{2}^-$ states ($\theta_S = 31°$). The angle θ_D is the mixing angle associated with the $N\frac{3}{2}^-$ states ($\theta_D = 6°$)

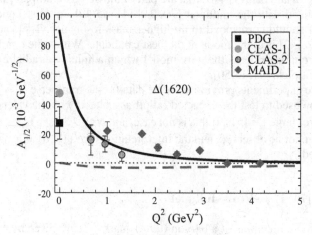

Fig. 5 Results for $N(1650)$. The models 1 and 2 gave the same result (*solid line*)

proton targets (average on the SQTM coefficients). Based on the amplitudes used in the calibration we expect the estimates to be accurate for $Q^2 \gtrsim 2\,\text{GeV}^2$ [16].

From the study of the $\gamma^* N \to N(1520)$ transition within the covariant spectator quark model it is possible to conclude that the contributions for the amplitude $A_{3/2}$ due to valence quarks are very small [15,16]. This conclusion in consistent with others estimates from quark models, where the valence quark component is about 20–40 % [15,58]. An accurate description of the $\gamma^* N \to N(1520)$ transition requires, then, a significant meson contribution for the amplitude $A_{3/2}$, as suggested also by the EBAC/Argonne-Osaka model [59]. Therefore, in Ref. [15], we developed an effective parametrization of the amplitude $A_{3/2}$, inspired on the meson cloud contributions for the $\gamma^* N \to \Delta$ transition [15,16,28,29].

From the relations (2), we can conclude in the limit where no meson cloud is considered ($A_{3/2}^{D13} = 0$), one has $C = -A$. The last condition defines the model 1, dependent only of the parameters A and B. Since, as discussed, the description of the $\gamma^* N \to N(1520)$ is limited when we neglect the meson cloud contributions for $A_{3/2}$, we consider a second model (model 2) where in addition to the valence quark contributions for the amplitudes $A_{1/2}^{S11}$ and $A_{1/2}^{D13}$, we include a parametrization for $A_{3/2}^{D13}$ derived in Ref. [15].

The models are compared with the available data for the amplitudes associated with the $[70, 1^-]$ $SU(6)$-multiplet. The data available for those resonances, from CLAS [3,33], MAID [4,34–36], PDG [42] and others [52], is very scarce particularly for $Q^2 > 1.5\,\text{GeV}^2$. Nevertheless, we conclude that for the cases $N(1650)$ and $\Delta(1620)$, the model 2 gives a good description of the data (see Figs. 5, 6).

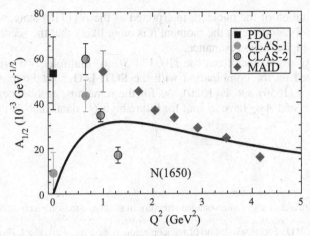

Fig. 6 Results for $\Delta(1620)$. Model 1 (*dashed line*) and model 2 (*solid-line*)

Table 2 Parameters from the high Q^2 parametrization given by Eq. (3)

State	Amplitude	$D(10^{-3}\,\text{GeV}^{-1/2})$	$\Lambda^2\,(\text{GeV}^2)$
$N(1650)$	$A_{1/2}$	68.90	3.35
$\Delta(1620)$	$A_{1/2}$	...	...
$N(1700)$	$A_{1/2}$	-8.51	2.82
	$A_{3/2}$	4.36	3.61
$\Delta(1700)$	$A_{1/2}$	39.22	2.69
	$A_{3/2}$	42.15	8.42

Based on the expected behavior for large Q^2 given by $A_{1/2} \propto 1/Q^3$ and $A_{1/2} \propto 1/Q^5$ in accordance with perturbative QCD arguments [51], we parametrize the amplitudes as

$$A_{1/2}(Q^2) = D\left(\frac{\Lambda^2}{\Lambda^2 + Q^2}\right)^{3/2}, \qquad A_{3/2}(Q^2) = D\left(\frac{\Lambda^2}{\Lambda^2 + Q^2}\right)^{5/2}. \qquad (3)$$

The coefficients D and the cutoff Λ are determined in order to be exact for $Q^2 = 5\,\text{GeV}^2$. The results of the parametrizations are in Table 2. Those parametrizations may be useful to compare with future experiments at large Q^2, as the ones predicted for the Jlab-12 GeV upgrade [1].

For the amplitude $A_{1/2}$ associated with the $\Delta(1620)$ state, it is not possible to find a parametrization consistent the power 3/2 for the amplitude $A_{1/2}$. This is because, for that particular amplitude, there is a partial cancellation between the leading terms (on $1/Q^3$) of our A, B and C parametrization, due to the difference of sign between the amplitudes $A_{1/2}^{S11}$ and $A_{1/2}^{D13}$ used in the determination of the SQTM coefficients (see dashed line in Fig. 6). As consequence, the amplitude $A_{1/2}$ for the state $\Delta(1620)$ is dominated by next leading terms (on $1/Q^5$) or contributions due to meson cloud effects ($A_{3/2}^{D13}$). It is clear in Fig. 6, that, when we neglect the contributions from $A_{3/2}^{D13}$ the result is almost zero (model 1; dashed line). This result shows that in the $\gamma^* N \to \Delta(1620)$ transition, contrarily to what is usually expected, there is a strong suppression of the valence quark effects for $Q^2 = 1\text{--}2\,\text{GeV}^2$. A better representation of the $\gamma^* N \to \Delta(1620)$ data is obtained using $A_{1/2} \propto \left(\frac{\Lambda^2}{\Lambda^2 + Q^2}\right)^{5/2}$, where $\Lambda^2 = 1\,\text{GeV}^2$ (note the power 5/2, instead of the expected 3/2). For a more detailed discussion see Ref. [16].

5 Summary and Conclusions

Answering to the challenge raised by recent experimental results and by the experiments planed for a near future, for resonances in the region $W = 1.6\text{--}1.8\,\text{GeV}$, with high photon virtualities, we present the more recent results from the covariant spectator quark model. The model is covariant, it is based on the valence quark degrees of freedom, and therefore potentially applicable in the region of the large energies and momenta. The estimates for the second radial excitation of the nucleon, interpreted as the $N(1710)$ state, are still under discussion, both theoretically and experimentally. At the moment it is more likely that the second radial excitation of the nucleon correspond to an higher mass resonance.

For the negative parity resonances from the $[70, 1^-]$ $SU(6)$-multiplet, we provide predictions for the transition amplitudes based on the combination with the SQTM. Our predictions compare well with the $Q^2 > 2$ GeV^2 data, for $N(1650)$ and $\Delta(1620)$. As for the remaining resonances, the predictions for the transverse amplitudes $A_{1/2}$ and $A_{3/2}$ have to wait for future high Q^2 data, such as the data expected from the Jlab-12 GeV upgrade.

References

1. Aznauryan, I.G., et al.: Studies of nucleon resonance structure in exclusive meson electroproduction. Int. J. Mod. Phys. E **22**, 1330015 (2013)
2. Aznauryan, I.G., Burkert, V.D.: Electroexcitation of nucleon resonances. Prog. Part. Nucl. Phys. **67**, 1 (2012)
3. Aznauryan, I.G., et al.: [CLAS collaboration]: Electroexcitation of nucleon resonances from CLAS data on single pion electroproduction. Phys. Rev. C **80**, 055203 (2009)
4. Tiator, L., Drechsel, D., Kamalov, S.S., Vanderhaeghen, M.: Electromagnetic excitation of nucleon resonances. Eur. Phys. J. Spec. Top. **198**, 141 (2011)
5. Tiator, L., Drechsel, D., Kamalov, S., Giannini, M.M., Santopinto, E., Vassallo, A.: Electroproduction of nucleon resonances. Eur. Phys. J. A **19**, 55 (2004)
6. Gross, F., Ramalho, G., Peña, M.T.: A pure S-wave covariant model for the nucleon. Phys. Rev. C **77**, 015202 (2008)
7. Covariant nucleon wave function with: S, D, and P-state components. Phys. Rev. D **85**, 093005 (2012)
8. Ramalho, G., Tsushima, K.: Octet baryon electromagnetic form factors in a relativistic quark model. Phys. Rev. D **84**, 054014 (2011)
9. Ramalho, G., Tsushima, K., Thomas, A.W.: Octet baryon electromagnetic form factors in nuclear medium. J. Phys. G **40**, 015102 (2013)
10. Ramalho, G., Tsushima, K., Gross, F.: A relativistic quark model for the Ω^- electromagnetic form factors. Phys. Rev. D **80**, 033004 (2009)
11. Ramalho, G., Tsushima, K.: Valence quark contributions for the $\gamma N \to P_{11}(1440)$ form factors. Phys. Rev. D **81**, 074020 (2010)
12. Ramalho, G., Tsushima, K.: $\gamma^* N \to N(1710)$ transition at high momentum transfer. Phys. Rev. D **89**, 073010 (2014)
13. Ramalho, G., Peña, M.T.: A covariant model for the $\gamma N \to N(1535)$ transition at high momentum transfer. Phys. Rev. D **84**, 033007 (2011)
14. Ramalho, G., Tsushima, K.: A simple relation between the $\gamma N \to N(1535)$ helicity amplitudes. Phys. Rev. D **84**, 051301 (2011)
15. Ramalho, G., Peña, M.T.: $\gamma^* N \to N^*(1520)$ form factors in the spacelike region. Phys. Rev. D **89**, 094016 (2014)
16. Ramalho, G.: Using the single quark transition model to predict nucleon resonance amplitudes. Phys. Rev. D **90**, 033010 (2014)
17. Ramalho, G., Peña, M.T.: Nucleon and $\gamma N \to \Delta$ lattice form factors in a constituent quark model. J. Phys. G **36**, 115011 (2009)
18. Ramalho, G., Peña, M.T.: Valence quark contribution for the $\gamma N \to \Delta$ quadrupole transition extracted from lattice QCD. Phys. Rev. D **80**, 013008 (2009)
19. Ramalho, G., Peña, M.T., Gross, F.: A covariant model for the nucleon and the Δ. Eur. Phys. J. A **36**, 329 (2008)
20. Ramalho, G., Peña, M.T., Gross, F.: D-state effects in the electromagnetic $N\Delta$ transition. Phys. Rev. D **78**, 114017 (2008)
21. Ramalho, G., Peña, M.T., Gross, F.: Electromagnetic form factors of the: Δ with D-waves. Phys. Rev. D **81**, 113011 (2010)
22. Ramalho, G., Peña, M.T., Stadler, A.: The shape of the Δ baryon in a covariant spectator quark model. Phys. Rev. D **86**, 093022 (2012)
23. Ramalho, G., Tsushima, K.: A model for the $\Delta(1600)$ resonance and $\gamma N \to \Delta(1600)$ transition. Phys. Rev. D **82**, 073007 (2010)
24. Ramalho, G., Tsushima, K.: Octet to decuplet electromagnetic transition in a relativistic quark model. Phys. Rev. D **87**, 093011 (2013)
25. Ramalho, G., Tsushima, K.: What is the role of the meson cloud in the $\Sigma^{*0} \to \gamma\Lambda$ and $\Sigma^* \to \gamma\Sigma$ decays? Phys. Rev. D **88**, 053002 (2013)
26. Ramalho, G., Tsushima, K.: Covariant spectator quark model description of the $\gamma^*\Lambda \to \Sigma^0$ transition. Phys. Rev. D **86**, 114030 (2012)
27. Ramalho, G., Peña, M.T.: Extracting the Ω^- electric quadrupole moment from lattice QCD data. Phys. Rev. D **83**, 054011 (2011)
28. Ramalho, G., Peña, M.T.: Timelike $\gamma^* N \to \Delta$ form factors and Delta Dalitz decay. Phys. Rev. D **85**, 113014 (2012)
29. Ramalho, G., Peña, M.T., Weil, J., van Hees, H., Mosel, U.: Role of the pion electromagnetic form factor in the $\Delta(1232) \to \gamma^* N$ timelike transition. Phys. Rev. D **93**, 033004 (2016)
30. Gross, F., Ramalho, G., Peña, M.T.: Spin and angular momentum in the nucleon. Phys. Rev. D **85**, 093006 (2012)
31. Ramalho, G., Tsushima, K.: Axial form factors of the octet baryons in a covariant quark model. Phys. Rev. D **94**, 014001 (2016)
32. Ramalho, G., Jido, D., Tsushima, K.: Valence quark and meson cloud contributions for the $\gamma^*\Lambda \to \Lambda^*$ and $\gamma^*\Sigma^0 \to \Lambda^*$ reactions. Phys. Rev. D **85**, 093014 (2012)

33. Mokeev, V.I., et al.: [CLAS collaboration]: Experimental study of the $P_{11}(1440)$ and $D_{13}(1520)$ resonances from CLAS data on $ep \rightarrow e'\pi^+\pi^-p'$. Phys. Rev. C **86**, 035203 (2012)
34. Drechsel, D., Kamalov, S.S., Tiator, L.: Unitary Isobar Model—MAID2007. Eur. Phys. J. A **34**, 69 (2007)
35. Tiator, L., Drechsel, D., Kamalov, S.S., Vanderhaeghen, M.: Baryon resonance analysis from MAID. Chin. Phys. C **33**, 1069 (2009)
36. Tiator, L., Drechsel, D., Kamalov, S.S., Vanderhaeghen, M.: Electromagnetic excitation of nucleon resonances. Eur. Phys. J. Spec. Top. **198**, 141 (2011)
37. Arrington, J., Melnitchouk, W., Tjon, J.A.: Global analysis of proton elastic form factor data with two-photon exchange corrections. Phys. Rev. C **76**, 035205 (2007)
38. Park, K., et al.: [CLAS collaboration]: Measurements of $ep \rightarrow e'\pi^+n$ at W = 1.6 - 2.0 GeV and extraction of nucleon resonance electrocouplings at CLAS. Phys. Rev. C **91**, 045203 (2015)
39. Melde, T., Plessas, W., Sengl, B.: Quark-model identification of baryon ground and resonant states. Phys. Rev. D **77**, 114002 (2008)
40. Ronniger, M., Metsch, B.C.: Effects of a spin-flavour dependent interaction on light-flavoured baryon helicity amplitudes. Eur. Phys. J. A **49**, 8 (2013)
41. Santopinto, E., Giannini, M.M.: Systematic study of longitudinal and transverse helicity amplitudes in the hypercentral constituent quark model. Phys. Rev. C **86**, 065202 (2012)
42. Beringer, J., et al.: [Particle data group collaboration]: Review of particle physics (RPP). Phys. Rev. D **86**, 010001 (2012)
43. Isgur, N., Karl, G.: P wave baryons in the quark model. Phys. Rev. D **18**, 4187 (1978)
44. Isgur, N., Karl, G.: Positive parity excited baryons in a quark model with hyperfine interactions. Phys. Rev. D **19**, 2653 (1979)
45. Isgur, N., Karl, G.: Ground state baryons in a quark model with hyperfine interactions. Phys. Rev. D **20**, 1191 (1979)
46. Suzuki, N., Julia-Diaz, B., Kamano, H., Lee, T.-S.H., Matsuyama, A., Sato, T.: Disentangling the dynamical origin of P_{11} nucleon resonances. Phys. Rev. Lett. **104**, 042302 (2010)
47. Santopinto, E.: An interacting quark–diquark model of baryons. Phys. Rev. C **72**, 022201 (2005)
48. Santopinto, E., Ferretti, J.: Strange and nonstrange baryon spectra in the relativistic interacting quark–diquark model with a Gürsey and Radicati-inspired exchange interaction. Phys. Rev. C **92**, 025202 (2015)
49. Giannini, M.M., Santopinto, E.: The hypercentral constituent quark model and its application to baryon properties. Chin. J. Phys. **53**, 020301 (2015)
50. Ferraris, M., Giannini, M.M., Pizzo, M., Santopinto, E., Tiator, L.: A three body force model for the baryon spectrum. Phys. Lett. B **364**, 231 (1995)
51. Carlson, C.E., Poor, J.L.: Distribution amplitudes and electroproduction of the delta and other low lying resonances. Phys. Rev. D **38**, 2758 (1988)
52. Burkert, V.D., De Vita, R., Battaglieri, M., Ripani, M., Mokeev, V.: Single quark transition model analysis of electromagnetic nucleon resonance transitions in the $[70, 1^-]$ supermultiplet. Phys. Rev. C **67**, 035204 (2003)
53. Hey, A.J.G., Weyers, J.: Quarks and the helicity structure of photoproduction amplitudes. Phys. Lett. B **48**, 69 (1974)
54. Cottingham, W.N., Dunbar, I.H.: Baryon multipole moments in the single quark transition model. Z. Phys. C **2**, 41 (1979)
55. Ramalho, G.: Improved empirical parametrizations of the $\gamma^*N \rightarrow N(1535)$ transition amplitudes and the Siegert's theorem. Phys. Lett. B **579**, 126 (2016)
56. Ramalho, G.: Improved empirical parametrizations of the $\gamma^*N \rightarrow \Delta(1232)$ and $\gamma^*N \rightarrow N(1520)$ transition amplitudes and Siegert's theorem. Phys. Rev. D **93**, 113012 (2016)
57. Ramalho, G.: Improved large N_c parametrizations of the $\gamma^*N \rightarrow \Delta(1232)$ quadrupole form factors and the Siegert's theorem. arXiv:1606.03042
58. Aiello, A., Giannini, M.M., Santopinto, E.: Electromagnetic transition form-factors of negative parity nucleon resonances. J. Phys. G **24**, 753 (1998)
59. Sato, T., Lee, T.-S.H.: Dynamical models of the excitations of nucleon resonances. J. Phys. G **36**, 073001 (2009)

Printed in the United States
By Bookmasters